City Planning for Civil Engineers, Environmental Engineers, and Surveyors

This thoroughly updated second edition focuses on the significant advances and shifts in urban and city planning and sustainable development. It includes new contents on sustainability and resilience in urban planning, climate change impact, technological advancements for urban planning, and electrification and decarbonization of urban environments. This comprehensive coverage balances theory and practice, including regulatory and policy insights, ensuring that the book remains a valuable resource with a unique perspective and technical depth that engineers and surveyors need. It also provides real-world examples and the use of state-of-the-art technologies adopted in urban planning.

New Features

- Bridges the gap between theoretical knowledge and practical application, offering insights into the complexities of urban development, infrastructure planning, zoning laws, and the effective implementation of city plans.
- Provides real-world examples related to the impact of climate change and explains the concept of resilient cities.
- Includes the new features of urban planning such as urban-scale electrification and decarbonization.
- Covers the state-of-the-art technologies adopted in urban planning and the concept of smart cities.
- Discusses the energy aspect in urban planning and the concept of sustainable cities.
- Explains cybersecurity in urban planning and considers a city a cyber-physical system.

This textbook is for senior undergraduate and graduate students in Civil and Environmental Engineering, Surveying, and those taking courses in land development and planning or urban planning. It's also an excellent text for civil engineers and professionals who work in zoning and planning commissions or in land development agencies.

City Planning for Civil Engineers, Environmental Engineers, and Surveyors

Second Edition

Yunyang Ye and Kurt W. Bauer

CRC Press
Taylor & Francis Group
Boca Raton London New York

CRC Press is an imprint of the
Taylor & Francis Group, an **informa** business

Second edition published 2026
by CRC Press
2385 NW Executive Center Drive, Suite 320, Boca Raton FL 33431

and by CRC Press
4 Park Square, Milton Park, Abingdon, Oxon, OX14 4RN

CRC Press is an imprint of Taylor & Francis Group, LLC

First edition published by CRC Press 2009

ISBN: 978-1-032-81562-6 (hbk)
ISBN: 978-1-032-81563-3 (pbk)
ISBN: 978-1-003-50038-4 (ebk)

DOI: 10.1201/9781003500384

Typeset in Times
by KnowledgeWorks Global Ltd.

Contents

Scope of the Book

As indicated in Chapter 1, this book is intended to introduce civil engineering students and practicing civil engineers to the concepts, principles, and practices underlying urban planning in the United States. It is not intended to make professional urban planners out of civil engineers, but rather to help practicing civil engineers become more knowledgeable participants in the urban planning process and more effective members of urban planning teams and governmental and consulting agency staffs.

As such, the book is, in effect, a primer on urban planning. It focuses on those areas of urban planning with which civil engineers are most likely to come into contact or conflict, in which civil engineers may be required to participate, and for which civil engineers may be required to provide necessary leadership. It stresses the basic concepts and principles of practice involved in urban planning as most widely practiced, particularly in small and medium-sized communities. It does not address, except in passing, the latest state-of-the-art approaches to the practice of planning and, indeed, seeks to avoid the always present tendency to follow the latest fads while neglecting time-tested and proven practices. It neglects some areas of urban planning—such as urban design—entirely, as it does some plan implementation devices such as building, housing, and sanitation codes. Nevertheless, it is hoped that this book provides an adequate introduction to the fascinating field of urban planning as practiced in the United States.

For those civil engineers interested in delving more deeply into the field of urban planning, further readings are suggested on some, but not all, topics. Any civil engineer so interested would do well to regularly read the quarterly *Journal of the American Planning Association* and the *Journal of Urban Planning and Development* of the American Society of Civil Engineers.

As noted in Chapter 1, for students and practicing planners, the book also may serve as an introduction to the engineering problems that are encountered in any urban planning effort. The quality and effectiveness of urban planning depend to a considerable degree upon successful solutions to those engineering problems. Good comprehensive planning and plan implementation require the close cooperation of the civil engineering and planning professions. It is hoped that this book may in some small way contribute to furthering that much needed cooperation.

Preface

As noted in Chapter 1, this book is intended to introduce civil engineers to the concepts, principles, and practices underlying urban planning as currently practiced in the United States and to introduce planners to the engineering problems encountered in any urban planning effort. The book is based on a popular senior and graduate level course in planning for civil engineers taught at Marquette University for over 40 years. The Marquette course was in turn modeled on a similar pioneering course taught at the University of Wisconsin–Madison since the early twentieth century, a course that may indeed have been one of the very first courses in city planning taught anywhere in the United States. It was taught at a time when civil engineers, to a considerable extent, constituted the city planning profession, and introduced the "city efficient" movement in planning. In 2025, the second edition of this book removed outdated content and added new features of city planning, making it a suitable textbook for continuing to teach engineering students about city planning. This second edition reflects the evolution of urban planning practice in the twenty-first century. It incorporates recent advances in spatial data infrastructure, sustainable and resilient city design, and the increasing emphasis on equity, climate adaptation, and community participation. Several chapters have been updated to include contemporary examples, digital mapping tools, and new regulatory approaches that better represent current U.S. planning practice.

The book is arranged in 21 chapters. Following a brief introductory chapter, Chapter 2 sets forth definitions of some of the terminology used in planning and the concepts that this terminology represents. The chapter describes the social and economic functions of cities, describes pertinent census geography, defines the terms planning and city planning, identifies core planning functions, presents criteria for good city planning, and considers the need for and value of city planning. Chapter 3 provides a brief account of the historical development of city planning in the United States.

Chapter 4 provides an introduction to the basic data-collection function of city planning. Chapter 5 describes the map requirements for city planning, stressing the importance of a common map projection, common horizontal and vertical datums, and a monumented survey control network for the preparation of the needed topographic and cadastral maps and the creation of computerized parcel-based land information and public works management systems. Chapters 6 and 7 deal with demographic and economic studies and the preparation of population and employment projections and forecasts. Chapter 8 focuses on land use and infrastructure inventories, while Chapter 9 examines needed natural resource base inventories.

Chapter 10 describes the institutional structure for city planning, the concept of the comprehensive plan, and staff organization for city planning. Chapter 11 deals with the formulation of objectives and supporting standards as a basis for plan design and evaluation. Chapter 12 discusses land use planning, including a description of the determinants of the urban land use pattern, the steps in land use planning, and the estimation of land requirements, with an example of an actual land use plan for a small community. Chapter 13 describes the neighborhood unit concept in city planning and includes an example of an actual neighborhood unit plan, while Chapter 14 deals with land subdivision design.

Chapter 15 deals with street patterns, street cross-sections, and functional and jurisdictional arterial street and highway system planning, and provides a brief introduction to mass transit system planning. Chapter 16 deals very briefly with other plan elements.

Chapter 17 discusses land subdivision control as a means of comprehensive plan implementation. An example of a land subdivision control ordinance is provided in an appendix to the book, and the salient provisions of the ordinance are commented upon in the chapter. Example preliminary and final land subdivision plats are provided. The required information for subdivision design is described. Importantly, required improvements are described.

Chapter 18 investigates zoning as a means of land use plan implementation. The techniques involved in the preparation of a zoning ordinance and attendant zoning district map are described. The problems inherent in zoning district boundary delineation related to natural resource protection are considered. An example of a zoning ordinance is provided in an appendix to the book, and the salient provisions of the ordinance are commented upon. Some common zoning problems are described.

Chapter 19 focuses on the official map as a means of arterial street, highway, park, and parkway plan implementation. Chapter 20 examines capital improvement programming as a means of comprehensive plan implementation. Finally, Chapter 21 introduces new features of urban planning for sustainable, resilient, and smart cities, with case studies.

The book emphasizes—perhaps far too subtly—the importance of good inventory data to sound plan preparation and implementation, and the importance of good mapping as an essential foundation for the collection, presentation, and application of the required data. The book also stresses that the land use element of a comprehensive plan is the essential foundation of all of the other elements of such a plan. Finally, the book explains the importance of the comprehensive plan as a foundation for the design and application of all plan implementation devices.

Acknowledgments from the First Edition

With the exception of the recommended system of survey control, none of the concepts, principles, or practices presented in this book represents original work on my part. I am indebted to many persons for the origins of these concepts, principles, and practices herein presented—primarily teachers and colleagues—with whom I studied, worked, and taught in the public and private sectors for over 50 years. I would be remiss, however, if I did not single out for acknowledgment Professor Lloyd F. Rader, who organized, and for many years taught, a course in city planning for civil engineers at the University of Wisconsin–Madison, and who inspired me to teach a similar course at Marquette University.

I would also be remiss if I did not gratefully acknowledge the invaluable secretarial help provided to me by Linette G. Heis and Marcia L. Hayd, and the equally invaluable drafting help provided to me by Nancee A. Nejedlo and, especially, by Donald P. Simon. As for the many others who contributed so much to this book, in the words of T. E. Lawrence: "The un-named rank and file miss their share of credit, as they must do, until they can write the dispatches." The truly complimentary nature of this apt quotation becomes apparent only when one remembers that it is the rank and file who fight the battles, do the hard work, and who, through their faithful devotion to duty, make our world, and especially our cities, livable.

Kurt W. Bauer, PE, RLS, AICP

Second Edition's Acknowledgments

Dr. Bauer's extraordinary effort makes this book's first edition successful. Well-organized, comprehensive content written by Dr. Bauer makes the revision process smoother. Over the past 15 years, the world has changed a lot. Advanced technologies are significantly changing city planning, and more interdisciplinary knowledge is needed to ensure suitable, sustainable urban design and planning. As a faculty member in civil engineering, I have expertise in my field but lack understanding of other areas required for editing this book. Luckily, experts in our community are hardworking. Many significant publications have been written and published, which have greatly helped me understand related knowledge in various fields and complete this edition. Thanks to all the experts' contributions in this book. Finally, I would also greatly acknowledge the support from my wife, Qingling Hu, which has enabled the completion of this edition.

About the Authors

Kurt W. Bauer received his B.S. degree in civil engineering from Marquette University and his M.S. and Ph.D. degrees in civil engineering from the University of Wisconsin–Madison. He was Executive Director of the Southeastern Wisconsin Regional Planning Commission from 1962 to 1997 and directed the preparation and implementation of that Commission's comprehensive plan for the physical development of a seven-county area of 1.8 million people. Dr. Bauer is a registered professional engineer and land surveyor, a charter member of the American Institute of Certified Planners, and a fellow of the American Society of Civil Engineers and the American Congress on Surveying and Mapping. He has served on the faculties of both Marquette University and the University of Wisconsin. Dr. Bauer has also served as a member of the Science Advisory Board of the International Joint Commission on the Great Lakes, as an advisor to the National Water Commission, and as a consultant to the U.S. Departments of Housing and Urban Development, Interior, and Transportation. Dr. Bauer is the recipient of awards from the U.S. Department of Transportation, the Great Lakes Commission, the American Society of Civil Engineers, the Soil Conservation Society of America, the University of Wisconsin–Madison, and Marquette University.

Dr. Yunyang Ye received his B.S. and M.S. degrees in mechanical engineering from Tongji University and his Ph.D. degree in civil engineering from the University of Colorado Boulder. He was a scientist at the Pacific Northwest National Laboratory and an assistant professor at the University of Alabama. Now, he is an assistant professor in the Civil and Environmental Engineering at Colorado School of Mines. His research focuses on sustainable and resilient buildings, sustainable and smart cities, and grid-interactive efficient buildings. His research spans from the system level to the urban scale, using system modeling methods that hybridize physics-based and cutting-edge artificial intelligence approaches. His spatial-temporal research is interdisciplinary and involves occupant behaviors (social), power systems, transportation (engineering), and climate disasters (science). He has led multiple projects and tasks sponsored by U.S. DOE, NSF, and ASHRAE, and published over 50 peer-reviewed articles. He is an Associate Editor for *Science and Technology for the Built Environment* and a Youth Editor for *Building Simulation: An International Journal*. He has also served as a guest editor for six special issues, as a member of the scientific committee organizing three international conferences, and as vice chair of ASHRAE Technical Committee 7.4. He has received multiple awards, including the ASHRAE New Investigator Award, IBPSA-USA Emerging Contributor Award, and the ASHRAE Homer Addams Award.

1 Introduction

For it is clear that in whatever it is our duty to act, these matters also it is our duty to study.

Thomas Arnold, DD
English educator
1822–1888

This book is intended to introduce civil engineering students—and practicing civil engineers—to the concepts, principles, and practices underlying urban planning in the United States. Emphasis is placed on the comprehensive plan, including its land use, transportation, public utility, and community facility elements, and on plan implementation devices such as zoning, land subdivision control, official mapping, and capital improvement programming. Additionally, this book also introduces the new technologies adopted in sustainable, resilient, and smart cities.

In this book, urban planning is considered a rational process that seeks the orderly, cost-effective development of the urban environment, oriented to the formulation and attainment of objectives and the application of standards that specify desirable physical arrangements. This view of urban planning assumes that the orderly physical development of urban areas is in the public interest and that public planning should be oriented to advancing public health, safety, and the general welfare rather than special interests. Other views of urban planning exist—such as user-oriented, participatory planning, advocacy planning, and incremental or market-oriented planning—but are not regarded as sound models in this book. Emphasis in this book is on physical, rather than social or economic, planning. It should be understood, however, that—properly conducted—physical planning is intended to help achieve the underlying social and economic objectives of a community.

Insofar as this book is intended for students and practicing civil engineers and surveyors, it is not meant to turn engineers and surveyors into planners, but rather to help them become more proficient practitioners of engineering and surveying. A high degree of competence in such civil engineering specialties as surveying and mapping, highway and transportation engineering, water resources engineering, environmental engineering, and, particularly, municipal engineering requires an understanding of urban development problems and of urban planning objectives, principles, and practices. This book is also intended to help practicing engineers and surveyors become more knowledgeable participants in the urban planning process and to make engineers, in particular, more effective members of urban planning teams and of governmental staffs, particularly at the county and municipal levels of government. At these levels of government, cooperative efforts and relationships between planning and engineering departments are particularly important to good

DOI: 10.1201/9781003500384-1

urban development and redevelopment. Cooperative efforts between planning and engineering staff are also crucial in the private sector, where, for example, consulting engineering firms may have urban planning divisions. Meaningful cooperation between planners and engineers in the public and private sectors requires, among other things, engineers who are knowledgeable about and appreciative of urban planning principles and practices.

This book may also serve as an introduction to what was, for a time, a highly specialized branch of civil engineering—namely, city and regional planning. Historically, courses in city and regional planning were often taught in civil engineering departments at major universities, and textbooks in the field were usually authored by practicing civil engineers. Indeed, the municipal engineers responsible for planning, design, construction, and maintenance of public infrastructure—the sewerage, water supply, drainage, and transportation facilities that make modern urban life possible and require a systematic approach to development—were, in many respects, the first modern city planners. The American Society of Civil Engineers, to this day, maintains an Urban Planning and Development Division, publishes an urban planning and development journal, and annually awards the Harland Bartholomew Prize for notable achievement in planning.

City and regional planning, however, has increasingly become a discipline separate from civil engineering, practiced by a now well-established planning profession. A close association and mutual understanding between these two professions is needed. Insofar as this book is addressed to students and practicing planners, it is intended to serve as a broad introduction to the engineering problems encountered in urban planning. The quality and effectiveness of urban planning depend, to a considerable degree, upon successful solutions to those engineering problems.

2 Definition of Terminology

One of the first facts which must be noted by those who would deal with problems of cities is that the political and institutional overlay of modern government does not describe or delimit the real city. Most cities grew beyond their statutory borders long ago, yet we who are responsible for the city and to its people must still operate within these obsolete and artificial boundaries. We are constrained by limits and institutions which are at best irrelevant to the structure and life of the real city and too often actually frustrate or strangle its natural vitality and development.

Phyllis Lamphere, President
National League of Cities, Nation's Cities
December 1976

INTRODUCTION

A good understanding of the principles and practices of urban planning requires the prefatory definition of certain terminology and an understanding of the basic concepts the terminology represents. In this respect, it should be noted that, for convenience, the older term *city planning* is sometimes used here in place of the newer term *urban planning*. Similarly, the term city is sometimes used interchangeably with the term *urban area*. The following definitions represent, in effect, very brief introductions to complex concepts that a variety of disciplines have developed concerning cities—and urban development. An awareness, if not a full understanding, of these concepts is essential to the practice of urban planning and to the truly professional practice of at least certain specialties of civil engineering.

DEFINITION OF THE TERM *CITY*

The rise of cities marked a great revolution in human culture and, unlike the agricultural revolution that preceded it, was predominantly a social process—a change in the way humans interacted with one another, rather than in the way humans interacted with the environment. Its essential element was the creation of a series of new institutions unlike the institutions created by earlier societies. The different forms that early urban societies took were the products of political and economic—that is, of human—forces, rather than the product of basic innovations in the pursuit of subsistence.

The term city may be defined as a relatively large, dense, and permanent settlement of people engaged in diverse economic activities. In addition, by definition, a city

DOI: 10.1201/9781003500384-2

must have a government and certain social institutions to serve the needs of the resident population, a population that must be agglomerated rather than isolated. This later phrase excludes from the definition such complexes as military cantonments and bases, monasteries, and prisons. As cities—that is, urban areas—have grown larger, the required governance of the urban area may be shared by different levels as well as by different individual units of government.

A more technically oriented definition of the term city is a complex arrangement of land uses, linked together by circulation systems and made viable by other systems of utilities such as sewerage, water supply, electric power, and telecommunications. It is the setting for economic activities and social intercourse, and provides certain important amenities. Cities have always attracted people seeking economic opportunity, improved housing conditions, better education, entertainment and, historically—at least in Europe—enhanced freedom.

In defining the term city, and in thinking about cities, a useful concept is that of "four cities," representing four different aspects of the city: the city as an economic entity, as a social organism, as a legal entity, and as a cyber-physical system.

The City as an Economic Entity

Economists regard the city as an economic entity existing to serve certain highly organized economic functions that act as an urbanizing force. These include four general functions that are performed by every city regardless of age, size, or location, and that remain basically unchanged over time. These four functions are production, distribution, consumption, and amenity service. More specifically, special urban economic functions include exchange, manufacturing, extraction, government, education and religion, and recreation and health. These functions will differ between cities and will determine, to a large extent, the characteristics of the city. In the contemporary context, cities increasingly serve as hubs for knowledge creation, digital services, advanced technologies, and platform economies. These functions do not rely on physical goods; instead, they rely more on ideas, data, and human capital. These functions also affect the characteristics of the city.

Historically, military considerations, which usually also had economic implications, often acted as an urbanizing force. Military considerations caused cities to grow around fortifications located either to dominate or to defend an area. Examples of cities that initially grew around fortifications include many European cities and such North American cities as Chicago, Illinois (Fort Dearborn); Green Bay, Wisconsin (Fort Howard); and Prairie DuChen, Wisconsin (Fort Crawford).

Multiply functions, instead of a single dominant function, are usually used to define a modern city. A modern city can also evolve across different economic roles over time. Thus, understanding a city's economic profile requires a dynamic and system-oriented perspective. Furthermore, issues of economic inequality, labor market exclusion, and spatial mismatch between jobs and housing increasingly shape the economic development agenda. Likewise, economic specialization can increase a city's vulnerability to technological disruption, market shifts, or climate-related risks. Thus, fostering resilient, diversified, and equitable local economies has become a critical aim of contemporary urban planning and policy.

Exchange

The special urban function of exchange—also known as trade and transport, or commerce—acts as an urbanizing force to cause cities to locate and grow at points of transshipment; at intersections and termini of trade routes; and at centers of agricultural regions. Examples of cities characterized by the exchange function include such major seaport cities as Boston, Massachusetts; New Orleans, Louisiana; New York, New York; Philadelphia, Pennsylvania; and London, England; and such railway cities as Chicago, Illinois; and Logansport, Indiana. Nowadays, the scope and mechanisms of exchange have significantly evolved. Modern cities grow around airports, inland logistics nodes, and digital infrastructure corridors. Examples of such modern cities include Atlanta, Georgia; Memphis, Tennessee; and Louisville, Kentucky.

Manufacturing

Manufacturing as an urbanizing force acts to cause cities to locate and grow at sources of raw materials, in proximity to markets, at sources of power and fuel, in areas marked by concentrations of labor and capital, and in areas of favorable climate. The momentum of an early application of a new technology, and even historic accident, may also be important factors determining the location of such cities. Examples of cities characterized, if not originally, then later, by the manufacturing function include Detroit, Michigan; Gary, Indiana; Milwaukee, Wisconsin; Peoria, Illinois; and Pittsburgh, Pennsylvania. In the twenty-first century, manufacturing has undergone a profound transformation. A new wave of advanced manufacturing has emerged, focused on technology-intensive, cleaner, and higher-value-added production. Examples include Austin, Texas; Phoenix, Arizona; and Detroit, Michigan.

Extraction

Extraction as an urbanizing force acts to cause cities to locate and grow at or in proximity to major mineral deposits, forests, and fisheries. Examples of cities originally characterized by the extraction function include Bessemer, Pennsylvania; Hibbing, Minnesota; Lewiston, Maine; and Gloucester, Massachusetts.

Government

Government as an urbanizing force is usually directed by political considerations and actions. Examples of cities originally characterized by the government function include Washington, D.C.; Madison, Wisconsin; Canberra, Australia; and St. Petersburg, Russia. Nowadays, the role of government extends far beyond the location of formal seats of power, and the government becomes a central actor in shaping urban growth, infrastructure, equity, and resilience.

Education and Religion

Education and religion as urbanizing forces cause cities to locate and grow in proximity to religious shrines, and in the case of major universities, similar to government, in response to political considerations and actions. Examples of cities originally characterized by education and religion include Mecca, Saudi Arabia; Oxford, England; and St. Andrews, Scotland.

Recreation and Health

Recreation and health as urbanizing forces cause cities to locate and grow in locations of favorable climate, at sites of mineral or hot springs, and at concentrations of medical facilities and highly skilled medical staff. Examples of cities originally characterized by the recreation and health functions include Lake Placid, New York; Hot Springs, Arkansas; Las Vegas, Nevada; Rochester, Minnesota; and Waukesha, Wisconsin. Today, planners and engineers must consider recreation and health as critical components not only of city identity but also of urban sustainability, equity, and resilience.

In considering these special urban functions, it must be remembered that as cities mature, their special functions may change, and they also become basically multi-functional. Waukesha, Wisconsin, may be cited as an example of such change. This city was originally founded as a market center in a growing agricultural area. Because of the presence of mineral springs, it became a health resort area. It later evolved into a railway and manufacturing center, and most recently into a satellite city in the greater Milwaukee area. Many examples could be cited of cities marked by multi-functionality, particularly including combinations of commerce and manufacturing and of government and education.

Quantitative Measures of Economic Functions

Economists and geographers have developed various quantitative measures of these general and special urban functions. Some of these measures are of particular interest to urban planners and useful in urban planning. These include the following.

Employment Structure

The employment structure of a city, or urban area, may be defined as the ratio of the employment in each major economic endeavor—manufacturing, construction, trade, government, education, and other services—to total employment. Today, additional layers such as employment in green jobs, creative sectors, and digital services can also be included as major economic endeavors. These ratios for individual cities can be compared to national averages to characterize the economic specialization of the individual cities. For example, in 2006, Janesville, Wisconsin, had 24 percent of its total employment engaged in manufacturing. The United States, as a whole, had 11 percent. The economy of Janesville was obviously concentrated in manufacturing, an important factor in planning for its development and redevelopment.

Industrial Structure

The industrial structure of a city or urban area may be defined as the ratio of the employment in each type of industry—for example, food processing, electronics, primary metals, fabricated metals, machinery, and transportation equipment—to total manufacturing employment. As with the employment structure, these ratios for individual cities can be compared to national averages to further characterize the economic specialization of the individual cities. For example, in 2006, Janesville, Wisconsin, had 42 percent of its total manufacturing employment engaged in the manufacture of transportation equipment. The United States, as a whole, had 12 percent of its manufacturing employment so engaged. The economy of Janesville was not only concentrated in manufacturing but also in transportation equipment—more specifically, automobile and light truck manufacturing. In fact, its economy was largely concentrated in one firm, again with important implications for planning. In the modern context, such insights are particularly useful for examining regional clusters, innovation hubs, and industry transition risks (e.g., coal-to-renewables).

Economic Base

The economic base of a city, or urban area, may be defined as those functions that bring a flow of purchasing power into the city—that is, that export goods and services. The economic base of a city is always associated with its special economic functions. The economic base is measured in terms of the number of employees engaged in basic economic activities as opposed to those engaged in service activities. The ratio of basic employment to service employment will vary with the age of the city from approximately 1:1 to 1:2; the ratio of basic employment to total employment will range from approximately 1:2 to 1:3; and the ratio of basic employment to total population will range from approximately 1:4 to 1:9. The economic base of an urban area is an important determinant of its overall size and rate of growth. The structure and diversification of the economic base significantly influence resilience, particularly in response to global supply chains, automation, or environmental change.

Sustainable Economy

Modern economic analysis increasingly incorporates sustainability metrics to assess whether an urban economy aligns with long-term environmental and social objectives. A sustainable economy considers carbon intensity, energy use efficiency, and the proportion of jobs in renewable energy, clean technology, and ecological services, while also accounting for the durability of the economic base—its ability to withstand shocks such as pandemics, climate disasters, or energy transitions. As cities adapt to climate change and resource constraints, the concept of a low-carbon economy—an economic system designed to reduce greenhouse gas emissions—has emerged as a key pathway to achieving sustainability. The low-carbon economy fosters innovation and resilience, creating new demands on urban infrastructure, from electric vehicle charging networks to green stormwater systems. It requires planners and engineers to rethink zoning, energy systems, transportation planning, and workforce development to align with sustainability goals. Additionally, equity remains an integral component of a sustainable economy, as coupling traditional economic indicators with climate- and justice-based data enables planners to support a financial strategy that is inclusive, low-carbon, and resilient.

The City as a Social Organism

Sociologists regard the city as a social organism, that is, as a complicated network of institutions that operate for the welfare of the residents. In this respect, the city may be regarded as a “community,” that is, a group of people with more or less common experiences, problems, and interests able to act in a corporate capacity. Although it may often be difficult to attach this definition of "community" to larger cities and urban areas, a strong sense of community by a substantial number of residents will make the city planning process both easier and more effective. Some observers would hold that the attainment of this sense of community should be a major objective of planning. The sociologist also regards the city as a form, or mode, of life wherein the associated beings acquire certain traits and create certain organizations for social action. It is within the framework of the social structure of a community that local government must operate, and within which the decision-making process relating to the planning, development, and redevelopment of the community must be conducted. Sociologists have also developed quantitative measures of some of these social characteristics.

Population Characteristics

One of the most useful contributions of the sociologist to urban planning deals with population characteristics: size, age, gender, and race. The statistical study of populations, particularly with reference to size, distribution, and vital statistics, is known as demography, a study of vital importance to urban planners and also to civil engineers, particularly those engaged in the planning and design of transportation systems, building infrastructure, public facilities and services, etc.

Institutional Structure

The institutional structure of a city may be defined as formal associational groupings that can be distinguished as either administrative agencies or organized groups, such as governmental agencies, nonprofit groups, religious institutions, civic associations, and advocacy networks. If church membership is excluded, probably less than one-half of the adult population of an urban area is normally a member of any organized, local group. Thus, when representatives of organized groups are brought into the planning process, it is important to remember that the people concerned are not necessarily representative of the community as a whole. Relatively small but dedicated organized groups can, in effect, capture the public decision-making process of an urban area and block the development of needed facilities, even though the majority of the population may support the proposed development.

Value System

The value system of a group consists of the morally binding customs and attitudes of the group—that is, of the things the members of the group hold dear. Many of the motivating forces that determine the characteristics of a community are rooted in the value system of its residents. Particularly important in this respect is how highly public enterprise, public goods, and public services are regarded. This aspect of a community's value system has subtle, yet important, implications for how planning is regarded within the community and how it can be effectively practiced. The values held, and the intensity with which they are held, will determine the preference and importance given, for example, to free-standing, single-family dwelling unit housing and lower density development, to the provision of public parks and open spaces, and to the aesthetic characteristics of the urban environment. The values held may also determine the intensity of opposition given to certain kinds of development, such as freeways, to public regulation of growth, and to what may be considered excessive taxation and public expenditure. Given the same climate and natural resources, the value systems held by the citizens may produce quite different communities.

In large cities and metropolitan areas, the relationship between value systems and communities becomes complex in that individuals may, in fact, formally and informally belong to a number of "layered" communities—from neighborhood organizations and local civic clubs to areawide industry associations and chambers of commerce. These communities may hold different values and take different positions with regard to some issues. For example, areawide business and industrial groups may favor an arterial street or freeway improvement or an airport expansion that neighborhood groups oppose—the latter giving rise to the "NIMBY" ("not in my back yard") syndrome, under which individuals, though generally appreciating the need for certain public improvements, may still oppose a specific improvement.

In this respect, it may be observed that the choice of elements comprising the value system of a community is not a task for the urban planner or sociologist. Sociologists can, however, assist the urban planner by conducting surveys identifying the values that the people comprising a community may hold, the intensity with which they may hold them, and the basis for the value choices. The increasing cultural pluralism of America complicates, at least in larger urban areas, the effort to deliberate and plan within a framework of common values. Hopefully, an agreement can be found on at least some values, such as striving for excellence, regard for tradition, and the frugal use of finite resources.

Social Stratification and Social Equity

Social stratification may be defined as the structure and process through which rights and privileges, and duties and obligations, are distributed unequally among socially designated, or ranked, grades of people. The basis for the ranking will vary from culture to culture and from community to community and is usually related to the value system. American communities are stratified into social classes, the rigidity of which is debatable, and also into racial and ethnic groups. These classes may hold different value systems, and some may be disadvantaged with respect to their position and role in society.

Social equity refers to fair and just access to opportunities, services, and environments for all residents, especially those historically excluded or marginalized. Planning for social equity requires urban planners to make efforts to reduce racial and economic segregation, promote inclusive public participation, distribute amenities equitably, and mitigate displacement and gentrification impacts.

Power Structure

The power structure of a community may be defined as the locus of the real power to make decisions concerning the development of the community. The power structure is not necessarily synonymous with the nominal leadership in a community that functions through the governmental structure. The power structure operates on the basis of informal but patterned relationships between individuals. Depending upon the community concerned, the locus of power may rest in the political organizations of the community, the economic organizations of the community, including labor, the church, or the media of mass communication. Sociologists have developed techniques for identifying members of the power structure of a community. Sociological methods such as power mapping, elite interviewing, and network analysis can illuminate informal but influential channels of urban governance. The power structure and the way in which that structure may be changing over time are—or should be—of interest to the practicing urban planner and civil engineer. Some observers would regard power structure as a negative characteristic, at least insofar as it may represent special interests as opposed to common interests.

Ecological Patterning

Ecological patterning can be defined as the processes that result in a division of a community into different types of land use areas, particularly various kinds of residential neighborhoods. When these types of areas invade each other, social and economic problems may be created. Some of these problems may relate to the physical development of the community, concerning issues such as the location of schools, libraries, police and fire stations, and other types of public infrastructure. When ecological zones collide or shift, conflicts can arise over issues such as health, noise, property values, or infrastructure needs. One of the contributions that sound physical planning can make to a community is to help the community foster desirable changes in the existing ecological patterns that promote community values.

The City as a Legal Entity

The city may also be regarded as a legal entity—a corporate being—existing in the United States as a creature and arm of state government. The concept of a city as a corporate entity is one that is apt to be used by the legal profession and is probably the only truly strict definition of the term city. More loosely, the system of local government may be shared with adjacent communities, as is the case in large metropolitan areas where counties and a multiplicity of municipalities, as well as special-purpose districts, may share in the governance of the area. It should also be noted that a city may, in fact, have several governments. It may have a general-purpose government: the corporate entity concerned, an overlying county government, and one or more overlying special-purpose governments. The rise of regional planning organizations, public-private partnerships, and metropolitan authorities for transportation, housing, and sustainability reflects the growing need to manage urbanization across jurisdictional boundaries. Cities are now engaging in digital governance, implementing smart city technologies, data-sharing platforms, and cybersecurity protocols, all of which require updated legal frameworks and interagency coordination.

For example, in Wisconsin, as in a number of states whose early settlers had roots in New York and New England, urban areas may be governed as cities, villages, or towns. In addition, counties and certain special-purpose districts may have important roles affecting the regulation of development and the provision of urban services. In other states, a simpler form of government may exist, composed only of cities, counties, and perhaps special-purpose districts. It is essential for practicing civil and environmental engineers engaged in urban planning and development to understand the structure of local government in the states in which they practice. Such understanding is important to the manner in which planning must be practiced and to achieving plan adoption and implementation. The following descriptions of the forms of local government in Wisconsin are provided as examples. These forms will differ in other states.

Cities

In Wisconsin, cities are governed by a mayor and common council. The members of the council, known as aldermen, are elected by ward or aldermanic district—a subdivision of the city. Cities are divided by statute into classes. Cities of the first class have a resident population of 150,000 or more; cities of the second class have a resident population of 39,000 to 149,999; cities of the third class have a population of 10,000 to 38,999; and cities of the fourth class have fewer than 10,000 inhabitants. These classes have implications with respect to the organization, functions, and duties of the cities concerned. With respect to planning, cities of the first, second, and third classes have extraterritorial planning powers that extend to a distance of three miles beyond their corporate limits. These extraterritorial powers, however, apply only to unincorporated areas—that is, to areas within towns—lying within the prescribed distance, and do not apply within other incorporated areas—that is, other cities and villages. Cities of the fourth class have extraterritorial planning powers that extend for a distance of 1.5 miles from their corporate limits.

Cities in Wisconsin must have a clerk, treasurer, health commissioner or board, police and fire commission, police and fire chiefs, park board, board of public works, city engineer, and assessor. A plan commission is optional, as is the exercise of the planning function.

Wisconsin law also provides for the use by cities and villages of the city manager form of government instead of the mayor and common council and president and trustee forms. Under this form of government, the manager—who is an appointed and not an elected official—is the chief executive officer, appoints all committees and boards, can appoint and remove municipal employees except police and fire personnel, and has the power to veto legislative actions. The manager is appointed by the legislative body. This is a very effective form of local government, but it is considered by some to be undemocratic. Wisconsin law also provides for the use by cities and villages of the commission form of government. Under this form, the governing body consists of a three-member council, one of whom serves as the mayor. Although available, this form of local government has not been used to date within Wisconsin.

Villages

In Wisconsin, villages are governed by a president and village board, the board consisting of six trustees who are elected at large and the president, who is also elected at large. To qualify for incorporation, a village needs a minimum resident population of 150 people. Villages, like fourth-class cities, have extraterritorial planning powers extending for a distance of 1.5 miles from their corporate limits. Villages in Wisconsin must have a clerk, treasurer, assessor, and marshal. A plan commission is optional, as is the exercise of the planning function.

Towns

In Wisconsin, civil towns are a legacy from New York and New England, and were originally envisioned as a rural form of government. The boundaries of the civil towns were generally made co-terminus with the boundaries of the U.S. Public Land Survey System townships. Towns are governed by a chairman and a board of supervisors elected at-large. The board usually consists of three (optionally five) supervisors, one of whom serves as chairman.

Towns have no explicit planning authority under the Wisconsin Statutes. Towns must adopt village powers in order to create a plan commission and prepare a comprehensive plan. The adoption of village powers does not change the form and structure of town government. This form of government approximates a true democracy in that the annual budget, work program, and tax levy must be approved at an annual meeting of the electors, that is, of all of the citizens of the town. As urban development patterns have changed, a number of towns in Wisconsin have become, in effect, urban, with the resident population of some towns exceeding the population of many villages and even of some cities of the second class. Unlike cities or villages, towns do not have annexation powers. This makes their area—subject to annexation by bordering cities and villages—and their tax base unstable. Towns are also subject to the extraterritorial planning powers of cities and villages. The state statutes concerned clearly envisioned that cities and villages would expand by growing into and annexing adjacent town areas.

Counties

In Wisconsin, counties are administrative districts originally formed to carry out state functions at the local level. Counties are governed by a chairman and a board of supervisors elected by supervisory districts. Counties may also choose to have an elected county executive or an appointed county administrator. As urban development patterns have changed, some counties have become important providers of certain urban facilities and services, including mass transit, airports, and parks. In Wisconsin, counties exercise joint zoning powers with towns.

Special Purpose Districts

Special purpose districts may be superimposed upon the general-purpose units of government. Such districts have separate governing bodies and have the power to levy taxes. Such districts may include school, metropolitan sewerage, sanitary, farm drainage, soil conservation, and fire protection districts. The existence of these special-purpose districts may complicate the work of the urban planner and engineer.

The City as a Cyber-Physical System

The city may also be regarded as a large and complex cyber-physical system. While the city continues to function as an economic entity, a social organism, and a legal construct, it is also now understood to be an interconnected system of physical infrastructure integrated with digital technologies. This cyber-physical perspective acknowledges that the city is not only a costly physical plant but also a dynamic platform for sensing, communication, automation, and data-driven decision-making. That system comprises the (1) buildings and structures that house various land uses; (2) the utility networks that deliver power, lighting, heating and cooling, communications, sewerage, water supply, and drainage; and (3) the transportation networks (i.e., streets, highways, transit, and logistics systems) that facilitate the movement of people, goods, and information within the city and beyond. Increasingly, each of these elements is embedded with sensors, connected devices, and intelligent controls that enable real-time monitoring, predictive maintenance, automated operations, and responsive services. The performance of a city for its social, economic, and environmental functions will depend not only on the design and operation of its physical infrastructure but also on the resilience, interoperability, cybersecurity, and ethical governance of the digital systems that now underlie it. In this framework, urban planning and civil engineering must integrate physical design with information architecture, data analytics, and the deployment of human-centered technology.

SOME PRACTICAL DEFINITIONS

Although the foregoing concepts are interesting and are intended to help one think about and better understand the city, urban planners and practicing engineers are confronted with the problem of operationally defining the "city." In this respect, the U.S. Department of Commerce, Census Bureau, uses the following definitions, which are useful in any operational definition of the term city and in the practice of urban planning and engineering. These definitions relate to what may be termed local census geography.

Places

Places are defined simply as concentrations of population or as densely settled areas.

Census-Designated Places

Census-designated places, formerly known as unincorporated places, are areas delineated by the Bureau of the Census for the purpose of providing place-level data to users. The boundaries of Census-designated places are intended to encompass densely settled population centers that lack legal corporate limits. A Census-designated place must have a locally recognized name and contain a mix of residential and commercial areas. There are no strict population size or density thresholds for Census-designated places, though guidance may consider housing unit density alongside population where appropriate.

Incorporated Places

Incorporated places are defined as political units incorporated as cities, villages, and boroughs and having legal boundaries called corporate limits. Civil towns are not considered incorporated places, but may contain Census-designated places.

Urban Areas

Urban areas are defined as densely developed residential, commercial, and other non-residential areas. Beginning with the 2020 Census, the Bureau of the Census simplified its classifications and no longer distinguishes between "urbanized areas" and "urban clusters." Urban areas should meet at least one of the following criteria: (1) at least 2,500 people in population or (2) at least 2,000 housing units. All remaining areas are classified as rural.

Metropolitan and Micropolitan Areas

For the 1990 census, the Bureau of the Census delineated metropolitan statistical areas. Such areas were defined as areas normally consisting of an urbanized area and the county in which the urbanized area is located. In addition, such adjacent counties, which according to certain criteria were considered to be metropolitan in character, being economically and socially linked to the central city, were included in the delineated metropolitan area. The criteria concerned were related to the percentage of the workers residing in an outlying county who commute to the central county and the percentage of workers residing in a central county who commute to an outlying county. If a metropolitan statistical area had a population of more than one million, primary metropolitan statistical areas may have been defined within it. Such primary areas comprised a county or counties that evidenced very strong internal social and economic linkages separate from such linkages to other parts of the metropolitan complex. When primary metropolitan statistical areas were defined, the original metropolitan statistical area was redesignated as a consolidated metropolitan statistical area.

As delineated by the Bureau of the Census, in 1990 there were 21 consolidated metropolitan statistical areas in the United States, 73 primary metropolitan statistical areas, and 268 metropolitan statistical areas. By way of comparison, in 1960 there were 212 standard metropolitan statistical areas in the United States.

For the 2000 census, the federal Office of Management and Budget identified "core-based statistical areas," which are defined as metropolitan areas based on densely settled population concentrations to be known as "cores." The building blocks of the core-based statistical areas are counties that contain either an urbanized area of 50,000 or more persons—resulting in a "metropolitan core-based statistical area"—or smaller urban clusters of 10,000 to 50,000 persons—resulting in a "micropolitan core-based statistical area."

As of OMB Bulletin No. 23-01, issued in July 2023, there are 392 metropolitan core-based statistical areas and 546 micropolitan core-based statistical areas in the United States, totaling 938 core-based statistical areas. In addition, 175 combined statistical areas have been identified, which are groupings of adjacent core-based statistical areas that have substantial employment interchange, and reflect broader labor and economic integration. These delineations serve a vital role in shaping regional planning, infrastructure investment, and the allocation of federal resources for transportation, housing, health, and economic development.

Census Tracts

Census tracts are delineated by the Bureau of the Census based on various geographic features and are intended to consist of areas having 2,500 to 8,000 inhabitants, and averaging about 4,000 inhabitants. The tracts are intended to encompass relatively homogeneous areas with respect to population characteristics. The census tracts may be further broken down into census block groups and census blocks, which provide more granular data used in demographic, housing, and economic analyses.

Concepts Represented by the Census Definitions

The census of population and housing conducted decennially by the Census Bureau provides an invaluable source of information for urban planners and civil engineers. The various areas delineated and used by the Census Bureau in conducting the census and in the presentation of the census data provide a disciplined basis for quantitative comparisons between delineated areas in a given census year, and between comparable areas in different historical census years.

Conceptually, the urban area may be characterized as the true physical city, distinguished from both the legal city and the metropolitan area. The latter may be characterized as an identifying area influenced socially and economically by a central city. Generally, the urbanized areas represent the densely settled cores of the metropolitan areas. These two definitions have particularly important implications for planning and engineering. For example, federal law requires the preparation of transportation system plans—highway and transit—for all metropolitan areas as a prerequisite for the provision of federal aid for facility planning, design, and construction. The system plans must cover at a minimum the delineated urbanized areas within the metropolitan areas concerned. Some salient features of the previously described census geography are illustrated in Figure 2.1. This figure shows the four-county Bureau of Census–defined Milwaukee metropolitan area together with the boundaries of the four counties and the 90 municipalities—cities, villages, and towns—making up the area, illustrating the governmental complexity and challenge to planning that metropolitan areas present. The metropolitan area may be viewed as the socio-economic city since it contains within its boundaries the commuter sheds of the urban core, the core area newspaper primary circulation area, the extent of core area telecommunications systems, and the associated commercial and labor market areas. The Census Bureau–delineated urban areas define the outer extent of urban development within the metropolitan area that is contiguous to the central city of the metropolitan area. It represents the true physical city.

PLANNING

The term *planning* is defined and used in different ways by different professions. Planning is often defined as one of the five basic functions of management, those being planning, organizing, directing, coordinating, and controlling. Used in this way, planning may be defined as an analytical and creative process that involves, first, the

Milwaukee Metropolitan Area

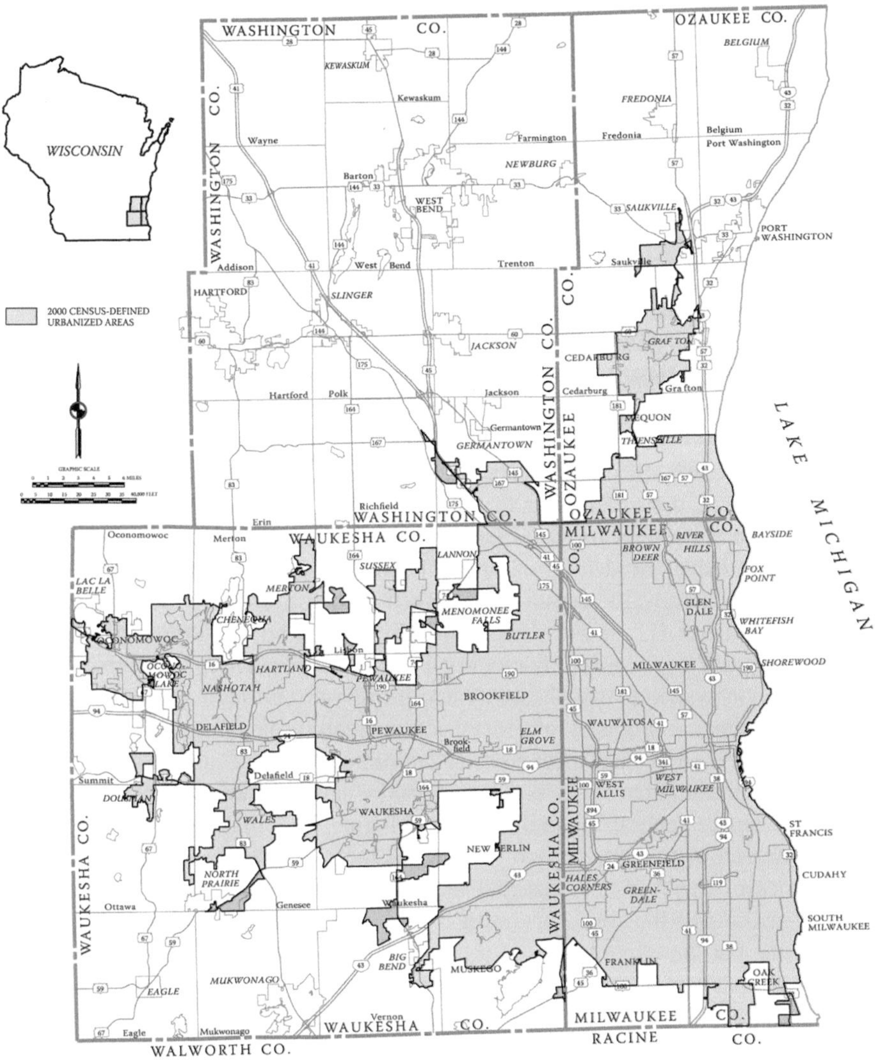

FIGURE 2.1 This figure shows the U.S. Census Bureau-defined Milwaukee Metropolitan Area, and the Census-delineated urban area within that metropolitan area. Within the context of this chapter, the metropolitan area may be thought of as defining the socioeconomic city, while the urban area may be thought of as defining the physical city. The figure also illustrates the complexity of the governmental structure of the metropolitan areas—there being four counties and 90 municipalities in the Milwaukee Metropolitan Area. In addition, there are a large number of special-purpose districts, such as school and sewerage districts, that are not shown on the figure.

Source: U.S. Census Bureau and SEWRPC.

establishment of objectives, and second, the establishment of a systematic means, or course of action, for the attainment of those objectives over time. Used in this way, planning is a rational process concerned with identifying needs or problems, gathering and analyzing relevant data concerning the problems, developing and testing alternative solutions to the identified problems, and selecting for adoption and implementation the solution that is judged best. It should be noted that, in uncertain situations, there may not be one best, or right, solution, but also that there are always poor, or wrong, solutions. These are incompatible with the attainment of the objectives, or grossly improbable in light of the assumptions and available data, and should be rejected.

Although planning used in this broadest sense is concerned with the future, the nature of this concern is also frequently misunderstood. The planner can neither predict nor control the future. The planner must, however, seek to understand the probable nature of the future, estimating the probabilities of existing trends continuing, and must attempt to identify possible occurrences that would have a significant effect upon attainment of the objectives. It should also be understood that planning does not deal with future decisions. Decisions exist only in the present. The issue that faces the planner is not what should be done tomorrow, but what must be done today to get ready for an uncertain tomorrow. Planning is necessary just because decisions can only be made in the present, and yet cannot be made for the present alone. The most expedient, the most opportunistic decisions, let alone the decision not to decide, may commit an organization in the long term to an undesirable course of action, often permanently and irrevocably.

Used in this general sense, planning is an activity of people who decide matters within an organization, and each unit in the organization may do some planning. Planning usually does not appear as a specialized function in an organization until the cooperative efforts of a substantial number of units are required to carry out the mission of the organization. Such appearance does not eliminate the need for planning by operating units, but simply enables the organization to more effectively carry out certain aspects of the planning function and to coordinate the planning efforts of the individual units.

Core Functions

Used in this broadest sense, planning encompasses three core tasks or functions: research, goal formulation, and plan design. The research function consists of the gathering, analyzing, and reporting of pertinent factual data about the existing situation and any needs or problems concerned. This function is intended to answer the question "Where are we now?" This function defines the existing situation, and may identify problems that need to be addressed. The goal formulation function is intended to identify objectives. This function is intended to answer the question "Where do we want to go?" This function identifies objectives. The plan design function is intended to identify and describe the actions that must be taken to attain the identified objectives. This function, which involves the consideration and evaluation of alternatives, is intended to answer the question "How can we best get to where we want to go?" Plan design is the pivotal function of any planning effort. It involves developing and comparatively evaluating alternative means for attaining the desired objectives. The alternatives to be considered should always include a "no action" alternative. The objectives need to be supported by measurable standards to facilitate evaluation of how well each alternative meets the desired objectives—in effect, an evaluation of the costs and benefits entailed.

In addition to these three core tasks, planning used in its broadest sense is often also defined to include coordination and the extension of assistance and advice. These two functions, however, may also be considered prerogatives of management and may be carried out independently of the planning function.

When considered one of the planning functions, coordination is achieved by reporting the findings of factual research, that is, by disseminating information, by maintaining a center of liaison with other concerned individuals and groups, and by participation in the formulation of objectives. Coordination requires negotiation, compromise, and adjustment in plans and, in some organizations, is the responsibility of the chief executive officer. The extension of assistance and advice is based on the recognition that planners possess special knowledge that in an organization may be extended upward to more central levels or downward to line operations.

Techniques aside, planning considered in this broadest sense is the same process whether conducted in the private or public sectors. That the afore-listed three functions are central to any definition of planning may be illustrated by the concept of the military staff study. Such a study is a formal method of developing a military plan. The steps in such a study may be defined as:

1. Determine objectives
2. Analyze existing situation
3. Consider alternative courses of action
4. Select the best course of action
5. Devise detailed procedures for carrying out the selected course of action

The concept of planning in this broadest sense may be thought of as "policy planning," that is, a process involving the continuing establishment of objectives for an organization as a whole, and the direction of its affairs to maximize the attainment of those objectives over time.

Classification of Planning

Planning can be more specifically classified in several ways. It may be classified according to the realm within which the planning decisions lie, as public or private sector planning; according to the resources or disciplines involved in the planning, as economic, social, military, or physical planning; according to the planning jurisdiction concerned, as city, county, state, or national planning; or according to the facility or service being planned, as school, hospital, highway, sewerage, water supply, park, or land use planning.

Often, several kinds of resources may be affected by a planning effort, so that planning for the development of any one of these resources may affect one or more others. When this occurs, then one of the resources concerned will be of primary concern, with the others secondary insofar as the planning process is concerned.

Thus, planning a cultural activity may be defined as social planning, while planning a building for the conduct of such activity may be defined as physical planning. In this case, physical planning is secondary to primary social planning. Planning the development of physical resources is almost always intended to serve broader social and economic objectives. Increasingly, planning efforts are becoming multidisciplinary and integrated, reflecting the interdependencies between different domains (e.g., housing, climate, and transportation).

Limitations of Planning

There is a trend within most organizations to rely less on intuitive reaction and more on planning as guides to decision-making. For some kinds of organizations, the propriety of this increased reliance on planning may have no logical limit. This is so for organizations whose objectives are reasonably simple and clearly understood by their members, such as private sector organizations whose primary objective is profit maximization. But society as a whole presents a quite different situation. Society as a whole consists of many diverse and often conflicting interests. This makes it challenging to formulate agreed-upon objectives and to adopt consistent means for meeting those objectives over time.

Free, capitalist societies have developed three parallel ways for making development decisions that shape communities:

1. Some decisions are left to the marketplace, where major changes can occur through the accretion of the results of myriad small transactions between the buyers and sellers of land, goods, services, and ideas.
2. Some decisions are made as a deliberate act of policy adoption through the political process.
3. Some decisions are left to professionals, usually licensed by the state, in whom society places confidence—for example, medical doctors, attorneys, and engineers.

Where the dividing lines between these three ways of making development decisions are drawn is a matter of vital importance. Planning plays quite different roles in the way it contributes to decision-making in these three ways. For the public sector planner, it is important that the line be properly drawn between technical decisions that should be left to professionals and political decisions that should be made by elected officials. It is even more important that the line be properly drawn between decisions that are to be left to the market and decisions that are to be made by public policy adoption. Additionally, planners today also have to navigate complex ethical, technological, and ecological considerations, making cross-sector engagement and inclusive participation more essential than ever.

City Planning

Planners operating within municipal government have appropriated the common word *planning*, as heretofore considered in its broadest sense, for their own particular use, giving it a more narrow and restricted meaning. This will, however, be the meaning used in the remainder of this text.

Sir Thomas Adams, a native of Scotland, was a British lawyer and city manager who advised the Canadian government on planning, who worked on the planning of London and on the preparation of the first regional plan for New York, and who later taught at Harvard. He gave perhaps one of the most succinct and better definitions of the term *city planning*.

> Town planning is a science, an art, and a movement of policy concerned with the shaping and guiding of the physical growth and arrangement of towns in harmony with their social and economic needs.

In this definition, Adams utilizes the English term *town planning*, which is synonymous with the term *city planning*—or *urban planning*—as used in this text. There are no superfluous words or phrases in this thoughtful and still sound definition. It should be noted that the definition recognizes that city planning is both a science and an art: that it is, in the first case, a branch of systematic knowledge demonstrating the relationship of causes to effects; and in the second case, a skill acquired by experience and marked by an ability to apply principles of practice. The definition recognizes that planning must be concerned with plan implementation—that is, with the movement of policy—as well as with plan design. And the definition recognizes that city planning deals with shaping the physical growth and development of cities to meet economic and social objectives. Civil engineers have contributed to city planning as a science, particularly in recent years, through the application of mathematical simulation modeling in transportation and water resource planning. More recently, engineers and planners have also adopted data-driven planning tools, such as digital twins, AI-assisted modeling, and integrated geospatial platforms, to enhance scenario testing, optimize resource allocation, and monitor urban performance in real-time.

Said another way, city planning is concerned with guiding the amount, rate, nature, and quality of urban change. As used in this text, it is concerned primarily with the arrangement of uses of land and with the configuration of the supporting transportation, utility, and community facilities. As used in this text, city planning is directly concerned only with physical development. It does not attempt to establish the basic policies to be pursued with respect to the social and economic development of the community. It does, however, provide the facilities for such development and thereby greatly influences the quality of the activities concerned. Thus, under the viewpoint adopted in this text, the core functions of the urban planner consist of the preparation of plans for the physical development of urban areas, including the land use, transportation, utility, and community facility elements of such plans, and the preparation and administration of plan implementation devices, including zoning, official mapping, land subdivision control, and capital improvement programming.

City planning is nevertheless aimed at fulfilling social and economic objectives that go beyond the physical form and arrangement of buildings, streets, parks, utilities, and other parts of the urban environment. City planning takes effect largely through the operations of government and requires the application of specialized techniques of survey, analysis, forecasting, and design. It has therefore also been described as a social movement, as a governmental function, and as a technical profession. Each such view of city planning has its own concepts, history, and theories. Recent shifts in governance models have emphasized participatory planning, equity-focused frameworks, and systems thinking, which makes new concepts and theories appear.

CITY PLANNING AS A TEAM EFFORT

Even when conceptually limited to dealing with the physical development of urban areas, city planning is so broad in scope that few individuals can master all of the disciplines and professional skills required for its successful execution. Urban planning, at least for metropolitan areas and for larger cities, is therefore best practiced through interdisciplinary staffs that can operate as teams to perform the functions and address the issues concerned.

Such interdisciplinary staffs, in the largest metropolitan and city planning organizations, may, in addition to urban planners, include municipal engineers, traffic and transportation engineers, environmental engineers, architects, landscape architects, geographers, economists, demographers, attorneys, and ecologists. The individual members of the interdisciplinary teams may be employed by the planning agency or planning department itself, or by line agencies or departments with the needed staff members concerned assigned to the planning agency or department as needed. Metropolitan planning agencies may have staff assigned to them by other governmental agencies such as state departments of transportation, state departments of natural resources, and federal highway and transit administrations. Even medium- and smaller-sized planning organizations can benefit by the assembly of interdisciplinary teams to conduct particular planning programs or address particular planning problems.

CRITERIA FOR GOOD CITY PLANNING

Good city planning should meet at least three criteria:

1. It should be comprehensive, considering all aspects of the physical development of the community and relating these to common unifying objectives. A comprehensive approach is essential to the making of intelligent decisions concerning relative needs and the effective allocation of resources to areas of greatest need. Contemporary planning practice also calls for integrating social, economic, and environmental dimensions into comprehensive planning, particularly in light of sustainability, climate resilience, and equity concerns.
2. It should be relatively long range, looking well beyond the obvious needs of the moment and attendant expedient solutions. Modern planning frameworks also emphasize adaptive planning and scenario analysis to address

uncertainty and allow flexibility in response to emerging conditions and technologies.

3. It should encompass a geographic area that permits a sound technical approach to the issues and problems concerned. Thus, for example, drainage and flood control planning must consider natural watersheds, while transportation planning must consider commuter sheds as rational planning areas. In addition, planning boundaries are now often informed by ecological regions, emissions zones, and digital infrastructure networks, reflecting the broader interconnectedness of urban systems.

Consideration of these criteria leads to the conclusion that some aspects of local urban planning can be properly conducted only within the framework of broader area-wide planning, such as regional or metropolitan planning. Moreover, regional or metropolitan planning can best be conducted within a framework provided by formally adopted state development objectives, and should serve to make those objectives operational at the multi-county regional or metropolitan level. Thus, an adopted state policy might be to maintain flood lands in essentially natural open uses. The hydrologic and hydraulic studies required to establish flood stages and to map the attendant flood hazard areas, and the conduct of the control surveys and large-scale topographic mapping required, can best be accomplished at the regional or metropolitan level based on area-wide land use plans. The regional or metropolitan plans should then be made operational through the preparation of local plans that are consistent with the regional or metropolitan plans. In the example used, the local land use and community facility plans may recommend the preservation of the mapped flood lands as park and open space. Local zoning, official mapping, and land subdivision control can then be used to carry out the local plans in a way consistent with the regional or metropolitan plans and the state development objectives.

NEED FOR CITY PLANNING

Communities have always had a need to be concerned with planning: that is, with preparing for change, whether this change is in the form of potential growth or in the form of decline. The need is particularly pressing in the current era due to rapid population growth, urbanization, and emerging environmental, technological, and socioeconomic challenges in the United States. Urbanized areas are becoming increasingly important in America. Consider the following:

1. From 1990 to 2000, the population of the United States increased from 249 million to 281 million, or by 32 million people. By 2020, the population reached approximately 331 million, and is projected by the U.S. Census Bureau to exceed 370 million by the 2040s. This rate of growth will require continued development of infrastructure and services to meet demand.
2. The population growth is largely urban. From 1990 to 2000, the rural farm population of the United States decreased while the urban population increased by 19 percent. By 2020, over 83 percent of the U.S. population

lived in urban areas. Only about one percent of the total population of the United States remains on farms, yet this small percentage provides much of the food and fiber required to sustain not only the United States but some of the rest of the world as well.

3. The population growth is largely metropolitan. From 1990 to 2000, 84 percent of the total national population increase occurred in metropolitan areas. This pattern has persisted through 2020, with growth continuing to concentrate in large metropolitan regions.
4. The population growth is occurring largely in the suburban and rural-urban fringe areas of the metropolitan areas. From 1990 to 2000, suburban and exurban growth outpaced that of central cities, contributing to decentralization within metropolitan regions.

For example, in many older metropolitan areas such as Milwaukee, the population within the central city has continued to decline, while suburban areas experience modest growth. In Milwaukee, the city population peaked in 1960 and steadily declined thereafter. This decentralization trend has reduced urban density and increased the complexity and cost of providing municipal services.

Population growth is only one measure of urban expansion. An increase in land devoted to urban uses is another. Although national land use statistics have evolved, recent trends confirm continued urban sprawl and land consumption exceeding population growth rates. Lower urban densities have important implications for the provision of such services as sanitary sewerage, public water supply, and bus transit, which generally require minimum densities for cost-effective service delivery.

The current pattern of urban growth within the United States is widely dispersed and of relatively low density. Industrial and commercial land uses, as well as residential land uses, are following this trend toward decentralization. If this decentralization—or recentralization along newer lines—is poorly planned from a physical development standpoint, major, costly developmental and environmental problems will be created. Large areas of urban development may be left without even the minimum level of municipal services that the older cities provided.

The attendant undesirable effects of poorly planned or unplanned urban growth include:

1. High capital and operating costs related to the provision of municipal services, with the customary tax revenues being inadequate
2. Unstable tax bases for both older central city and first-ring suburban areas and newer outer-ring suburban and rural-urban fringe areas
3. Pollution of streams, lakes, and watercourses
4. Drainage and flooding problems
5. Air pollution
6. Water supply problems
7. Power failures
8. Overcrowded and inadequate transportation facilities and overcrowded and inadequate community facilities, such as schools
9. Inadequate park and open space areas

10. Often irreparable misuse of the land
11. Exacerbation of the already serious problems of the central business district and surrounding core areas of the older central cities together with the appearance of blight in the older first-ring suburbs
12. Substandard areas of housing lacking essential public facilities and services, and inadequate housing for the needed labor force in proximity to employment concentrations

The planning problems of central cities and older suburbs are quite different from those of the newer suburbs and exurban areas. Older central cities of the United States grew out of the Industrial Revolution. Steam power and railway transportation required highly centralized urban development for efficient industrial location. The widespread availability of electric power and electronic communication and the highly developed state of motor vehicle transportation and accompanying highway networks make such centralization now unnecessary. The following factors have tended to promote decentralization of residential, commercial, and industrial land uses:

1. Widespread ownership and use of the automobile for commuting and other personal use, and widespread use of the motor truck for the transportation of goods and provision of services
2. Widespread availability of electric power
3. Introduction of electronic communication: telephone, radio, television, the personal computer, and the Internet
4. Onsite sewage disposal systems and private water supply wells
5. Public policies that make suburban and rural urban fringe areas attractive economically, such as state and federal grants-in-aid and low-interest loans for the development of highway, sewerage, and water supply systems
6. Affluence and increased leisure time for working people
7. The continuing strong preference of American consumers for free-standing, single-family homes on large lots with the attendant privacy and exclusionism, and of manufacturers for large, efficient, single-story plants in a campus-like setting

In addition, working remotely has shifted housing preferences post-COVID-19, as well as the increasing impact of climate resilience and energy efficiency concerns on location decisions.

The planning problems of the newer suburbs and rural-urban fringe areas relate primarily to shaping the new growth in accordance with agreed-upon objectives to bring about a more orderly, efficient, and environmentally sound development pattern, thereby creating a more desirable environment in which to live and work.

The planning problems of the central cities and older suburbs include congestion of population with attendant lack of light, air, and privacy; congestion of traffic and inadequate transportation facilities for both inter- and intra-city movements; congestion of industry with attendant inefficiencies, noise, glare, and air pollution; misuse of land, particularly incompatible mixed land uses; lack of adequate space; obsolescence of existing residential, commercial, and industrial facilities; lack of amenities, particularly beauty in surroundings; and social breakdown, particularly crime and poor elementary and secondary education facilities and services.

VALUE OF CITY PLANNING

Urban growth entails high capital and operation and maintenance costs. Good fiscal impact studies related to growth are difficult to conduct, involving as they must marginal—instead of average—costs attendant to new growth. Moreover, such costs are highly sensitive to the capacity of the existing infrastructure. The impacts are far-reaching, affecting every function of local government, including schools, police and fire protection, sewage treatment and conveyance, water treatment and transmission, stormwater management, solid waste collection, street maintenance and traffic management, and library facilities. Such costs are not borne by the newly developing communities alone, but must be shared by other communities in a developing region. In the twenty-first century, these challenges are further complicated by climate change, aging infrastructure, technological disruption, and equity concerns, which elevate the importance of effective planning.

Good planning can reduce the costs attendant to urban growth and change in at least seven ways:

1. Through advance reservation and acquisition of land for needed facilities. Good planning permits the acquisition of sites and rights-of-way in advance of needed construction and thereby avoids increased costs of land acquisition. Good planning can forestall the construction of buildings that must later be acquired and razed. Good planning may permit the acquisition by dedication—and, therefore, at no cost to the governmental unit or agency concerned—of park and school sites and of rights-of-way for arterial streets and urban drainage facilities as development or redevelopment takes place. This, however, requires plans definitively showing the proposed location of future parks, school sites, and drainage ways, as well as arterial streets and highways.
2. Through elimination of conflicting public uses. Good planning can, for example, keep new schools and other community facilities out of the right-of-way of needed future arterial streets and highways and avoid locating such facilities where their service areas will be cut by the future arterial streets and highways.
3. Through elimination of weak and uneconomic projects. Good planning may show certain proposed projects to be in improper locations because of otherwise unforeseen population or land use changes; or may show such projects to be unnecessary or to become obsolete before their costs can be amortized. Savings of this type often apply to fire stations and schools, whose locations should be determined on the basis of long-range land use plans rather than on existing land use patterns.

4. Through scheduling projects for construction in advance of apparent need. Projects undertaken in advance of actual need, but in time to meet such need, can effect savings by lowering construction costs, avoiding the need for costly emergency measures and expensive enlargements later. Long-term rising labor and material costs may make early construction economical. The advance construction of sanitary sewer and stormwater drainage facilities and water supply mains may effect a double economy if coordinated with street and highway construction. Trunk sewers and water transmission mains should be sized to serve ultimate service areas rather than immediate land development projects.
5. Through advantageous financing. Good planning may make it possible to effect savings by selling bonds at more favorable times and under the most favorable conditions. Such planning may also permit more flexibility in the choice of the methods of financing to be used—current revenue as opposed to debt financing.
6. Through providing factual data and rational forecasts that can be used by the private sector as well as by the public sector and by government agencies other than the one for which the data were originally developed. This is particularly true of regional or metropolitan planning, where much of the data collected in base mapping, aerial photography, and demographic, economic, land use, natural resource, and traffic studies will be useful in local and in private sector planning. Today, planners increasingly leverage GIS, real-time sensor data, and digital twins to support more informed, data-driven decision making.
7. Through encouraging cooperative action by two or more units or agencies of government. A significant potential exists for reducing costs of government through cooperative action by the various units and agencies of government, particularly in metropolitan areas. Good planning makes the joint or shared use of facilities—such as joint school and park sites—possible, avoiding needless duplication. Modern intergovernmental collaborations may also include shared climate resilience planning, regional transit investment, or coordinated affordable housing strategies.

Important as monetary savings may be, the most important value of planning lies not in such savings but in the objectives good planning makes possible to attain. Monetary savings aside, good planning is its own best justification because it enables communities to achieve agreed-upon objectives, as, for example, a new elementary school located within walking distance of all pupil residences in contrast to a school attended in half-day shifts and reached mainly by school bus, or trunk sewer and water transmission lines in place when a residential area is to be developed, thereby eliminating the need for interim onsite sewage disposal systems and private wells. Planning is thus important primarily for what it achieves: a good environment in which to live and work. Importantly, good planning can protect, preserve, and enhance the natural amenities of an area—the floodplains, woodlands and wetlands, wildlife habitat, groundwater recharge areas, mineral deposits, and prime agricultural lands—and can avoid the creation of serious and costly developmental and environmental problems such as flooding and poor drainage, surface and groundwater pollution, and air pollution. Planning can also advance climate mitigation and adaptation goals, social equity, walkability, and access to essential services—critical considerations in an era of increasing urban complexity. Indeed, for monetary savings to be meaningful, it must first be shown that the savings involved have been attended by the achievement of social value. Otherwise, the planning becomes pointless.

FURTHER READING

Cronon, William. *Nature's Metropolis—Chicago and the Great West.* W. W. Norton, 1991.
Pacione, Michael. *Urban Geography: A Global Perspective.* Routledge, 2001.

3 The Historical Context of Urban Planning in the United States

To understand a science it is necessary to know its history.

Auguste Comte
French philosopher
1798–1857

INTRODUCTION

Forms of city planning were practiced by ancient civilizations, including the Egyptian, Greek, Roman, and Chinese. Forms of city planning were also practiced in medieval Europe and during the Renaissance. City planning was widely practiced in colonial North America from the time of its settlement by Europeans to the American Revolution in 1776. To understand urban planning in colonial North America, it is essential to contextualize it within the broader European planning traditions that preceded it. At that time, a number of treatises were published by utopians, architects, and military engineers on the principles of city planning, on urban reconstruction and extension, and on the design of new towns. Importantly, a number of new towns were planned and actually constructed. These efforts laid the foundation for modern urban design principles that continue to influence planning practices today.

EUROPEAN NEW TOWNS

The European new towns were built primarily for military purposes. Plans for these towns were marked by rectangular, or sometimes radial, street patterns and by the provision of public squares or plazas. The towns usually had a fortified perimeter and, therefore, were of restricted size. Water supply and drainage were carefully considered in the planning. The towns usually were intended to militarily dominate an occupied or hostile territory or to defend the territory of a principality or nation-state.

Architects tended to like certain features of these plans, such as the straight streets and well-proportioned plazas for the monumental character given to the town. Military engineers, of course, deliberately designed the plans to facilitate the exercise and movement of troops and the use of artillery. Such new towns were constructed in France, Spain, Holland, and Ireland.

Many of the new towns reflected Renaissance ideals of geometric harmony and axial planning. A number were founded by royal decree, and construction often

DOI: 10.1201/9781003500384-3

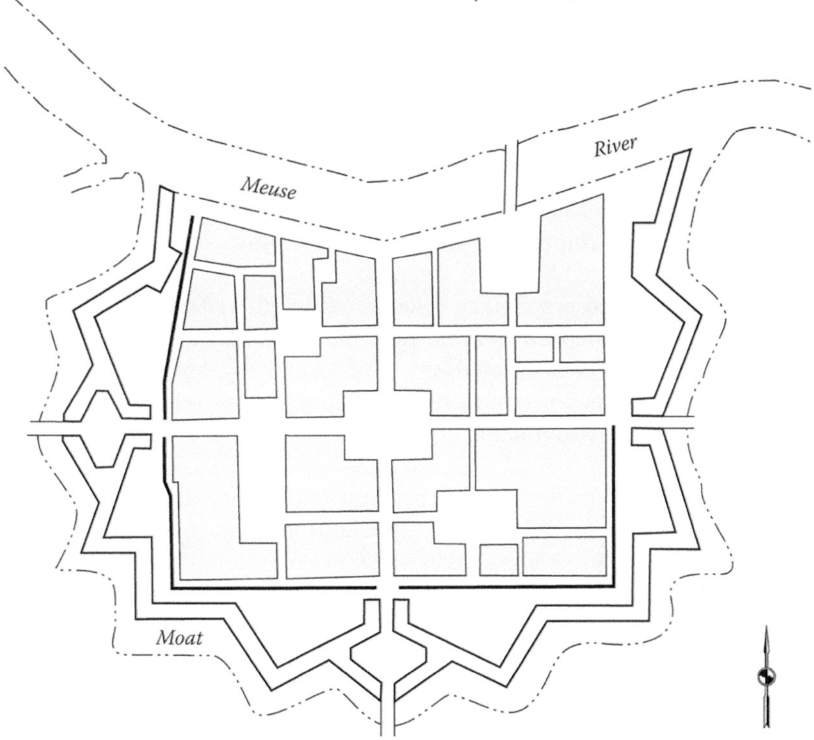

FIGURE 3.1 The historic plan for Charleville in the Ardennes region of France is an example of the type of planned new towns developed in Europe primarily for military purposes that influenced city planning in colonial America.

occurred over a short time span, resulting in unusually uniform building styles. Although originally designed for defense, their formal layouts and spatial order came to influence broader urban design principles, including those later adopted in colonial North America.

Figure 3.1 illustrates the plan for one of these European new towns, that for Charleville in the Ardennes region of France. The plan was prepared in 1656 and is marked by a fortified perimeter, a rectangular street pattern interrupted by public squares, and sites specified for the location of principal buildings.

Contributions of Colonial Planning

Colonial planning made a number of significant contributions to the principles and practices of city planning and urban design, some of which are still valid, namely:

1. Differentiation of major and minor streets.
2. Reservation of adequate and well-located open spaces.

3. Use of formal axes and harmonious architecture to achieve an aesthetically pleasing and pleasant environment.
4. The plans were genuinely regional, subjecting both rural and urban development to an integrated plan. The plans often extended well beyond the limits of the towns proper to include subsistence garden plots, grazing areas, and woodlots, as well as agricultural lands in the surrounding area with holdings for individual and communal uses.
5. The plans accommodated the technology of the time and emphasized the important foci of community life.
6. The plans managed well the forces by which cities were established. The corporate and proprietary colonies existed to make money from the distribution and improvement of land, but profit-making was secondary to considerations of the establishment of sovereignty, promotion of trade, development of raw materials, religious freedom, and other social objectives such as the rehabilitation of debtors.

These foundational principles—particularly attention to the street hierarchy, open space, and integrated land use—continue to inform contemporary planning frameworks.

Planning of the National Capital

In 1790, Congress authorized the President to select a site of 10 square miles on the Potomac River for a national capital. President Washington personally selected the site and appointed Andrew Ellicott to survey the exterior boundaries and topography of the site and to lay out the plat for what was to become the city of Washington. Major Pierre Charles L'Enfant was appointed to prepare the plan. L'Enfant was a French artist-architect who had served as a military engineer in Washington's Continental Army during the Revolutionary War. President Washington referred to L'Enfant as the planner and Ellicott as the surveyor of the work.

A Board of District Commissioners consisting of Daniel Carroll, Thomas Jefferson, Thomas Johnson, and David Steward was appointed to oversee the preparation of the plan. Jefferson personally proposed a rectangular grid plan for the site with a peculiar block and lot arrangement that had no side lot lines.

Major L'Enfant based his 1791 plan for Washington, D.C.—illustrated in Figure 3.2—on the topography of the site, identifying locations for major buildings, such as the Capitol, which he sited on a central hill, and the President's House (now the White House), which he sited on another prominent hill about three miles from the Capitol, and a number of plazas and malls that he sited on other high points. The plazas became focal points of the major street system, which was designed to connect the focal points with radial avenues having right-of-way widths ranging from 90 to 160 feet. These major streets were carefully oriented to provide impressive and commanding views of the city. A rectangular grid of minor streets was superimposed upon the focal point and radial street layout and provided land access to individual building sites.

Plan for Washington, D.C.: 1791

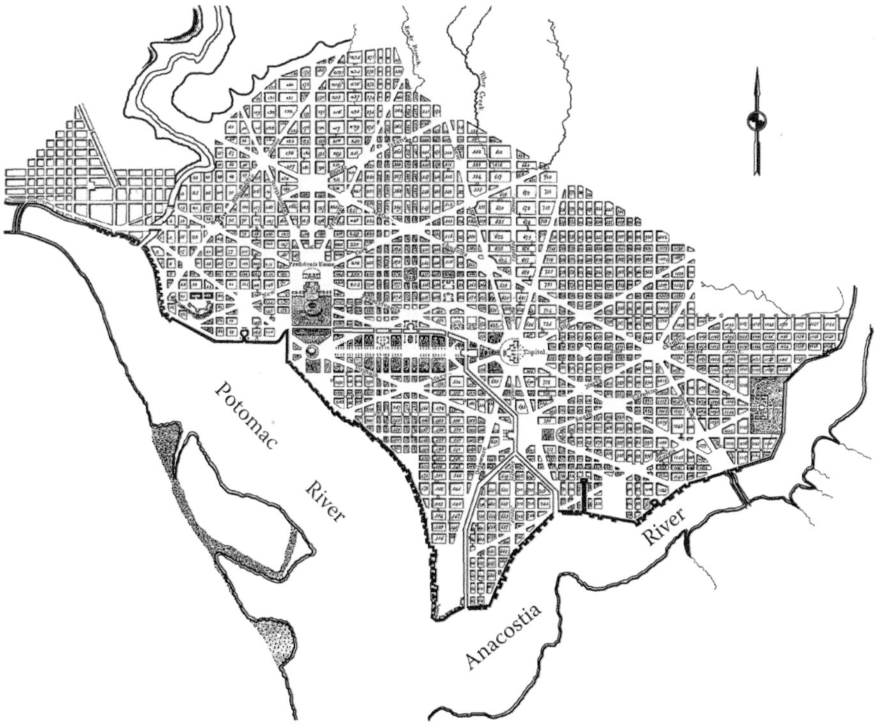

FIGURE 3.2 The plan for Washington, D.C., as prepared by Pierre L'Enfant represents a remarkable accomplishment in city planning, having admirably served as the basis for the development of the national capital for more than 200 years. L'Enfant's plan was slightly changed by the surveyor, Andrew Ellicott, and the plan shown is Ellicott's modification of L'Enfant's original plan.

The plan provided a number of squares for the location of monuments, academies, and churches. It also identified locations for key land uses such as a central marketplace and civic functions. The plan in its visionary magnitude, in its cleverly fitting of a symmetrical design to the topography of the site, and in the provision of a variety of spectacular open spaces represents one of the great city planning achievements of all time. The plan was essentially completed by L'Enfant in August 1791, after only six months of intensive work.

L'Enfant came into conflict with the District Commissioners over such matters as the manner in which lots were to be sold, construction of needed improvements, and publication of the plan map. When he refused to make a copy of his plan available for publication, he was finally dismissed by the Commissioners in February 1792. He died a pauper in 1825 and was initially buried at a friend's plantation. He was reburied in Arlington National Cemetery in 1909. His sarcophagus is located at the front entrance to the Lee mansion on the hilltop overlooking the city across the Potomac. The sarcophagus is engraved with his plan for the city and is a fitting tribute to this remarkable planner.

When L'Enfant refused to make his plan available for publication, Andrew Ellicott, a prominent colonial astronomer and surveyor, was commissioned to complete and publish the plan. As completed by Ellicott, the plan differs in some important details from L'Enfant's plan, but in most of its features faithfully follows it. The published plan sadly does not contain L'Enfant's name anywhere on its face. It should be noted that Ellicott was a highly regarded and respected professional whose work qualified him for membership in the Royal Society of Britain, a singular honor for a colonial. He was commissioned to survey several state boundaries, including that between New York and Pennsylvania, and the international boundaries between Spanish Florida and the United States and between Canada and the United States. He served as Geographer General of the United States and taught mathematics at West Point, the U.S. Military Academy.

Decline of Public Planning

After the Revolutionary War, new social and economic forces led to a decline of public planning. Four factors contributed to this decline:

1. The newly formed federal government did not provide national incentives or frameworks for city planning. The US Constitution does not mention cities, and powers not granted to the federal government were retained by the states. Cities became creatures of the states, and generally there was no state-enabling legislation for city planning or meaningful regulation of private land development.
2. An anti-urban bias existed among American intellectuals, exemplified by Thomas Jefferson's strong anti-urban bias.
3. The Industrial Revolution created the factory system. The steam-powered railway was developed, and an intense competition developed between cities for economic development.
4. Given the laissez-faire approach to development, speculation in land became widespread, particularly during the conduct and completion of the U.S. Public Land Survey across newly acquired territories.

The new social and economic forces brought unfortunate changes to the plans for Philadelphia, Washington, D.C., and New York. In Philadelphia, Penn's vision of a city dominated by single-family residences with ample open space was lost with the introduction of dense row housing, mid-block alleys, and lot subdivision to accommodate growing population and commercial needs.

New York, which was originally founded as the Dutch colonial city of New Amsterdam, featured a colonial layout including a fort and defensive wall—now remembered in the name Wall Street—with planned open space. In 1811, a plan for the further growth of New York was prepared by a commission created for this purpose. That plan imposed a rigid grid of streets on the island of Manhattan north of Canal Street. The island, which is about two miles wide and 13 miles long, was to be developed with 12 north-south streets, 100 feet in width, and 155 east-west streets between the rivers, 60 feet in width with every tenth street 100 feet in width. One diagonal street was provided: Broadway. The blocks ranged in width from 600 to 900 feet with a depth of 200 feet. Only 500 acres of open space were provided in the plan: a parade ground, three small parks, a public market area, and a water reservoir. The principal objective of the design was to facilitate economic growth. The plan illustrates the permanence of a street pattern once established. The legacy of the plan includes a lack of suitable sites for public buildings, traffic congestion due to frequent intersections and the lack of adequate north-south arterial capacity, and overcrowding of the land on small lots. Speculation in land was greatly enabled by the uniform grid layout and predictable lot sizes.

After 1830, new town plans, which were usually prepared in the private sector with little or no guidance from municipal government, were almost universally marked by a rigid rectangular street pattern. An example of this is the original plat of the city of Chicago. A civil engineer, James Thompson, in 1830 platted the original town site, utilizing a rectangular grid street pattern. The city was founded during the canal era of transportation system development, and the proceeds of the lot sales were to be used to build a Lake Michigan to Mississippi River canal. The plan is shown in Figure 3.3. Part of a military reservation east of the town site was later preserved as public land, becoming today's Grant Park along Lake Michigan. The rectangular grid of streets was subsequently extended on a large scale with a dreary and unstabilizing effect.

As the U.S. Public Land Surveys were completed, the hasty use of rectangular grid plans by speculators became widespread. Spurred on by changes in intercity transportation—from canals and river steamboats to railways—speculation in land became a virtual mania.

As railroads extended into the Midwest and West, railway companies became active town builders, reversing the earlier pattern of towns shaping rail routes. They chose sites based on operational needs—such as servicing locomotives or transshipping goods—and then platted and named the towns themselves. The Illinois Central Railway, for example, used a standard rectangular grid layout for 33 towns, specifying lot sizes, street widths, depot locations, and even commercial zones. This efficient, repetitive model prioritized railway traffic generation and land sales over urban variety or long-term livability.

Simultaneously, industrial corporations began founding new towns around factories, sometimes with distinct social objectives. In New England, textile towns like Lowell and Holyoke were planned to reflect local moral values and address labor shortages. Factory owners constructed supervised boarding houses for "mill girls"—young, unmarried women—alongside mills, churches, libraries, and homes, aiming to make factory work respectable and appealing to rural families.

Other corporate towns followed suit. Pullman, Illinois, developed by the Pullman Car Company, featured a cohesive architectural style, public squares, and retained corporate ownership. Similar efforts occurred in Gary, Indiana; Hershey, Pennsylvania; and Kohler, Wisconsin. Some religious groups, like the Mormons in Nauvoo and Salt Lake City, also employed standardized town plans for community development.

Plan of Chicago, Illinois: 1834

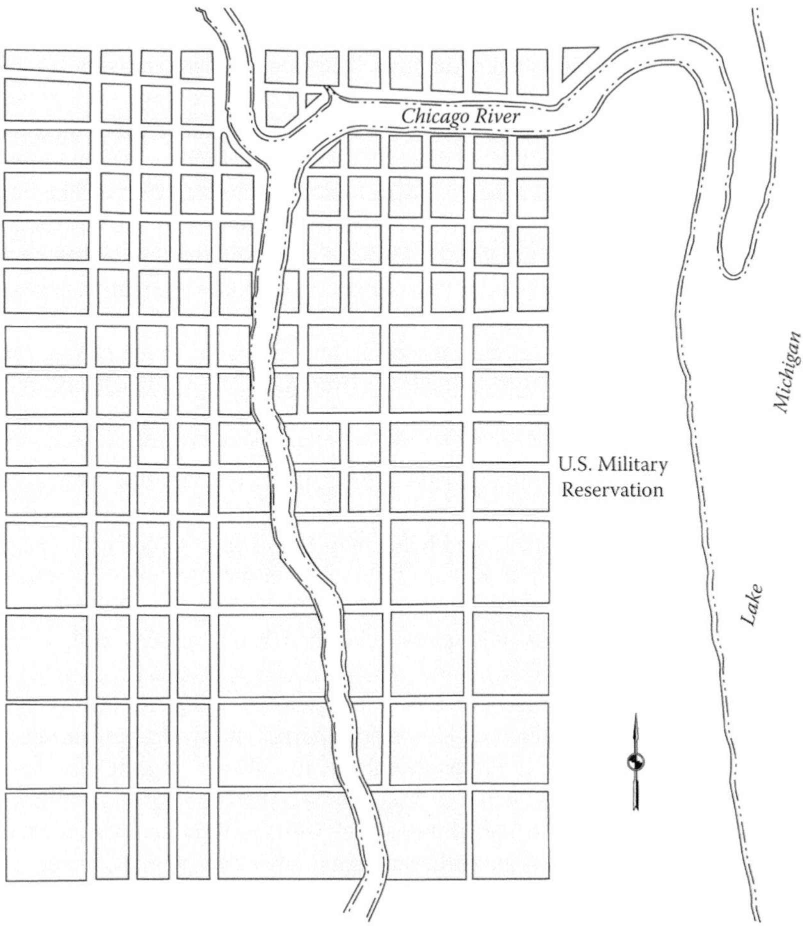

FIGURE 3.3 After 1830 most plans for new towns in the United States were marked by the use of rigidly rectangular street patterns, such as that for the original townsite of Chicago.

However, few of these industrial-era towns had lasting influence. As labor sources shifted—particularly after the Civil War toward European immigrants—the New England mill towns declined. Meanwhile, most urban growth during the Industrial Revolution occurred without planning, producing the squalid, overcrowded conditions famously documented in The Jungle by Upton Sinclair and Jacob Riis's photography.

Renaissance of Public Planning

The 1893 Columbian Exposition in Chicago marked a turning point in American urban consciousness, inspiring widespread discontent with industrial city conditions and sparking the City Beautiful movement. Led by Daniel Burnham and Frederick Law Olmsted, the fair's monumental design and technical innovations—such as electric power systems and the Ferris wheel—revived public interest in planning as both an art and a civic responsibility. Its legacy included the McMillan Plan for Washington, D.C. (1901) and Burnham's 1909 Plan of Chicago, both emphasizing grandeur, order, and civic pride.

The City Beautiful era (1890s–1910s) promoted monumental public buildings, parks, boulevards, and civic centers, embedding beauty and moral purpose into urban form. It also laid the groundwork for modern governance through zoning, subdivision control, and the establishment of professional planning institutions. Yet as engineers and administrators assumed greater roles, aesthetics gave way to efficiency, ushering in the City Efficient movement (1920s). This phase emphasized infrastructure, transportation, and utilities, supported by the 1927 Standard City Planning Enabling Act and early consulting practices that professionalized urban design.

The Great Depression revived federal leadership in planning through experimental "greenbelt towns" such as Greendale, Wisconsin (1936), designed by Elbert Peets under the Resettlement Administration to blend housing, open space, and self-sufficiency. Later, during and after World War II, federal and private initiatives expanded planned development—from war housing and garden cities like Radburn, New Jersey, to postwar new towns such as Reston, Virginia, and Columbia, Maryland. Together, these movements—the City Beautiful, City Efficient, and Greenbelt experiments—constituted a renaissance of public planning, transforming it from aesthetic vision to institutionalized civic enterprise rooted in both design and function.

CONCLUSION

By the late 1950s city planning was well established in the United States, and most progressive communities were engaged in planning efforts with the assistance of resident or consulting staffs. Initially, the resident staff were often provided by the city engineering departments, but as the planning profession became well established, such staffs were provided in separate planning departments. The planning principles and practices applied was largely those developed during the city efficient movement.

The science—as opposed to art—of planning was advanced through the techniques developed with federal funding, under the metropolitan transportation planning efforts required by the 1962 Federal Aid Highway Act. These planning efforts developed the quantitative relationships existing between land use and transportation and applied system engineering, including mathematical travel and traffic simulation modeling to the plan development process. These planning efforts also pioneered the application of digital computer technology to planning and helped foster the creation and application in urban planning of computerized geographic information systems, parcel-based land information systems, and public works management systems.

The art—as opposed to science—of planning saw the refinement of zoning and land subdivision design techniques and the development of the "new urbanism" and "smart growth" movements in reaction to the perceived disbenefits of urban sprawl. The new urbanism, however, was in reality a return to the old urbanism, emphasizing higher densities of development, mixed land uses, and modified grid street patterns considered to be more amenable to pedestrian movement.

FURTHER READING

Howard, Ebenezer. *Garden Cities of Tomorrow.* Swann Sonnenschein, 1898; republished by MIT Press, 1965; and by Routledge, 2003, *as Tomorrow: A Peaceful Path to Real Reform.*
Larson, Erik. *Devil in the White City.* Brown Publishers, 2003.
Reps, John W. *The Making of Urban America.* Princeton University Press, 1965.
Riis, Jacob. *How the Other Half Lives.* Dover, 1970 (reprint).

4 Compilation of Essential Data—A Brief Overview

> If we would first know where we are and whither we are tending, we could better judge what to do and how to do it.
>
> **Abraham Lincoln**
> *16th President of the United States*
> *1861–1865*

Reliable basic planning data are absolutely essential to the formulation of workable development plans. Consequently, inventory becomes the first operational step in the planning process. The crucial nature of factual information in the planning process should be evident since no intelligent forecasts can be made, or alternative courses of action evaluated, without knowledge of the current state of the system being planned. The inventories provide not only data describing the existing conditions but also a basis for identifying existing and potential problems in the planning area as well as opportunities for good development. The type and amount of data to be compiled will depend upon the size and character of the planning area, the type of planning concerned, and the degree of detail to which the plans concerned are to be carried. Usually factual data will be required on the existing land use development pattern, on the transportation, public utility, and natural resource bases, and on the ability of these bases to support development and redevelopment.

The collection, analysis, and dissemination of basic planning and engineering data are important not only to plan design but also to facilitate good development decisions in both the public and private sectors. In today's digital era, this includes integrating data from remote sensing, real-time sensor networks, mobile location data, and geographic information systems (GIS), which allow dynamic and interactive analysis. Cloud-based data platforms and open-data portals are also enabling better transparency and public engagement in planning processes. The provision of such data can, in and of itself, significantly contribute to shaping development and redevelopment decisions in the public interest.

Important as this "research" function may be, there is a real danger in any planning operation of devoting too much time and too many resources to the compilation of basic data. Data collation and collection operations should be carefully designed to insure that the resulting data are complete and accurate for the intended use. A good database facilitates the preparation of sound and effective plans, provides the basis for sound arguments in support of planning recommendations, familiarizes planning staff with the planning area, its problems, and its probable future requirements, and, should plans be challenged, supports the attainment of favorable administrative and court decisions. The planning data required include pertinent demographic,

DOI: 10.1201/9781003500384-4

economic, natural resource base, land use, housing, traffic and transportation, utility, and community facility data. However, the first and most basic data requirement is for good maps of the planning area. Accordingly, the next chapter deals with mapping for city planning. Subsequent chapters deal with the demographic, economic, land use, and natural base inventories needed for city planning and engineering. The inventories considered in these chapters are those that provide data essential to the preparation of all of the elements of a comprehensive city plan, and particularly for the preparation of the land use element of such a plan. Modern inventories often include spatial and temporal data collected via satellite imagery, drone-based mapping, LiDAR, and community-sourced platforms, which contribute to a more robust and adaptive planning process.

5 Map Requirements for City Planning

In the conduct of a city's business, two factors require constant consideration. These are the land itself, with its configuration and other physical characteristics, and the lines that form the boundaries of real property. Generally speaking, the first of these factors is practically unchangeable. Modifications in the topography of a region are only superficial; the surface is scratched, but the structure is unaffected. The second factor—the boundaries of the original land subdivisions—forms the basis of all subsequent property divisions, and is a matter of permanent importance and influence. The need for full information concerning these two factors is apparent. If they are ignored during the planning stage of a project, obstacles will be encountered during its construction.

American Society of Civil Engineers
Manual of Engineering Practice No. 10, 1957

INTRODUCTION

The most efficient and effective way of not only presenting but also of integrating information about a number of factors that must be considered in city planning and engineering is through maps. Maps serve at least two important purposes in city planning:

1. To provide a graphic representation of the planning area
2. To relate pertinent planning data to geographic location

The maps must, however, be designed and prepared specifically for planning and engineering applications. The design and proper preparation of such maps require proficiency in surveying and mapping.

Historically, surveying and mapping constituted a highly specialized branch of civil engineering, just as city and regional planning did at one time. And, historically, university civil engineering curricula included extensive coursework in surveying and mapping, and textbooks in surveying and mapping were often authored by practicing engineers. The American Society of Civil Engineers still maintains a Surveying and Mapping Division and annually awards a prize for notable achievements in the field. Surveying and mapping—now sometimes known as geomatics—have, however, increasingly become a discipline separate from civil engineering, and some branches of geomatics, such as land surveying, are now practiced by a separate licensed profession. Some universities no longer include any courses in surveying and mapping in the civil engineering curriculum. Nevertheless, the practice of civil

DOI: 10.1201/9781003500384-5

engineering will inevitably involve the use of maps and, in many cases, the conduct of surveying operations.

Surveying and mapping technology has undergone a period of rapid, indeed revolutionary, change. This change has been marked by the introduction of new instrumentation for the conduct of field surveys, such as electronic theodolites and distance-measuring devices (EDM), the development of software and the application of computers for the reduction and storage of survey data, and the development of new photogrammetric plotters and digitizers to permit the development and use of maps in digital form. Perhaps the most revolutionary has been the development of Global Positioning System (GPS) technology. Today, GPS is typically integrated with real-time kinematic (RTK) systems and drone-based aerial surveys, enabling centimeter-level accuracy and high-resolution topographic mapping.

Another major development is the widespread adoption of Geographic Information Systems (GIS), which allow planners to visualize, analyze, and model complex datasets spatially. GIS has become a standard tool for integrating land use, environmental, transportation, and demographic data into a single platform. Cloud-based GIS and open data sharing initiatives now allow multiple stakeholders to collaborate on spatial planning in real-time, greatly increasing transparency and efficiency.

Emerging technologies, such as LiDAR (Light Detection and Ranging), 3D city modeling, and augmented reality (AR) applications, are also enhancing the documentation and visualization of urban environments. These tools allow planners to create highly detailed simulations and predictive models for future development and environmental impact assessments.

In spite of these technological advances, many municipalities in the United States remain inadequately mapped for municipal planning and engineering purposes. This fact, combined with the nature of the technological changes taking place in surveying and mapping, warrants a restatement of some of the basic principles that should underlie the preparation of maps for municipal planning and engineering applications. If these basic principles are ignored, the application of the new technologies to urban surveying and mapping practice may prove to be an unmitigated disaster. Indeed, for many communities the application of the newer technologies may not be as cost-effective as the application of more conventional techniques based on sound principles of practice. The old warning concerning the development of costly computerized data banks expressed as "garbage in—garbage out" certainly applies to maps and survey data. Therefore, this chapter reiterates certain basic principles of practice that should be followed in the preparation of maps for municipal planning and engineering purposes. These principles of practice apply equally to maps that are prepared and used in a more conventional manner through hard copy and to maps prepared and used in the newer computer-manipulatable digital forms.

BASIC DEFINITIONS AND CONCEPTS

A map may be defined as a flat, true-scale, graphical representation of a portion of the earth's surface. In this respect, it should be recognized that the spherical surface of the earth cannot be presented on a flat surface without distortion. The map scale is defined as the relationship between a distance on a map and the corresponding distance on the surface of the earth. A map scale may be expressed as an equivalence, as a ratio, or as a graph. The creation of a map requires three foundational elements: (1) a system for accurately locating features on the surface of the earth, (2) a projection to reduce the spherical surface of the ellipsoid to a plane surface, and (3) a system of survey control to manifest the projection used on the surface of the earth.

Spherical Coordinate System

The system used for accurately locating features on the surface of the earth is illustrated in Figure 5.1. It consists of spherical coordinates expressed as latitude and longitude.

The surface of the physical earth is not shown in Figure 5.1, except conceptually by the location of the point P, shown above the ellipsoid. Figure 5.2 illustrates the relationship between the topographic surface of the earth, the geoid, and the ellipsoid. The relationships shown are particularly important in understanding the use of GPS instrumentation in determining elevations. The physical surface of the earth is the surface on which measurements are made. Due to topographic configuration and the effects of gravity and centrifugal force, the physical earth is an irregular, nonmathematical surface, oblate, that is, flattened at the poles and elongated through the equatorial plane.

The geoid is defined as an equipotential surface, a surface on which the gravity and centrifugal forces are balanced. It is a surface everywhere perpendicular to the direction of gravity, a level surface conceptually equivalent to the surface that would

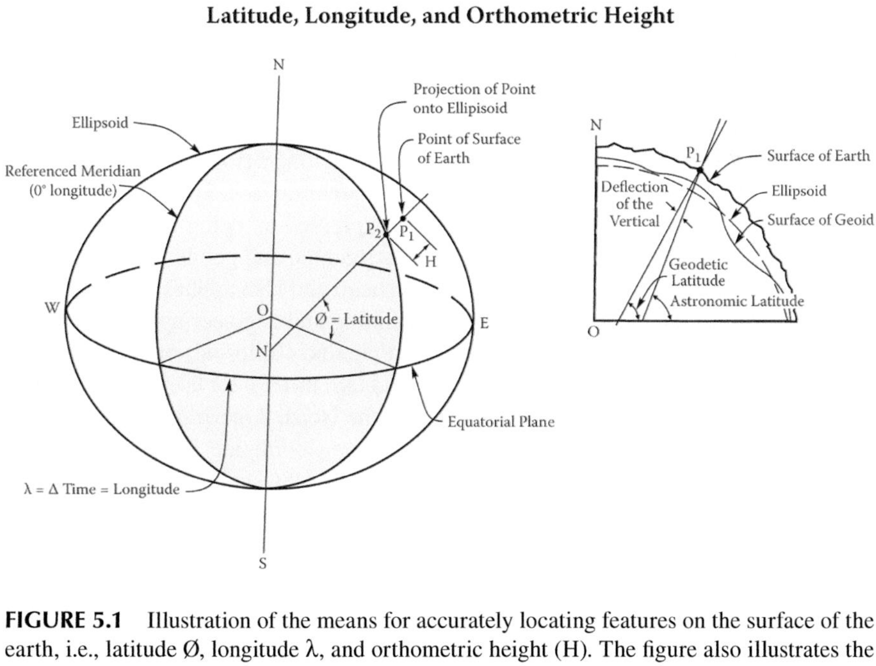

FIGURE 5.1 Illustration of the means for accurately locating features on the surface of the earth, i.e., latitude Ø, longitude λ, and orthometric height (H). The figure also illustrates the difference between astronomic and geodetic latitude by means of the meridian section shown.

Orthometric and Ellipsoid Heights

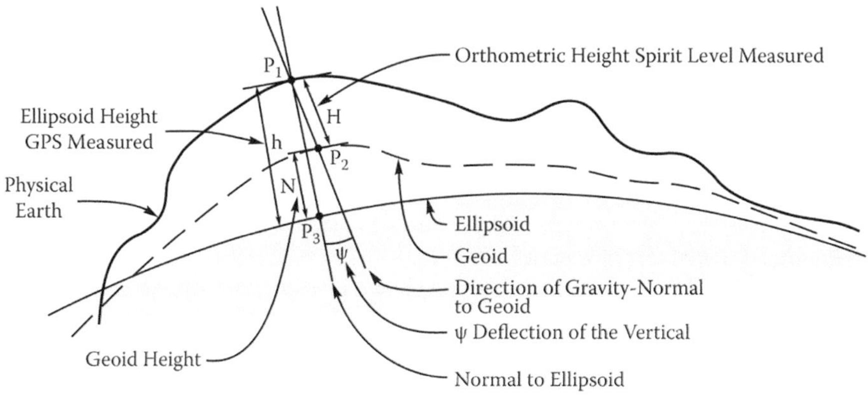

FIGURE 5.2 Illustration of the relationships between the topographic surface of the earth, the geoid, and a referenced ellipsoid. These relationships are important to understanding the use of GPS instrumentation in determining elevations. (From Nemerow, Nelson L., Agardy, Franklin J., Sullivan, Patrick, and Salvato, Joseph A. *Environmental Health and Safety for Municipal Infrastructure, Land Use and Planning, and Industry*, 6th ed., John Wiley & Sons, 2009.)

Source: Reprinted with permission from John Wiley & Sons.

be assumed by mean sea level if the seas extended under the continents. The geoid is the figure to which all physical measurements are referred. Surveying instruments are oriented to the geoid through the use of optical or physical plumb lines or spirit level vials. The geoid is, however, an irregular, undulating, non-mathematical surface.

Because it is impossible to make survey computations on an irregular surface such as the geoid, a mathematical surface—an ellipsoid—is chosen that closely approximates the geoid. This surface is created by rotating a two-dimensional ellipse about its semi-minor axis to create a three-dimensional surface. Spatial locations and relationships—distances and directions—are expressed as being on the ellipsoid. Figure 5.2 illustrates the difference between astronomic and geodetic latitude. This difference is known as the deflection of the vertical, and also applies to longitude.

Two ellipsoids are of importance to city planning and engineering practice in the United States. The first of these is the Clarke Spheroid (Ellipsoid) of 1866, adopted by the U.S. Coast and Geodetic Survey (USC and GS) in 1880 for use in the continental United States. This ellipsoid is the basis for the North American Datum of 1927 (NAD-27) and the basis for the original State Plane Coordinate Systems within the various states of the United States as developed by the USC and GS. The coordinates on the original State Plane Coordinate Systems were expressed in U.S. Survey feet.

The second of the two ellipsoids of importance is the World Reference Spheroid (Ellipsoid) (WRS-80) of 1980, also known as the Geodetic Reference System of 1980 (GRS-80), a global reference system. This ellipsoid is the basis for the North American Datum of 1983 (NAD-83) and the basis for a revised State Plane Coordinate System. The grid coordinates for this revised system are given in meters.

In recent years, the National Geodetic Survey (NGS) has begun transitioning to the new geometric reference frame known as the North American Terrestrial Reference Frame of 2022 (NATRF2022), which replaces NAD 83 and aligns more closely with the global ITRF system. This change improves GPS accuracy and consistency with global positioning systems.

The term *datum* may be defined as any numeric or geometric quantity that serves as a reference or base for other quantities. A geodetic horizontal datum consists of four defining quantities: its ellipsoid, the latitude and the longitude of an initial point, and the azimuth of a line from this point. A geodetic datum forms the basis for the conduct of horizontal control surveys in which the curvature of the earth is considered to determine the accurate latitude and longitude of points on the surface of the earth.

MAP PROJECTIONS

The second foundational element required for the creation of a map is a map projection. A map projection may be defined as a set of mathematical equations for converting the curved surface of the ellipsoid to a flat surface upon which computations can be made and maps constructed. Map projections are thus used to convert the spherical geometry of the mapping ellipsoid to the plane geometry of the flat mapping surface.

Three projection systems are of particular interest for city planning and engineering: the tangent plane projection, the Lambert Conformal Conic projection, and the Transverse Mercator projection.

The tangent plane projection is not a projection in the sense of the other two projections of interest in that in its most common form, this projection is related to the geoid and not the ellipsoid, as shown in Figure 5.3. It is applicable only to surveys, primarily land surveys, of small areas and is being replaced by more sophisticated

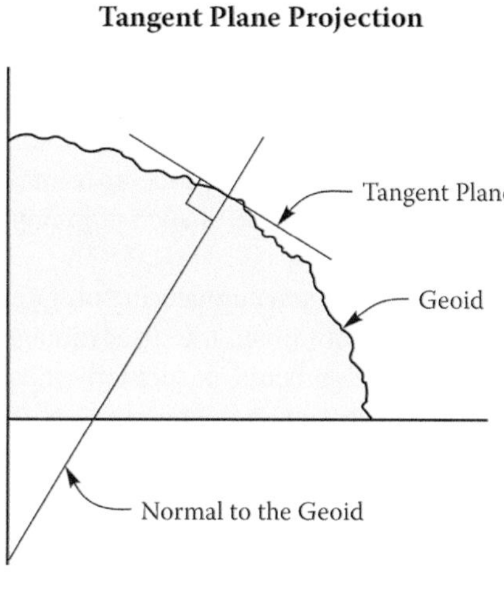

FIGURE 5.3 Illustration of the concepts underlying the use of a tangent plane projection such as was, in effect, historically done in land surveying.

projections. Most land surveyors are probably not aware that they are using this type of projection in their work. In application, the surveyor selects a point in or near the area to be surveyed and mapped where the survey can be oriented to some form of directional control that is recoverable. Conceptually, a tangent to the geoid is assumed to exist at this point. The directional control may be by magnetic observation, by celestial observation, or by the direction of a monumented line such as a U.S. Public Land Survey System one-quarter section line. The surveyor then measures the angles and distances formed by the lines of the survey, the angles, usually in the form of bearings, being referred to the central directional control. This procedure is different from independently measuring a magnetic or astronomic direction—or bearing—for each line, as is the case in projectionless surveying mapping. Bearings shown on tangent projection maps do not represent astronomic or geodetic bearings of the survey lines. The curvature of the earth and convergence of the meridians are ignored. The measured distances are measured as, or reduced to, horizontal distances, and are assumed to be measured at the mean elevation of the area surveyed. The map derived from the measured angles and distances is, in effect, a projection of the curved surface of the earth onto a flat plane.

The principal advantage of this system is its simplicity. Straight lines are considered to have a constant bearing; parallel straight lines are considered to have the same bearing; level surfaces are considered to be flat planes; and plumb lines are considered to be parallel. The errors introduced by these assumptions become noticeable when the areas concerned exceed about 75 square miles and approximate 0.05 foot and 0.1 second of arc. Individually conducted survey maps cannot be coordinated, and maps based on such surveys cannot be combined. Other surveys made in the same way will disagree in dimensions and directions of common lines, and directions between identical points on adjacent parcels will have different values. This means that discrepancies, such as gaps and overlaps, may not be apparent from mere review of survey maps of adjacent parcels. Resurveys are entirely dependent upon recovery of survey markers or monuments set during the original work.

Tangent plane projection surveys are of limited use to comprehensive urban planners and to engineers concerned with areawide projects. The lack of a common reference makes the task of relating several parcels to one another difficult or impossible. Indeed, existing municipal maps compiled from plats of survey are often no more than compilations of paper records so poorly done as to make use in planning and engineering difficult and costly, and the use of such plan implementation devices as official mapping questionable.

The conduct of areawide planning and engineering programs has always required the use of projection systems that eliminate the disadvantages of the tangent plane system. Recent technological developments, particularly the use of GPS instrumentation and techniques, require an understanding and use of such projections and the datums on which they are based. Two such projections are of importance to the practice of city planning and engineering in the United States: the Lambert Conformal Conic projection and the Transverse Mercator projection.

The Lambert Conformal Conic projection, as shown in Figure 5.4, conceptually uses a cone passed through two parallels of latitude of the ellipsoid—known as standard parallels—to develop the spherical surface into a plane surface. Meridians of

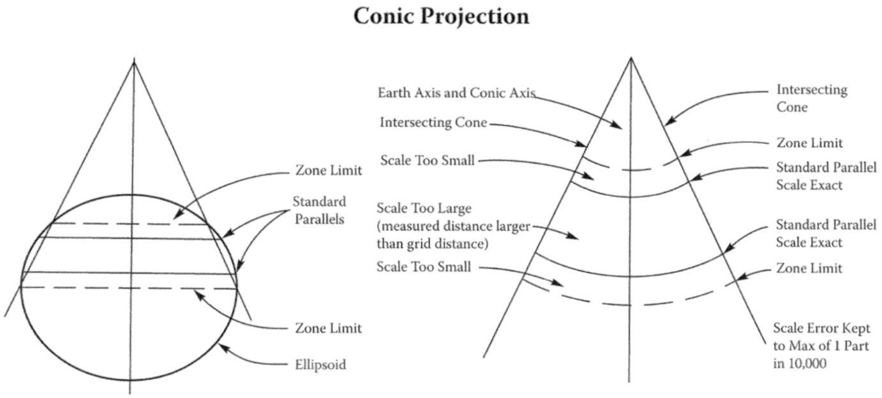

FIGURE 5.4 Illustration the concepts underlying the creation of a conic projection such as the Lambert conformal conic projection used for some State Plane Coordinate Systems.

longitude are represented on the projection by converging straight lines, and parallels of latitude are represented as arcs of circles with a common center. Angles are correctly represented, but the scale is exact only along the standard parallels, continuously changing along the meridians, being too large between the standard parallels and too small beyond them.

This projection is generally used as the basis for the State Plane Coordinate System in states having their greatest dimension in an east-west direction. The State Plane Coordinate System converts the spherical coordinates of latitude and longitude of the projection, expressed in spherical coordinates, to rectangular grid coordinates expressed in feet: the x-axis values are known as "eastings," and the y-axis values are known as "northings." This permits the conduct of surveys using plane surveying methods while accounting for the curvature of the earth and maintaining geodetic survey accuracies within specified limits, generally a maximum error of one part in 10,000. When used for the original State Plane Coordinate System, the projection was based on the Clarke Spheroid of 1866 and the North American Datum of 1927 (NAD-27). When used for the revised State Plane Coordinate System, the projection is based on the Geodetic Reference System of 1980 (GRS-80) and the North American Datum of 1983 (NAD-83).

The Mercator projection is a conformed projection that conceptually uses a cylinder tangent to the ellipsoid at the equator to develop the spherical surface into a plane surface. The equator is represented as a straight line true to scale, meridians of longitude are represented as straight lines perpendicular to the equator, and parallels of latitude are represented by straight lines parallel to the equator. The Transverse Mercator projection may be thought of as a Mercator projection rotated 90 degrees. Neither meridians of longitude—except for the central meridian—nor parallels of latitude appear as straight lines. This projection is used as a basis for the State Plane Coordinate systems in states having their greatest dimension in a north-south direction. Like the Lambert Conformal Conic projection, when used for the original State Plane Coordinate System, the Transverse Mercator projection was based on the

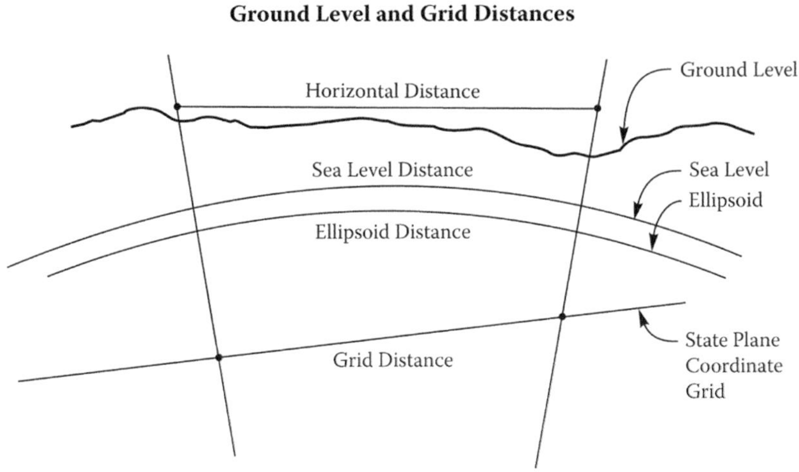

FIGURE 5.5 Illustration of the relationship of ground-level distances to State Plane Grid Distances. For the original State Plane Coordinate System the reduction of the horizontal ground-level distance is to sea-level distance and then grid distance. For the revised State Plane Coordinate System, the reduction is to ellipsoid distance and then grid distance.

Clarke Spheroid of 1866 and the North American Datum of 1927 (NAD-27). When used for the revised State Plane Coordinate System, the projection is based on the Geodetic Reference System of 1980 (GRS-80) and the North American Datum of 1983 (NAD-83).

Use of the original State Plane Coordinate System requires knowledge of the sea level or ellipsoid and scale reduction factors. These factors are illustrated in Figure 5.5. For applications where geodetic rather than grid bearings are needed, knowledge of the theta angle, the angle between the central meridian and the meridian through a given survey point or area, is also required. The sea level elevation factor is applied in conjunction with the orthometric heights of the points concerned. Use of the revised State Plane Coordinate System requires knowledge of the elevation and scale reduction factors and of the theta angle. The elevation factor is applied in conjunction with the ellipsoid heights of the points concerned. Because the two projections concerned are based on different ellipsoids, the scale factors are different.

With respect to projections, the term conformal has a special meaning. Since it is impossible to develop a spherical surface onto a plane, all maps will contain distortions according to the projection used in their compilation. The distortions may relate to scale, to area, to angles, or to the shape of figures. The unique properties of a conformal projection include the aspects that all figures on the surface of the earth retain their shape on the map, angles measured on the map approximate their true values on the surface of the earth, and the map scale at any point is uniform in all directions.

These properties are important to all who use grid coordinates in their work, including surveyors, engineers, and city planners.

Survey Control

The third foundational element required for the creation of a map is a system of survey control. A system of survey control is necessary to manifest the map projection used on the surface of the earth and to make it possible to relate the measurements involved in surveying and mapping to the map projection. A survey control network consists of a framework of monumented points whose locations on the surface of the earth are accurately known, either relatively or absolutely. The monumented stations are used to locate, orient, and adjust local surveys, and provide a means of verification and check for such surveys, and as a framework for the preparation of maps. Survey control networks may be either horizontal, in which case the locations of the monumented stations are given either in latitude and longitude or in State Plane coordinates, or vertical, in which case the elevations—orthometric heights—of the monumented stations, called benchmarks, are given in feet above the geoid or mean sea level configuration of the earth.

There are, in effect, two survey control networks in place within the United States: (1) the geodetic control survey network created by the federal government through the National Geodetic Survey—formerly the U.S. Coast and Geodetic Survey (USC and GS) and its predecessor agencies and (2) the U.S. Public Land Survey System (USPLS), also created by the federal government through the Bureau of Land Management and its predecessor agencies. The geodetic control survey system provides the basis for all accurate topographic mapping and for the preparation of all nautical and aeronautical charts. The Public Land Survey System provides the basis for all real property boundary surveys and mapping in much, but not all, of the United States. With the modernization of the National Spatial Reference System (NSRS), the NGS is replacing traditional vertical datums such as NAVD 88 with new geopotential datums based on gravimetric geoid models. This will improve elevation data accuracy and consistency across regions.

NATIONAL GEODETIC SURVEY CONTROL SYSTEM

The national geodetic survey control system actually consists of two networks: a horizontal survey control network and a complementary vertical survey control network. The system is scientific in nature and is designed to provide the basic control for all federal topographic and hydrographic mapping operations.

The horizontal survey control network consists of thousands of monumented stations, or points, whose positions on the surface of the earth, expressed in terms of latitude and longitude, were established to known orders of accuracy by the U.S. Coast and Geodetic Survey (USC and GS), now the NGS. The positions of the stations were established by high-order triangulation and traverse surveys. The stations are relatively widely separated, generally located on high points, and generally relatively inaccessible. The use of the stations requires geodetic survey techniques and equipment not historically available to public works engineers and land surveyors.

In order to make the national horizontal survey control network more readily available to and useable by local surveyors, the USC and GS in 1933 devised the State Plane Coordinate System. As already noted, the State Plane Coordinate System, as originally developed, was based on the Clark spheroid of 1866 and the attendant North American Datum of 1927 (NAD-27). The State Plane Coordinate grid values are expressed in U.S. Survey feet. Also, as already noted, the system translates the spherical coordinates—latitude and longitude—of the primary federal survey control stations into rectangular coordinates on a plane surface mathematically related to the ellipsoid on which the spherical coordinates have been determined. The original State Plane Coordinate System was designed so that the effect of the distortion inherent in the projection of the curved surface of the ellipsoid concerned onto the plane used for the rectangular grid is always less than one part in 10,000. The State Plane Coordinate System, thus, permits local engineers and surveyors to connect surveys by simple, well-established plane surveying techniques to the extensive network of precisely located triangulation and traverse stations of the national geodetic control survey network.

If the location of a point is known by stated grid coordinates on the State Plane Coordinate System grid, then the location of that point is also known by its corresponding latitude and longitude. So, too, with respect to the lengths and bearings of lines. Conversely, if the location of a point is known by stated spherical coordinates of latitude and longitude, then the location of that point is known in grid coordinates of the State Plane Coordinate System. Thus, the precise location on the surface of the earth of all survey stations and landmarks established in local engineering and land surveys can be accurately described by stating their coordinates, referring to the common origin of the State Plane Coordinate System grid. Once plane coordinates are established for any survey station or landmark, these coordinates may become the best available evidence for the original positions concerned, should the physical monuments marking the positions be lost. Computations using the State Plane Coordinate system are simple, being made with the well-established formulas of plane surveying.

The NGS readjusted the national horizontal survey control network in 1983, creating the North American Datum of 1983 (NAD-83), since further refined as NAD-83(91). As already noted, NAD-83 is based on the Geodetic Reference Spheroid of 1980 (GRS-80). The attendant State Plane Coordinate values are given in meters. By way of example, within southeastern Wisconsin, the maximum shift in latitude between NAD-27 and NAD-83(91) approximates 11 feet, while maximum shift in longitude approximates 39 feet. While these shifts are important globally, affecting courses and distances between intercontinental locations, the shifts do not affect the relative bearings and distances between monumented survey control stations within southeastern Wisconsin.

The vertical survey control network of the national geodetic control system consists of a network of monumented benchmarks, the elevations—orthometric heights—of which have been determined by the USC and GS through high-order, differential level surveys. The original vertical datum is known as the National Geodetic Vertical Datum of 1929 (NGVD-29). This datum was also known as Mean Sea Level Datum. The national level net was based on 26 tide stations located along the coasts of the United States and Canada.

The NGS readjusted the national-level network in 1988 to produce a new vertical datum known as the North American Vertical Datum of 1988 (NAVD-88). By way of example, the difference between NGVD and NAVD within southeastern Wisconsin ranges from 0.08 to 0.32 foot, NGVD being the "higher" datum.

Currently, as part of the modernization of the NSRS, the NGS is replacing traditional vertical datums such as NAVD-88 with new geopotential datums based on gravimetric geoid models. This update aims to improve the accuracy, consistency, and integration of elevation data, especially when using GPS-derived coordinates. The upcoming geopotential datum will eliminate reliance on physical benchmarks and enhance alignment with global positioning technologies.

Local vertical control survey datums are still in use throughout the United States. For example, the city of Milwaukee uses a local datum based on the level of the Milwaukee River in March 1836. To reduce elevations referred to either NGVD-29 or NAVD-88 to the city of Milwaukee datum, equations between the datums must be determined by differential leveling.

U.S. PUBLIC LAND SURVEY SYSTEM

The present economic society is based, quite fundamentally, on the concept of property or sense of ownership, more precisely defined as the exclusive right to control an economic good. One of the most important forms of property is that of real property, which takes the form of rights in land—the commodity traded on the real estate market. The first step in enforcing this privilege of ownership in land, along with concomitant duties, is to identify the property object. Exactly that task is the object of land surveying and of the U.S. Public Land Survey System. Incidental to a necessary technical description of the system, the creation and application of that system constitutes an interesting example of effective planning designed to carry out an economic development policy.

After the Revolutionary War and the Louisiana Purchase, the newly created federal government found itself to be the owner of a vast wilderness stretching westward from the original 13 eastern seaboard colonies. Under the Constitution, the states had no original rights in such land. It became a national policy to sell this public domain to private owners for development in the belief that the wide extension of land ownership would result in the highest productivity for the nation. Much of the public land was sold at $1.25 per acre, regardless of quality. After 1862, it was granted free to any settler who would occupy and improve the homestead. The deed, or patent, from the US government became the basis for all subsequent ownership in the claim of title.

To facilitate the sale of the public lands, it was necessary to provide a straightforward and convenient method of land description and identification. As a result, the rectangular land survey system, the U.S. Public Land Survey System (USPLSS), was devised, and federal law required that the land be surveyed and monumented by the federal government before sale. The USPLSS is a brilliantly devised system providing a simple yet unambiguous means of writing unique legal descriptions for each and every parcel of land in the area covered. The system depends—as a survey control system—upon being able to identify, in the field, the location of the original monuments established by the government surveyors prior to the transfer of the land from federal to private ownership. This dependence upon the perpetuation of the monuments marking the system on the surface of the earth is the only significant weakness in an otherwise virtually ideal system.

The scheme of subdivision is shown in Figure 5.6. The primary unit of division is the survey township, a nominally 36-square-mile area. The secondary unit of division is the section, a nominally one-square-mile, or 640-acre, area. These units are located with respect to a set of coordinate axes consisting of an initial point, a prime meridian through that point, and a baseline surveyed as a parallel of latitude through the initial point, and to an implementing set of standard parallels as correction lines

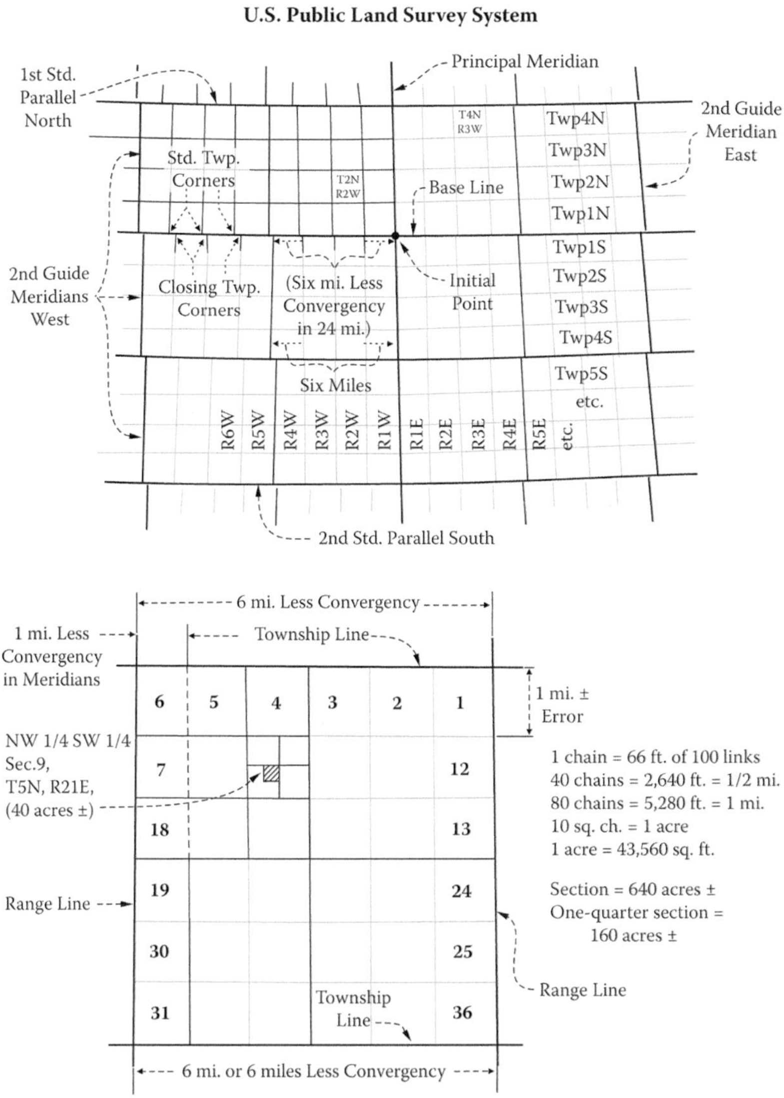

FIGURE 5.6 Illustration of the construct of the U.S. Public Land Survey System. This system provides the basis for all land surveys in much of the United States. Its perpetuation and use depend upon the preservation of the monuments marking the section and one-quarter section corners.

and guide meridians. The approximately 24-mile by 24-mile areas created by the standard parallels and guide meridians are subdivided into 16 nominally 36-square-mile townships by meridional range lines and latitudinal and township lines. The townships are further divided into 36 nominally one-square-mile, or 640-acre, sections numbered in boustrophedonic fashion. The sections can be readily further subdivided into one-quarter sections—nominally 160 acres in area, and into still smaller areas such as 80-, 40-, 20-, 10- and 5-acre areas by simple description.

A typical legal description, illustrated in Figure 5.6, might read: the NW ¼ of the SW ¼ of Section 9, T5N, R21-E, 4th P.M. Often, the reference to the principal meridian concerned is dropped, with county and state location substituted. In this example, a nominally 40-acre tract is simply and unambiguously identified.

The U.S. Public Land Survey System provided not only a simple, unambiguous means for describing and locating real property boundaries, but also the first maps and natural resource inventories of the vast areas covered by the survey. The instructions to the government surveyors provided that as the township, range, and section lines were run, notes were to be kept indicating where such lines crossed streams and watercourses, together with the width and direction of flow, intersected lake shores, which were then meandered, entered and left wetlands, and entered and left woodlands. The government surveyors also had to make notations concerning the principal species of the woodlands, the types and potential fertility of the soils traversed, and the potential presence of gravel, stone, and other mineral deposits. The survey data, when assembled on township plats, provided, as already noted, the first maps and the first accurate natural resource inventories of the areas surveyed. The survey notes, together with the township plats, can be and have been used, for example, to create accurate maps of the surface water system, wetlands, woodlands, and prairies of a planning area as these features existed prior to settlement by Europeans.

While the U.S. Public Land Survey System remains foundational for legal land description and property boundaries across much of the United States, modern surveying increasingly incorporates satellite-based technologies such as the GPS and RTK positioning. These technologies enable sub-centimeter accuracy and rapid data collection. Federal initiatives—particularly the modernization of the NSRS by the NGS—are enhancing compatibility between geodetic control systems and legacy frameworks such as the USPLSS. This integration supports improved land administration, cadastral mapping, and infrastructure planning.

MAP REQUIREMENTS FOR CITY PLANNING

Two basic types of maps are required to adequately meet city planning and engineering needs. The first of these is the large-scale topographic map. Such maps accurately show the configuration and elevation of the ground, the stream and watercourse lines, and other natural and cultural (man-made) features of the land and cityscapes. The parts of the map that present the relief, or elevation, of the land surface with reference to a vertical datum are known as hypsometry. The parts of the map that present the cultural and natural features—that is, that represent everything but relief—are known as planimetry. The hypsometry is usually portrayed by contour lines, that is, lines that represent lines of equal elevation on the ground. Hypsometry may also be virtually portrayed by computer-readable digital terrain models.

For city planning and engineering purposes, the topographic maps should be at scales ranging from one inch equals 50 feet to one inch equals 200 feet, and should have a vertical contour interval ranging from one to two feet. The topographic maps should be prepared on a map projection to specified accuracy standards, such as the National Map Accuracy Standards. The latter standards require that all horizontal survey control stations and all map projection tick marks should be plotted to an accuracy of 0.01 inch on the map sheet; that all well-defined planimetric features should be plotted to within 1/30th of an inch of their true position as referred to the map projection; that 90 percent of all contour lines should be accurate to within one-half of contour interval; that no contour lines should be in error by more than one contour interval; and that all spot elevations should be accurate to within one-quarter of a contour interval.

Such maps permit drainage areas to be accurately delineated and measured; distances between existing natural and cultural features and between such features and existing and proposed public works to be accurately scaled; profiles to be drawn; preliminary grade lines for proposed public works to be established and computed; flood-hazard areas to be delineated; and alternative locations and configurations for various types of proposed public works to be considered and evaluated.

The second type of map required for city planning and engineering purposes is the large-scale cadastral map. Such maps accurately depict the location, arrangement, and dimensions of all real property boundary lines, including all street and other public rights-of-way, and all existing land subdivisions. The cadastral maps should be prepared on the same map projection and at the same scale as the topographic maps. The two types of maps should be carefully designed so that the data presented on the cadastral maps can be readily and precisely correlated with the data presented on the topographic maps through modern Geographic Information System (GIS) platforms, which support digital overlay, spatial analysis, and real-time data integration. It should be noted that survey data and maps prepared using different datums and projections will not "overlay," that is, cannot be accurately correlated. With the adoption of updated datums, such as NAD 83 (2011), and the upcoming NSRS 2022 modernization by the NGS, new mapping projects must align their coordinate systems and vertical references accordingly. These updates will enhance the consistency and interoperability of map data across agencies and regions.

These two basic types of maps, if properly prepared, are not only useful in the conduct of day-to-day planning and engineering functions, but can also provide the basis for the economical preparation of a number of other maps required from time to time for city planning and engineering purposes, including planning base maps and maps for certain legal plan implementation devices, such as zoning district maps and official maps. Base maps show specified types of fundamental information, real property lines, for example, copies of which can then be utilized to display other information such as existing land use.

The topographic maps should be prepared by photogrammetric methods. The cadastral maps are typically prepared using conventional plotting techniques, which require the interpretation of legal descriptions of real property boundaries, including street and other public right-of-way boundaries. Modern cadastral mapping is increasingly incorporating technologies such as drone-based photogrammetry, mobile LiDAR, and high-resolution orthoimagery, which allow for rapid and accurate mapping of both topographic and cadastral features. The cadastral maps should be prepared using the same map projection as the companion topographic maps. Accuracy standards should be carefully specified for cadastral maps just as for topographic maps. These standards might, for example, require that all horizontal survey control stations and all map projection grid tick marks be shown with an accuracy of 0.01 inch, all real property boundary lines be plotted to an accuracy of 2.5 feet, and all gaps or overlaps of property boundary lines of 2.5 feet or more be shown.

A peculiar feature of cadastral maps that is not in accord with the basic definition of a map is that the dimensions and bearings of real property boundary lines are usually shown on the maps. However, the dimensions given are usually ground-level figures; that is, they have not been reduced to grid values. This practice introduces a usually small difference of about 0.01 feet per 100 feet in the values concerned.

SURVEY CONTROL FOR CITY PLANNING AND ENGINEERING

To make the topographic and cadastral maps the truly effective planning and engineering tools that they ought to be—and thereby save much duplication of survey efforts at a later date—the maps should be based on a carefully designed survey control network. This survey control network should meet two basic design criteria. First, it should permit the accurate correlation of real property boundary line information with topographic data. Second, it should be permanently monumented on the ground so that lines drawn on the maps can be accurately produced in the field. This facilitates field verification that the maps do indeed meet the specifications governing their preparation and the layout of planned public works projects when they reach the construction stage, and facilitate the enforcement of public land use controls such as zoning and official mapping. Importantly, the survey control network facilitates the collection of accurate "as-built" data for all constructed works, and particularly for subsurface works.

In areas covered by the U.S. Public Land Survey System, the control survey network should entail the location and monumentation of all U.S. Public Land Survey System section and quarter-section corners, including the centers of the sections, and the establishment of State Plane coordinates for the monumented corners. The coordinates of the monumented corners should be established by Third Order Class 1 or higher-order horizontal surveys. Although this level of accuracy may not be required for the topographic mapping work itself, such accuracy is required if the survey control network is to have permanent utility in all subsequent surveying and mapping operations within the area. Second-order or higher differential level lines should be run to establish the elevations to National Geodetic Vertical Datum of 1929 (NGVD-29) or North American Vertical Datum of 1988 (NAVD-88) monuments marking the U.S. Public Land Survey corners and attendant reference benchmarks. With the modernization of the NSRS, the NGS is replacing traditional vertical datums such as NAVD 88 with new geopotential datums based on gravimetric geoid models, such as the North American-Pacific Geopotential Datum of 2022 (NAPGD2022). This update will improve elevation accuracy and consistency nationwide, especially when combined with Global Navigation Satellite System

Portion of a Survey Control Summary Diagram

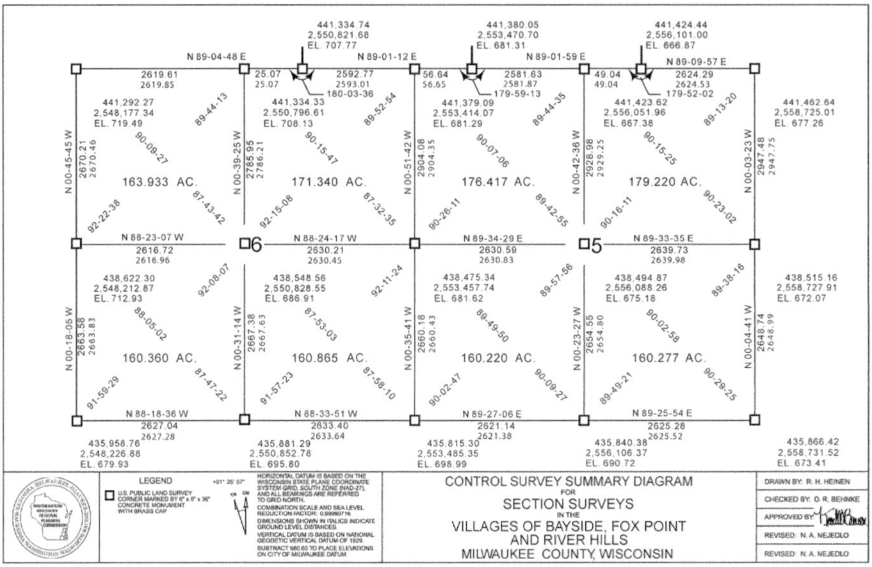

FIGURE 5.7 Illustration of the data given on a survey control diagram designed for use in city planning and engineering. The bearings are given as grid bearings, the lengths of the one-quarter section lines are given in both grid and ground level values even though the combination scale and sea level reduction factor are given in the title block data, and the areas of the one-quarter sections are ground level areas.

Source: SEWRPC.

(GNSS)-based methods. As such, future survey control networks should consider compatibility with these emerging datums to ensure long-term utility and seamless integration with geospatial datasets.

The resulting survey control network should be summarized for ready use in the form of a set of survey control summary diagrams covering the planning area. The diagrams should show, as illustrated in Figure 5.7, the State Plane coordinates of the monumented USPLSS corners; the ground level and grid level lengths and the grid bearings of the exterior boundaries of each one-quarter section; and the elevations—orthometric heights—of the corner monuments. The angle between geodetic and grid north (theta angle) should be given together with the combination scale and sea level reduction factor if the datum used is NAD-27, or the combination scale and elevation reduction factor if the datum used is NAD-83. These factors are used to convert grid distances to horizontal ground-level distances, and ground-level distances to grid distances.

In order to facilitate the recovery and use of the monumented survey control stations—the USPLSS corners—in subsequent land and engineering surveys, a data sheet, such as shown in Figure 5.8, should be prepared for each monumented corner. The sheet should identify the USPLSS corner concerned and give the State Plane

Typical Record of U.S. Public Land Survey Control Station

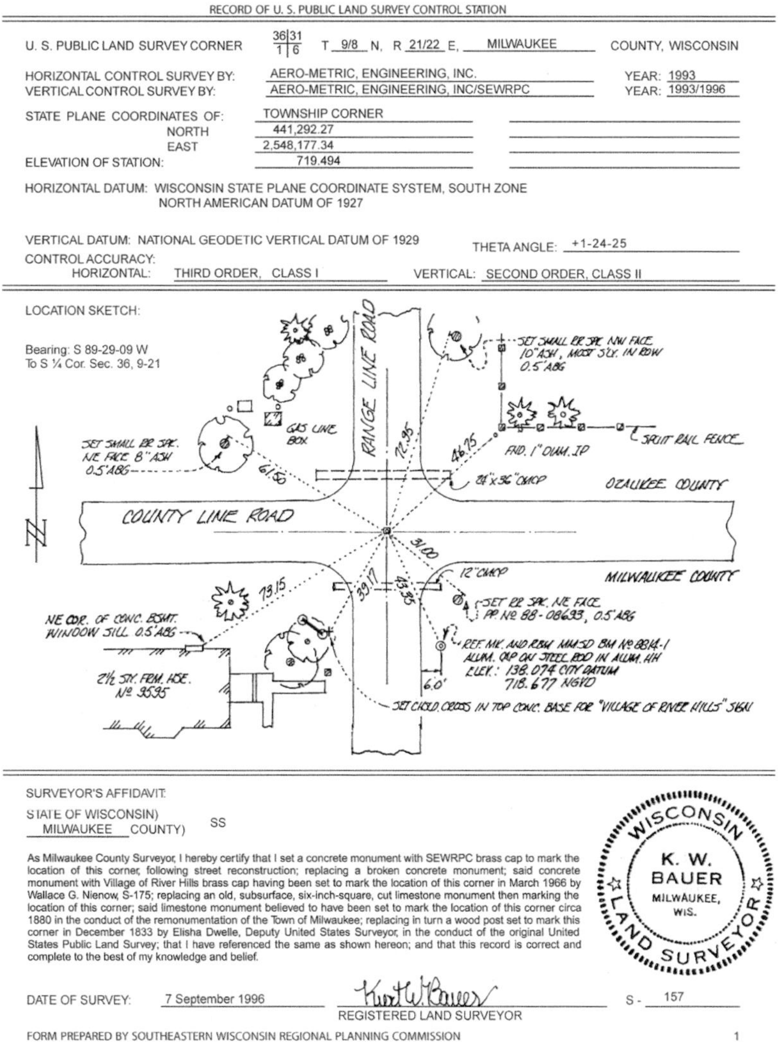

RECORD OF U. S. PUBLIC LAND SURVEY CONTROL STATION

U. S. PUBLIC LAND SURVEY CORNER 36|31 / 1|6 T 9/8 N, R 21/22 E, MILWAUKEE COUNTY, WISCONSIN

HORIZONTAL CONTROL SURVEY BY: AERO-METRIC, ENGINEERING, INC. YEAR: 1993
VERTICAL CONTROL SURVEY BY: AERO-METRIC, ENGINEERING, INC/SEWRPC YEAR: 1993/1996

STATE PLANE COORDINATES OF: TOWNSHIP CORNER
NORTH 441,292.27
EAST 2,548,177.34
ELEVATION OF STATION: 719.494

HORIZONTAL DATUM: WISCONSIN STATE PLANE COORDINATE SYSTEM, SOUTH ZONE
NORTH AMERICAN DATUM OF 1927

VERTICAL DATUM: NATIONAL GEODETIC VERTICAL DATUM OF 1929 THETA ANGLE: +1-24-25
CONTROL ACCURACY:
HORIZONTAL: THIRD ORDER, CLASS I VERTICAL: SECOND ORDER, CLASS II

LOCATION SKETCH:

Bearing: S 89-29-09 W
To S ¼ Cor. Sec. 36, 9-21

SURVEYOR'S AFFIDAVIT:
STATE OF WISCONSIN)
MILWAUKEE COUNTY) SS

As Milwaukee County Surveyor, I hereby certify that I set a concrete monument with SEWRPC brass cap to mark the location of this corner, following street reconstruction; replacing a broken concrete monument; said concrete monument with Village of River Hills brass cap having been set to mark the location of this corner in March 1966 by Wallace G. Nienow, S-175; replacing an old, subsurface, six-inch-square, cut limestone monument then marking the location of this corner; said limestone monument believed to have been set to mark the location of this corner circa 1880 in the conduct of the remonumentation of the Town of Milwaukee; replacing in turn a wood post set to mark this corner in December 1833 by Elisha Dwelle, Deputy United States Surveyor, in the conduct of the original United States Public Land Survey; that I have referenced the same as shown hereon; and that this record is correct and complete to the best of my knowledge and belief.

WISCONSIN K. W. BAUER MILWAUKEE, WIS. LAND SURVEYOR

DATE OF SURVEY: 7 September 1996 Kurt W. Bauer S- 157
REGISTERED LAND SURVEYOR

FORM PREPARED BY SOUTHEASTERN WISCONSIN REGIONAL PLANNING COMMISSION 1

FIGURE 5.8 Illustration of a survey control station "dossier" sheet intended for use in city surveying and mapping operations to recover and use monumented survey control stations. This sheet is for the station location at the extreme northwest corner of the survey control summary diagram shown in Figure 5.7.

coordinates of the corner, the elevation—orthometric height—of the corner monument and of one or more attendant reference benchmarks, and the order of accuracy of the horizontal and vertical surveys establishing the coordinate values and elevation of the station. The sheet should contain the county surveyor's affidavit certifying the validity of the corner locations for its use in the conduct of land surveys.

Since the boundaries of the original government land subdivision form the basis for all subsequent real property divisions and boundaries, and since the State Plane Coordinate System provides, in effect, the basis for all earth-science-related surveys, the survey control network described provides a common reference system for both topographic and real property boundary line mapping. The accurate reestablishment of the quarter-section corners and the accurate location of those corners on the State Plane Coordinate System permit the accurate compilation of both property boundary line and topographic maps. Indeed, the topographic maps will provide the "ground truth" required for the proper construction of the companion cadastral maps. The topographic and property boundary line maps can be readily and accurately updated and extended since all new surveys can be readily tied to the corners of the U.S. Public Land Survey System and through those corners to the State Plane Coordinate System. The survey control network permits the ready transfer of details supplied by aerial mapping, including contour lines, to property boundary line maps and the equally ready transfer of property boundary line data to topographic maps by simple overlay techniques. Significant savings in office research time are possible during the planning and design phases of municipal public works projects by having all available information accurately correlated. This makes possible the consideration and evaluation of many alternatives in the planning and design of public works, such as streets, trunk sewers, water transmission mains, and stormwater management facilities.

The survey control network described provides an extremely practical system that makes the State Plane Coordinate System conveniently available to public and private surveyors and engineers as a basis for all subsequent survey work within the mapped area. The survey control system places a monumented, recoverable control station accurately located upon both the U.S. Public Land Survey and State Plane Coordinate systems, and a point of known elevation at half-mile intervals throughout the mapped area. Thus the land and engineering surveyor is never more than one-quarter mile from a point of known horizontal location on two control systems and of known vertical location. Not only does this monumented control network expedite engineering surveys made by various public works agencies, but it also serves to coordinate all the land and engineering survey work throughout the mapped area. By using this survey control network, local land and engineering surveyors can, without changing their methods of operation or incurring any additional expense, automatically relate all surveys to the State Plane Coordinate System, reference all bearings used in land surveys, plats, and legal descriptions to grid north, and reference all elevations to the National Geodetic Vertical Datum of 1929 (NGVD-29) or the North American Vertical Datum of 1988 (NAVD-88). Importantly, the control network also permits boundaries drawn on the topographic and cadastral maps—regardless of whether these boundaries represent the limits of land to be reserved for immediate or future public use, the limits of districts to which public regulations are to be applied, or the location and alignment of proposed public works—to be accurately and precisely reproduced in the field.

The finished topographic and cadastral maps should show, in addition to the usual features contained on such maps, the monumented USPLSS corners and the section and quarter-section lines as established by the field surveys, the grid lengths and bearings of those lines, and the State Plane coordinates and elevations of the monuments marking the corners. The map sheets should also show the angle between geodetic and grid bearing, the combined sea level and scale reduction factor or combined ellipsoid and scale reduction factor applicable to the sheet, and the equation between the National Geodetic Vertical datum, and any local datum previously used in the area covered by the map, as shown in Figure 5.9. Depending upon scale, each map sheet should cover one U.S. Public Land Survey quarter section or one section.

Typical Topographic and Typical Cadastral Map

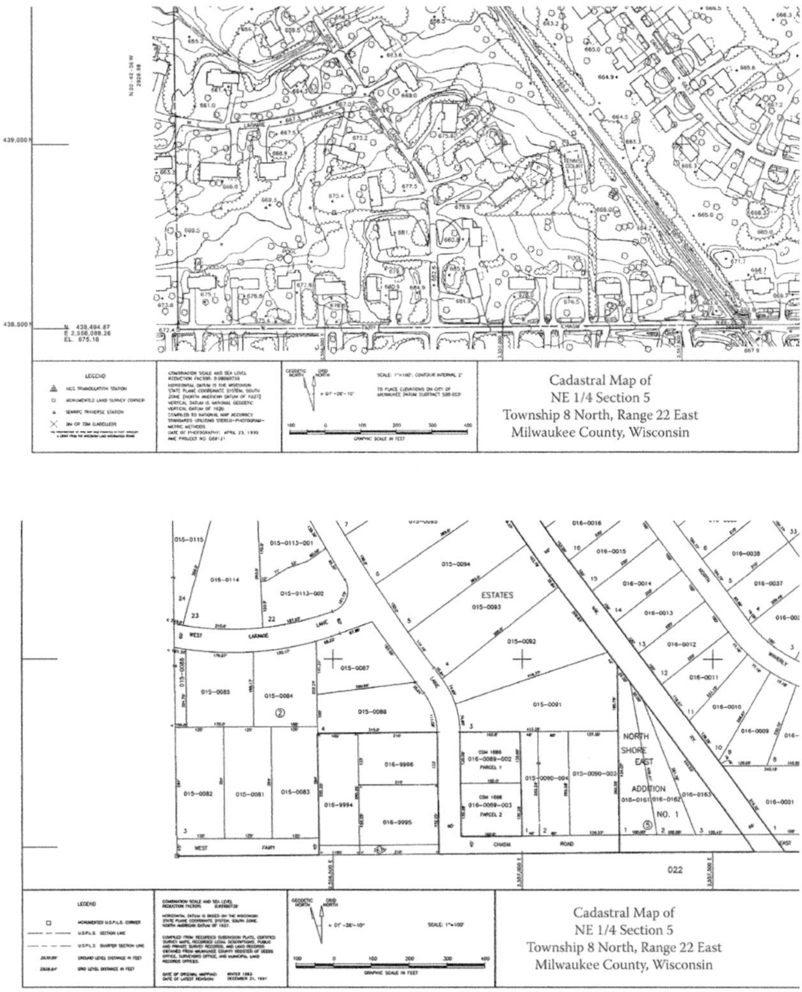

FIGURE 5.9 Illustration of the type of topographic and cadastral maps needed for city planning and engineering. Note that the maps are compiled on the same projection utilizing the same datum and survey control, and can therefore be accurately "overlaid" in hardcopy or digital format. This projection is manifest by the State Plane Coordinate System tick marks. The monumented survey control stations are also shown together with their State Plane Coordinate System values.

Source: SEWRPC.

STATE PLANE COORDINATE COMPUTATIONS

Both the simplicity of computations involving the State Plane Coordinate System and the utility of the control survey network herein described may be illustrated by the following example.

Refer to Figure 5.10, and assume that the State Plane coordinates of Point B within the Southeast one-quarter of USPLSS Section 5, Township 8 North, Range 22 East

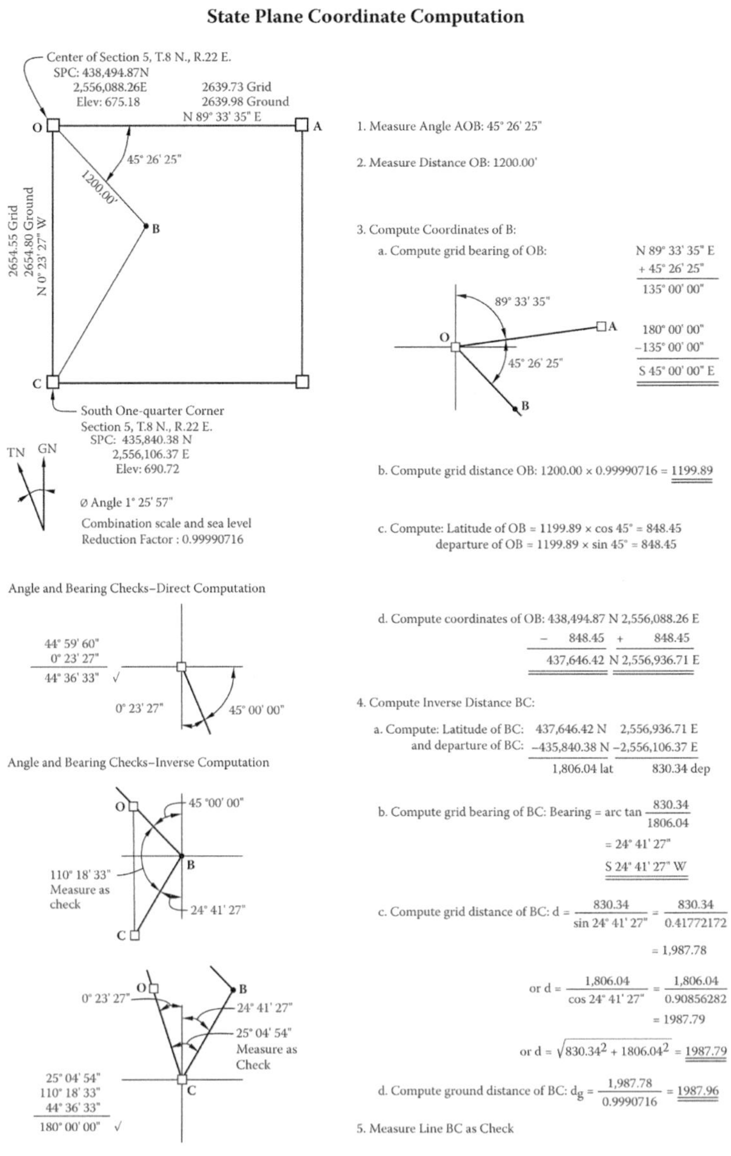

FIGURE 5.10 State plane coordinate computation.

are to be determined. The ground-level lengths and grid bearings of the north and west lines of the one-quarter section are obtained from the survey control network summary diagram, such as that shown in Figure 5.7. U.S. Public Land Survey corner recovery data sheets, such as that shown in Figure 5.8, are obtained and used to recover the monuments marking points A, O, and C. Point O is occupied with a total station, point A sighted, and the angle AOB and distance OB are measured as 45° 26′ 25″ and 1200.00 feet, respectively. The coordinates of point B are then computed as shown in Figure 5.10. To check the coordinate position of point B, the grid bearing and length of line BC are computed as shown in Figure 5.10, and the accuracy of the work is verified by field measurement of the bearing and distance. The latter verification illustrates the utility of the survey control network as a means of verification and check, eliminating the need to run traverses that close upon the point of beginning for such verification and check. A similar benefit would exist with respect to differential level surveys.

USE IN CREATION OF LAND INFORMATION SYSTEMS

The control survey network and the topographic and cadastral maps herein described also provide the essential foundational elements for sound, computer-manipulatable, parcel-based land information and public works management systems. Such systems are becoming essential to the efficient and effective conduct of city planning and engineering, and are profoundly influencing the scope and content and the techniques involved in such programs.

In this respect, the parcel identification numbers shown on the cadastral map in Figure 5.8 should be noted. These numbers provide the link between digital planning and engineering databases and the geographic location, configuration, and areal extent of the attribute data concerned. The data that can be so linked are virtually infinite, including parcel ownership, assessed valuation, street address, existing and planned land use, soil type and properties, vegetation, flood hazard, and zoning. The cadastral maps also provide the bases for the preparation of accurate sanitary sewerage, storm water drainage, water supply and other utility system and facility maps, and the linkage of engineering data to those maps for use in public works management.

If a community has a properly staffed engineering department, the basic mapping work should be the responsibility of that department, thereby freeing the city planning department for the conduct of current and long-range planning operations. The city planner and land information system technicians should, in any case, be familiar with the use and limitations of maps in the conduct of the necessary planning inventory operations.

AERIAL PHOTOGRAPHY

Aerial photography provides another useful tool for city planning and engineering. These corrections are now typically performed using photogrammetry software such as Agisoft Metashape, Pix4D, or Esri Drone2Map. The digital workflow includes image orientation, tie point extraction, terrain modeling, and orthorectification.

Although oblique photographs have use for general informational and display purposes in planning, the most useful photographs are vertical. Such vertical photographs may be provided in the form of simple contact prints made from aerial negatives, in the form of ratioed enlargements, ratioed and rectified enlargements, or in the form of orthophotographs. The basic relationships involved in aerial photography are illustrated in Figure 5.11. The relationships shown indicate that contact prints

Basic Relationship Involved in Photogrammetry

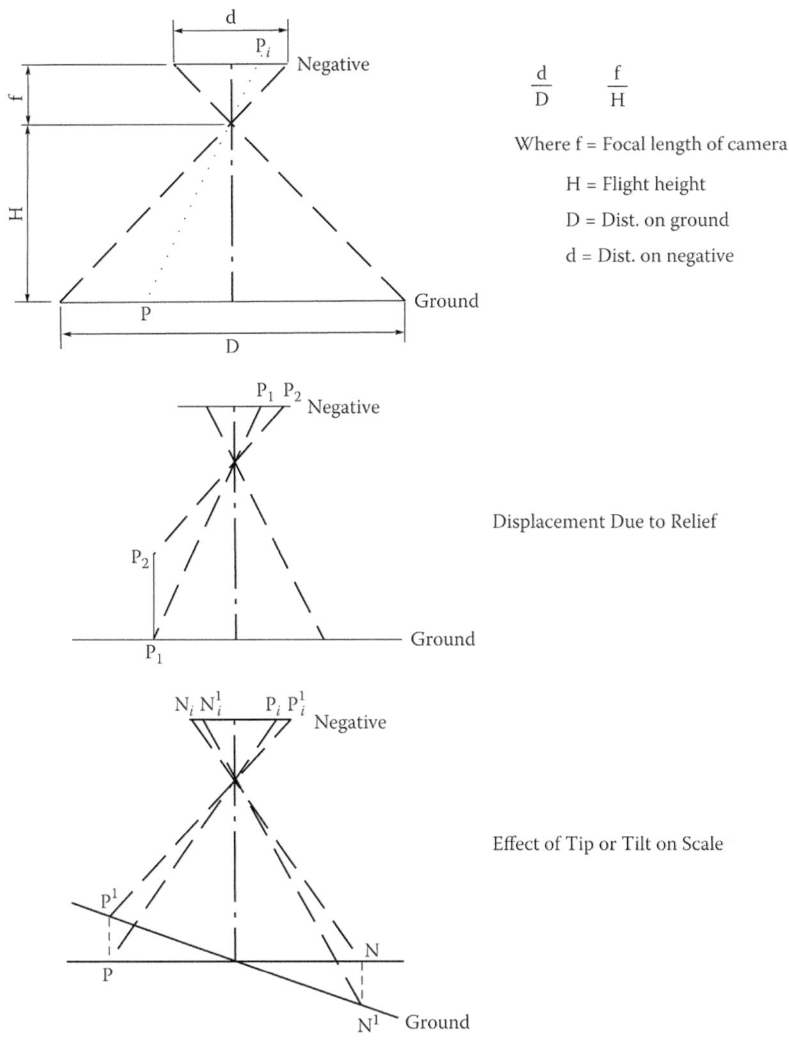

FIGURE 5.11 Illustration of the basic relationships involved in photogrammetry. Aerial photographs can be totally uncorrected, ratioed (that is, corrected for scale), rectified (that is, corrected for tip and tilt), or ratioed and rectified. Only orthophotographs which partially correct for displacement due to relief approximate true maps.

contain distortions due to the effects of variations in the flight altitude and in the tip and tilt of the aircraft during photography. Distance and areas cannot be accurately scaled from contact prints, and aerial mosaics assembled from such prints are primarily useful only in the preparation of graphic displays.

Ratioed enlargements made from aerial negatives correct for the changes in scale caused by variations in the flight altitude of the aircraft, but still contain distortions due to the tip and tilt of the aircraft during photography and due to relief of the ground and related features being photographed. Ratioed and rectified enlargements made from aerial negatives remove the effects of variations in flight altitude and tip and tilt of the aircraft during photography, leaving only displacement due to relief as a source of distortion in the photograph. In photography of areas marked by only moderate relief, the resulting enlargements may provide a basis for the measurement of distances and areas within acceptable limits of accuracy. Ratioed and ratioed-rectified aerial photographs can be prepared by optical projection or computer manipulation.

Orthophotographs are adjusted by computer manipulation not only for variations in flight altitude and in the tip and tilt of the aircraft during photography, but also for variations in relief. Such photographs approach—but are not—true maps, since the adjustment for relief is made only for the average ground level of the photography. True Orthophotos, a newer development, apply structure-specific corrections to eliminate distortion even from vertical surfaces and elevated features. These are particularly important for urban environments, where building outlines and infrastructure must be precisely mapped. Thus, major variations from the average ground elevation, such as the tops of tall buildings or other structures such as chimneys and electric power transmission line towers, will still have marked displacement. An example of an orthophoto is given in Figure 5.12. The orthophotograph covers the same area as the topographic and cadastral maps shown in Figure 5.9. Note that the orthophotograph contains the same survey control data as the maps. Today, orthophotos and associated mapping data are routinely overlaid and analyzed within Geographic Information Systems (GIS) like ArcGIS or QGIS. This integration supports advanced analysis such as land cover classification, change detection, flood modeling, and infrastructure inventory.

The preparation of ratioed, ratioed and rectified, and othophotographs requires survey control data, and such preparation is greatly facilitated by the survey control network described here.

MODERN MAPPING TECHNOLOGIES

Advances in surveying and mapping have introduced a range of modern technologies that enhance the accuracy, efficiency, and utility of maps used in city planning and engineering. These include photogrammetric mapping, digital terrain models, orthophotography, and computer-manipulatable formats such as those used in parcel-based land information systems. More recently, technologies such as LiDAR, GPS, GIS, and UAV (unmanned aerial vehicle, or drone) mapping have been integrated into municipal workflows, providing higher-resolution spatial data and facilitating real-time data updates. These tools build upon the foundational principles of

Portion of Orthophotograph

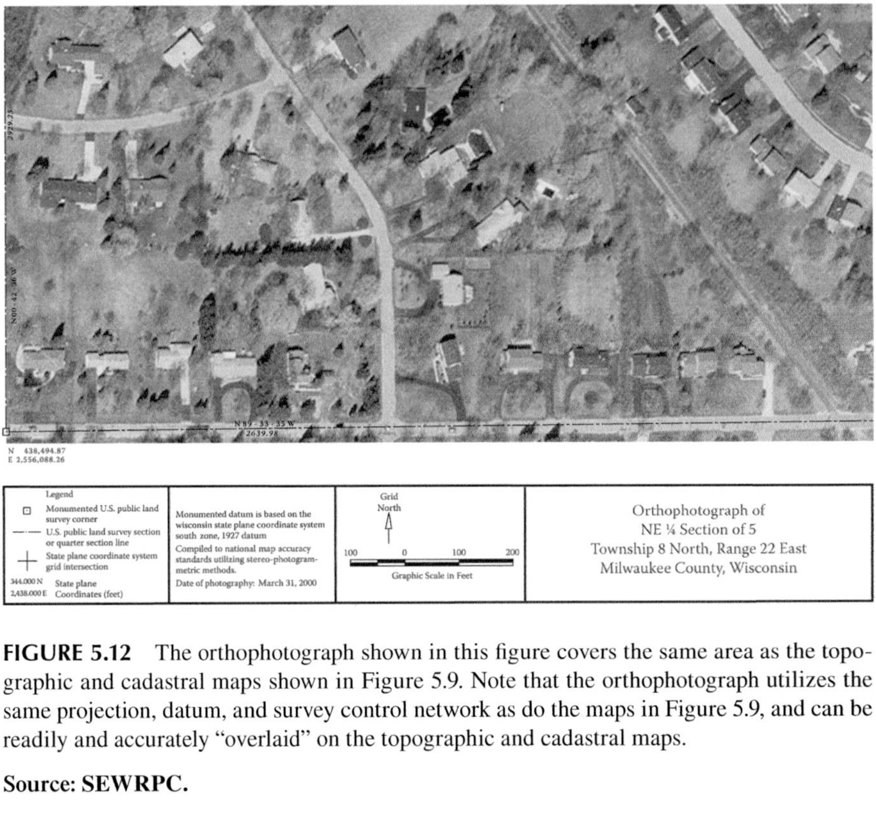

FIGURE 5.12 The orthophotograph shown in this figure covers the same area as the topographic and cadastral maps shown in Figure 5.9. Note that the orthophotograph utilizes the same projection, datum, and survey control network as do the maps in Figure 5.9, and can be readily and accurately "overlaid" on the topographic and cadastral maps.

Source: SEWRPC.

accurate control surveys and standardized datums and projections, enabling cities to support increasingly complex planning, public works, and infrastructure management needs.

FURTHER READING

Bauer, K. W. "A Control Survey and Mapping Project for an Urbanizing Region," *Journal of Survey and Land Information Science*, Volume 65, Number 2, 2005.

Blachut, Teodor J., Chrzanowski, Adam, and Saastamoinen, Janko H. *Urban Surveying and Mapping*. Springer-Verlag, 1979.

Burkholder, Earl F. *The 3-D Global Spatial Data Model—Foundation of the Spatial Data Infrastructure*. CRC Press, 2008.

Kissam, Philip. *Surveying for Civil Engineers*. McGraw-Hill, 1981.

Nemerow, Nelson L., Agardy, Franklin J., Sullivan, Patrick, and Salvato, Joseph A. *Environmental Health and Safety for Municipal Infrastructure, Land Use and Planning, and Industry*, 6th ed., John Wiley & Sons, 2009.

6 Population Data and Forecasts

Population projections are useful analytical tools: they show what would happen if certain demographic assumptions were met. They do not say, and should not be construed as saying, that these assumptions will be met. In other words, projections are not predictions, despite their frequent misinterpretation as such.

"Demographic Change and the Economy of the Nineties"
Report of the Joint Economic Committee of the Congress of the United States, December 1991

INTRODUCTION

The ultimate objective of city planning is to provide for the future needs of the community in an effective, efficient, and environmentally sound manner. These needs will be directly related to, among other factors, the resident population of the planning area. Therefore, a study of the past and present resident population and of the probable future growth or decline in that population is an essential part of any city planning effort.

The city planner and the municipal engineer are particularly interested in three aspects of the existing and probable future resident population of the planning area concerned: size, composition, and distribution. The size of the existing and probable future resident population of the planning area provides the most basic measure for necessary estimates of land and facility needs—present and probable future. The composition of the population in terms of such aspects as age, income, education, race, and household type and size provides additional information essential to transportation, housing, school, and park planning. The distribution of the population adds a geographic dimension, providing information about where land and facilities need to be provided. The distribution may also vary with time as measured by the daytime and nighttime population distributions. Sources of data on population include:

1. The decennial U.S. Census of Population and Housing
2. Local censuses, such as school censuses
3. Statistical sampling
4. Local utility agencies, such as water and power utilities
5. State and regional demographic agencies
6. American Community Survey (ACS)
7. Commercial demographic data providers

DOI: 10.1201/9781003500384-6

ESTIMATING CURRENT POPULATION LEVELS

All planning must begin with an analysis of the existing situation. Therefore, one of the first concerns in the conduct of population studies is data on the current size, composition, and distribution of the population. Information on the current population is essential to certain analyses critical to city planning and engineering, such as determining per capita water consumption, sewage contribution, and household trip generation.

The U.S. Census Bureau provides "hard count" population data only at 10-year intervals. Therefore, a need often exists to estimate current population data for intercensal years, particularly for a year that is to constitute the base year of the planning effort. The U.S. Bureau of the Census provides periodic estimates of the size of the population of the nation as a whole and for individual states. In addition, the American Community Survey (ACS), conducted annually, provides valuable ongoing estimates of population characteristics. Some state agencies, such as the Wisconsin Department of Administration, may provide annual estimates of the size of the population of counties and of some or all municipalities. Where such estimates are available in a timely fashion for the planning area concerned, and are found to be reliable, they should, of course, be used. Where such activities are not available, several techniques are available for estimating current population levels of planning areas. These techniques are essentially the same as those used for projecting and forecasting future population levels.

The distinctions between a population estimate, projection, and forecast are important. A projection is defined as a technique that uses facts about a population at a particular time to reach conclusions about that population at some other time. If the time for which a projection is made is in the present or past, the result of the technique is called an estimate; if in the future, it is a projection. The term projection implies a conditional assertion about future population based on a stated set of assumptions concerning the components of population change. The term forecast implies an unconditional assertion for use in plan preparation. A forecast should be selected from a range of projections.

PROJECTING AND FORECASTING FUTURE POPULATION LEVELS

City plans deal with public infrastructure and other improvements that are of considerable cost and permanence. In the interest of economy, such improvements must be designed to adequately serve the needs of the community for considerable lengths of time after their completion. The periods of cost amortization involved are often long. Therefore, city planning and engineering require the preparation of population forecasts. The selection of a forecast period requires consideration of a number of factors, including the physical and economic life of the various improvements concerned, the potential periods of cost amortization, and the limitations of the available forecasting techniques. Different forecast periods may be involved in, for example, the planning and design of trunk sewers and hydraulic channels—perhaps 50–80 years; pumps and electrical equipment—perhaps 10–20 years; street pavements—perhaps 20–25 years; and bridges—perhaps 50–80 years. Forecasts are of particular importance in the design of waterworks, sewerage facilities, and transportation facilities.

Large errors in population forecasts have different implications depending upon their direction. Under estimation may result in significant increases in costs entailed in the remediation of inadequate facility capacities. Overestimation may result in significant overinvestment attendant to the provision of larger than needed facility capacities. A large underestimation is likely to be more serious because of the costs associated with the construction of additional capacity. It might, therefore, be concluded that population forecasts should always be adjusted upward. But such a conclusion would be wrong. What should be concluded is that the process requires alternative population projections to be prepared so as to provide a realistic range of probable future conditions from which a forecast level can be selected, based on careful considerations of the implications of the alternative projections on the cost and performance of the facilities being planned.

Techniques

A number of techniques have been developed for projecting and forecasting population change. Some of these are quite simple, and others are quite complex. All are ultimately based on historical data and rely upon a combination of mathematical formulation and professional judgment to analyze these data and project them into the future.

In exercising the professional judgment inevitably involved, it must be recognized that local population growth will be influenced by a number of factors, including the structure and vitality of the urban economy, particularly the availability of employment opportunities; land available for growth; the location, configuration, and capacity of transportation facilities; climate; topography; the quality of the local school, park, police and fire protection, water supply, and sewerage facilities and services; and taxes. Changes in any of these factors should be a warning that past trends may not necessarily continue.

The principal difference between any of the available techniques is the degree of emphasis placed upon mathematical formulation as opposed to professional judgment. At one extreme, a technique may involve little or no mathematical formulation and depend almost entirely upon experienced judgment. Because the considerations entering into such judgment are often not clearly articulated, perhaps not even in the minds of the persons making the judgments, the projections concerned are generally not capable of replication by others. At the other extreme, a technique may depend almost entirely on mathematical formulation and may be readily replicated, but not on that account be necessarily more accurate or valid.

There are two basic types of population projection techniques: those that treat the population and changes in the population as a single aggregate, and those that deal with the disaggregate components of the population and of population change. Aggregate approaches include mathematical and graphical extrapolation, ratio, and analog techniques. Disaggregate approaches include cohort-survival and cohort-change techniques. The disaggregate approaches have two general advantages: (1) they permit explicit consideration of the components of population change, and (2) they provide more detailed projections by age and sex. Disaggregate techniques, however, are not necessarily more accurate than simpler aggregate techniques.

More specifically, the following techniques are commonly used in city planning and engineering for projecting population levels:

1. Cohort-survival method, applied age specifically
2. Cohort-change method
3. Mathematical or graphical projection methods
4. Projections based on independently prepared estimates of future employment
5. Estimates based on symptomatic data
6. Other methods, such as the ratio or shift-share method based on projections independently prepared for a larger area and projections based on the holding capacity of the planning area concerned

Cohort-Survival Method

A birth cohort is defined as a group of persons born in a specified year or period. A cohort can be traced through time. For example, survivors of the cohort born in the period April 1, 1980 to April 1, 1985, are 5–9 years old on April 1, 1990, 15–19 years old on April 1, 2000, and so on. The term cohort indicates that the computation is done by age group, retaining the identity of the group as it is carried forward in time. The term component indicates that the method explicitly considers the three components of population change: births, deaths, and in or out migration. Thus, the method recognizes that the number of people residing in an area can be changed in only three ways: by births, by deaths, and by in or out migration.

The basic formula used in the method may be stated as:

> Civilian population at the end of a projection period is equal to the population at the beginning of a period plus the natural increase (births minus deaths) plus or minus net migration minus the loss to the armed forces. Total population is equal to the civilian population plus the armed forces stationed in the area.

The steps in the application of the method as used to determine the civilian population of an area may be summarized as:

1. Establish the base population and the date from which the projection is to be carried forward.
2. Establish a schedule of vital rates—birth, death, migration—estimated to be in effect at the base date.
3. Establish an estimation cycle—usually five years—and an estimated schedule of age-specific vital rates to be assumed to apply during each cycle following the base date.
4. Multiply the base population cohorts by the vital rates to obtain estimates of the number of births, deaths, and migrations that will occur during the next cycle period.
5. Add these estimated numbers to, or subtract them from, the base population.

Recent advances in data availability have significantly streamlined this process. APIs and data portals provided by the U.S. Census Bureau, CDC, and state health departments now offer downloadable datasets and machine-readable formats that support automated workflows.

In addition, data on the armed forces stationed in the area must be obtained from cognizant military authorities, and the number added to the civilian population to compute the total population of the area.

The population cohorts age as the cycle progresses. This yields, in an orderly and methodical manner, an estimate or a projection of what the population may be expected to be at the end of each cycle, given the assumed vital rates. The validity of the projection depends entirely on how successfully fertility, mortality, and migration rates are selected.

Mortality and fertility rate data are now widely accessible through sources like the CDC WONDER database, the National Center for Health Statistics, and state open data platforms. The total fertility rate is defined as the expected number of children that 1,000 women would bear if they completed their reproductive lives at the age-specific rate concerned. The reproductive life is considered to be ages 15 through 44. A total fertility rate of 2.1 is referred to as "replacement fertility" because this rate will produce just enough births to sustain the population through natural increase. For high growth estimates or projections, moderate increases in the fertility rates may be assumed; for intermediate growth estimates or projections, stability in the fertility rate may be assumed; and for low growth estimates or projections, a moderate decline in the fertility rate may be assumed. It is worth noting that even small differences in the assumed rates will compound over the projection period.

Migration rates are not generally available from federal or state agencies and must be estimated. One method of estimation utilizes school enrollment data to calculate an estimated rate. In some states, such as Wisconsin, local school superintendents annually file school enrollment data with the state. Using these data, the local migration rate can be estimated. The local migration rate is estimated as equal to the change in local enrollment for the period concerned divided by the base year enrollment less the percent change in state enrollment.

Modern methods also use IRS migration statistics, USPS change-of-address data, and administrative datasets from Medicaid and driver's licensing records to estimate net migration flows at sub-county levels.

The cohort-survival method is generally recognized by demographics as the "gold standard" for preparing population estimates, projections, and forecasts. It provides estimates, projections, and forecasts by sex and age, component data that are particularly important for housing and transportation system planning. Importantly, it permits explicit consideration of the factors of change—birth rates, death rates, and net migration rates—and permits varying assumptions to be made in those factors and alternative projections to be prepared. The computations required for the application of the method are tedious, but can be programmed for computer use. Although the method may give the appearance of great accuracy, it should be noted that when used for projections and forecasts, assumptions must be made as to probable future birth, death, and migration rates that cannot be known. Therefore, the method is best applied by an experienced demographer. For practical planning applications, small and medium-sized jurisdictions often rely on projections developed by state demographers, university research centers, or regional planning commissions. The estimates, projections, and forecasts prepared by their agencies can be adapted for use in county and municipal planning operations. It is unlikely that planners employed by small and medium-sized counties or municipalities will ever have to employ the method directly. Nevertheless, planners should understand the assumptions and limitations underlying the projections they use.

Cohort-Change Method

A simplified component method of estimating, projecting, and forecasting population levels by age and by sex is provided in the cohort-change method. If access is not available to data supplied by the cohort-survival method, the cohort-change method can be used to obtain population estimates, projections, and forecasts by age and sex. The method is simple enough to be applied without the assistance of a demographer. The method requires data from two successive decennial censuses. Data on births are generally available from cognizant state agencies for counties and municipalities. While not as detailed as cohort-survival methods, it offers a practical and replicable approach for small or medium-sized jurisdictions.

Mathematical and Graphical Methods

Mathematical and graphical methods of preparing population estimates, projections, and forecasts are based on the assumption that future population levels can be predicted solely upon projections of past trends without separately analyzing component trends. These methods have several advantages for application in city planning and engineering and have long been described in engineering texts on sanitary sewerage and water supply planning and engineering. The methods are quite rational for use in the preparation of relatively short-term projections and forecasts since a high correlation usually exists between population changes in successive time periods. The methods require fewer data than other methods, are simple, and can be quickly applied by an experienced city planner or engineer. The principal disadvantage of these methods is that they can provide only aggregate estimates, projections, and forecasts, and therefore cannot provide data on characteristics such as age and sex.

The mathematical and graphical methods can be applied using a number of different trend lines, including arithmetic progression, geometric progression, decreasing rate of increase, and logistic curve trend lines. These are illustrated in Figure 6.1. The arithmetic progression variation assumes a uniform absolute rate of growth; consequently, the percentage rate of change will be different each year. This method has been used by the U.S. Bureau of Census to estimate component trends between decennial census years. Under this method, communities are assumed to gain approximately the same number of inhabitants each year. Commercial centers of prosperous agricultural regions often appear to follow this pattern. An example of the actual application of this method is provided in Figure 6.2. The geometric progression variation assumes a uniform percentage rate of growth. It is often used

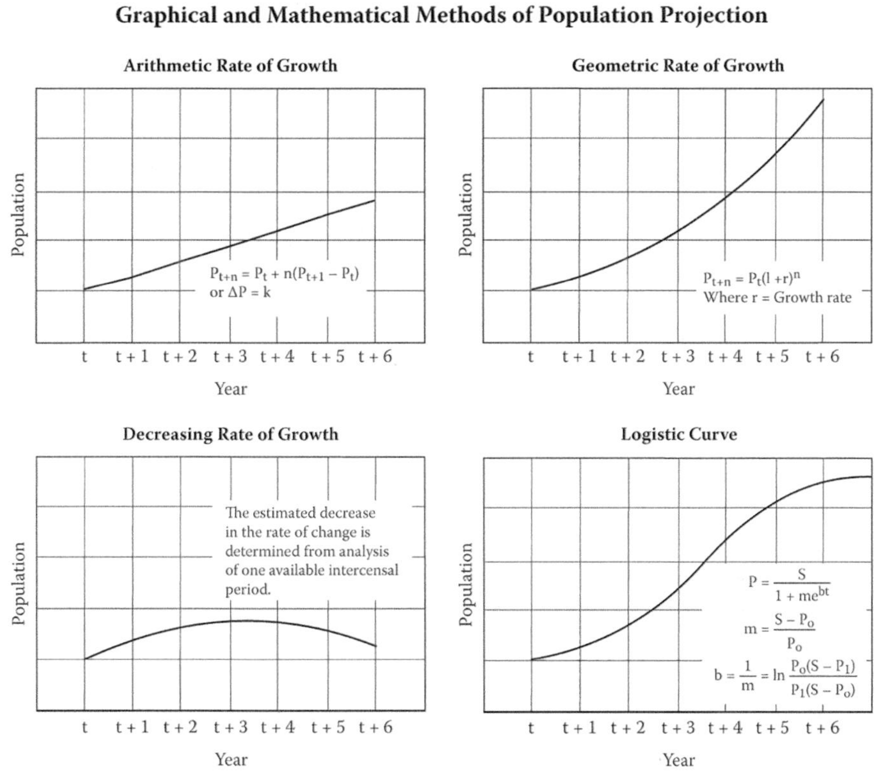

FIGURE 6.1 Illustration of the concepts and mathematical formulae underlying graphical and mathematical methods of population projection. These methods can provide only aggregate projections. Being based upon historic trends, all are conceptually valid even though simplistic.

to represent the type of growth experienced by young cities. Application for long-term projections or forecasts may lead to overestimation of the future levels. A decreasing percentage rate of growth may be applicable to larger, older cities. The decrease in the rate of growth should be based on an analysis of the experience over at least one intercensal period.

Modern forecasting often incorporates nonlinear regression, spline-based methods, or machine learning models, which can identify turning points or saturation without assuming symmetry. Additionally, techniques such as Bayesian time-series models allow planners to account for uncertainty and multiple future scenarios. These approaches are particularly useful when integrated with spatial datasets or infrastructure capacity models. Furthermore, modern planning departments often implement these methods through statistical tools such as R or Python, or through platforms like Power BI, Tableau, and ArcGIS Pro. These tools allow the graphical presentation of trends and simulation of alternative futures. In small communities or resource-limited settings, traditional trend-based projections remain valid but should

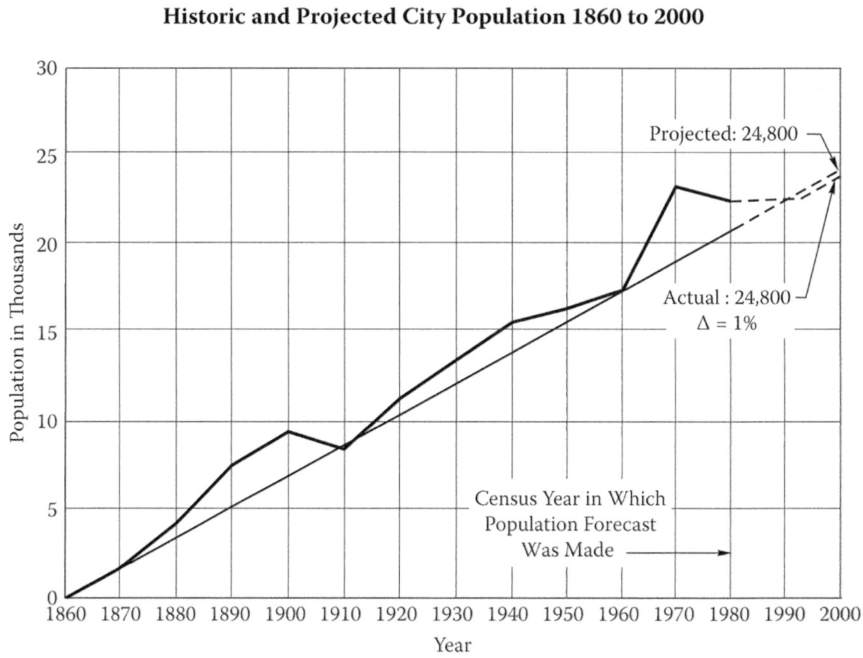

FIGURE 6.2 Illustration of the actual application of the arithmetic method of population projection. The successful application illustrated is largely fortuitous and must be regarded with a certain sense of humor.

ideally be supplemented with sensitivity analysis or scenario comparisons to reflect demographic volatility.

Employment Relationship Method

The employment relationship method of population projection and forecast is based on the concept that a determinable ratio exists between the employment levels in a planning area and the resident population. Data on annual employment levels are available from the U.S. Bureau of Economic Analysis (BEA), the Bureau of Labor Statistics (BLS), and state labor departments, often disaggregated by sector and geography. These "place-of-work" data indicate the number and types of jobs located within a defined area. By correlating these data with available population and labor force participation data, planners can establish the current employment–population ratio and examine its trends over time. Population projections and forecasts can then be derived using regional or state employment forecasts, adjusting for expected changes in commuting behavior, remote work trends, and economic diversification. This method remains valuable for metropolitan and regional analysis but should be used with caution in areas experiencing significant economic restructuring, industrial transition, or shifts toward telecommuting. Integrating spatial employment models or regional input–output forecasts can strengthen its reliability.

Symptomatic Data Method

Symptomatic data can serve as practical indicators for estimating intercensal or post-censal population changes within a planning area. One such type of symptomatic data is electric utility meter connections, for which historical records are typically maintained by local or regional power providers. By correlating these data with known census counts, planners can estimate a relationship between service connections and population size and apply it to recent connection growth to estimate incremental population change. While useful for interim updates, symptomatic indicators should not be used for long-range projections without supporting demographic analysis. Other useful symptomatic indicators include housing construction, occupancy permits, water service connections, school enrollment, or even mobile device activity. These data, when used carefully, can help estimate short-term demographic shifts. Modern applications often integrate multiple indicators—such as utility usage, postal change-of-address data, or building permits—into regression or machine-learning models to produce more robust intercensal estimates.

Other Methods

Several additional methods for population projection and forecasting are occasionally used in city planning, though they are generally less robust or applicable under specific conditions. The composite diagram method involves graphically comparing the growth trends of a smaller city with historically similar but larger cities by shifting trend curves to match a common base year. While intuitive, this method heavily relies on judgment and can mislead if the cities are at different stages of development. The ratio method—also known as the shift-share method—projects a smaller area's population based on its historical share of a larger region's population, assuming the share remains stable; this can falter when external factors like infrastructure changes alter growth patterns. The holding capacity method is useful for municipalities with fixed boundaries, estimating future population based on land use plans, zoning, and parcel-level data, assuming full build-out and local development rates. Though these methods offer simplicity or localized control, they are highly sensitive to assumptions and less reliable than more data-driven approaches. When used, they should be supported with sensitivity testing and integrated into a broader modeling strategy.

Area Considerations

Historical population data are usually presented for geographic areas fixed at the time of the census: states, counties, metropolitan areas, urbanized areas, and municipalities, down to the census tract level. While the geographic boundaries of states and counties remain fixed, the boundaries of the other kinds of areas utilized may change between census years. Knowledge of these changes is necessary for an understanding and proper use of the census data.

Population projections and forecasts made for city planning and engineering purposes implicitly assume that the urban area concerned will expand as necessary to accommodate the projected population increases. Municipal boundary changes and annexation policies can significantly affect planning assumptions, especially in rapidly urbanizing regions or in areas undergoing political restructuring.

In metropolitan areas, and particularly within the urbanized areas of such metropolitan areas, local planning agencies cannot limit their consideration to the population of a single municipality—some of which may have fixed boundaries and others of which may not—but must consider the trends within larger areas. Consequently, population projections and forecasts are best made for entire metropolitan areas and then allocated successively to the constituent counties and then to subareas of the counties. Local planning agencies also need to consider population estimates, projections, and forecasts for smaller subdivisions of their formal jurisdictions such as neighborhood units. School district, transit service, and utility service boundaries also form critical geographic units for population analysis, especially in integrated regional planning frameworks.

Accuracy

City planners and engineers, as well as demographers, have long been concerned about the accuracy of population estimates, projections, and forecasts. To date, no single method of population estimation, projection, and forecasting has proven to be more accurate than any other, though advances in statistical modeling and scenario-based planning have improved the quantification of uncertainty. Once a particular projection has been selected for use as a forecast, it is appropriate to assess its accuracy by a subsequent comparison with a census count. Attention should be focused in such a comparison on the accuracy of the projected net change and on the components of that change—births, deaths, and migration—since it is these components that are actually projected, at least in the cohort-survival method.

Lacking objective tests, projection accuracy requirements are largely a function of the use to be made of the projections or forecasts. As applied to city planning and engineering, the critical question is what effect any inaccuracies may have on the basic structure of the plans concerned. The projection tolerances should be within the range wherein only the timing and not the structure of the plans will be affected. Experience has indicated that this tolerance approximates 10 percent per decade.

Finally, it must be recognized that no one can "predict" the future and that all forecasts, however made, involve uncertainty and, therefore, must always be used with great caution. Forecasts cannot take into account events that are unpredictable but that may have a major effect upon future conditions. Such events with respect to population forecasting may include wars, epidemics, major social, political, and economic upheavals, and radical institutional changes. Moreover, both public and private decisions of a less radical nature than the foregoing can be made that may significantly affect the ultimate accuracy of any forecast. The very act of preparing forecasts that present a distasteful situation to society may lead to actions that will negate those forecasts. For example, climate-related migration, emerging technologies (e.g., remote work), and housing policy shifts have introduced new volatility in population dynamics. For these reasons, forecasting, like planning, must be a continuing process. As otherwise unforeseen events unfold, forecast results must be revised, and plans that are based on such forecasts must be reviewed and revised accordingly.

Policy Impacts

Urban policies, whether at the municipal, regional, or national level, can have substantial and often rapid impacts on population trends, land use, and infrastructure demand. Zoning reforms, investments in public transit, housing subsidies, and changes in immigration or taxation policy can either accelerate or dampen population growth in targeted areas. While forecasts are often based on demographic trends and historical data, they may not fully capture the effects of recent or forthcoming policy decisions. For this reason, planners must remain aware of evolving policy landscapes and incorporate scenario-based approaches when considering long-term plans. In some cases, the planning process itself may influence policy direction, creating feedback loops that complicate predictive efforts but highlight the need for flexibility and continuous reevaluation.

Concluding Comments on Population Projection and Forecast

All of the methods describe base population projections and forecasts on the extrapolation of historical trends. Each method has its own particular shortcomings. The planner or engineer must choose the method that fits the individual community concerned least unsatisfactorily. All projections should be accepted with reservations as indications of trends rather than absolute quantities. Whenever possible, more than one method should be applied, and a range of growth proposed. Projections and forecasts for smaller communities are much more uncertain than for larger ones, since economic change will have a greater relative demographic effect. Effects of war, restriction of immigration, and industrial change all disturb historical population curves. Recent events such as global pandemics and climate-induced disasters have also introduced new forms of demographic disruption.

With respect to estimates, projections, and forecasts, three techniques are particularly applicable in city planning and engineering. The first and best of these is the cohort-survival method, which permits explicit consideration of the factors of population change—births, deaths, and migration—and which permits population estimates, projections, and forecasts to be made by age, race, and sex. By varying the assumptions concerning birth, death, and migration rates, alternative projections can be readily made and interpreted for planning purposes.

The second of these methods consists of a variety of mathematical and graphical extrapolations to project and forecast the aggregate population of an area by identifying and extending historical trends. While not recommended for wide application, the mathematical and graphic projection methods and related mathematical curve-fitting techniques may still be useful in preparing relatively short-range population projections and forecasts for small cities and villages in largely isolated, rural settings, and for applications that require only aggregate projections and forecasts, such as projections and forecasts for the design of small sewage treatment plants or water supply facilities. These methods are generally considered too simplistic for application to communities located within metropolitan areas, and in any case their use may be questionable as a basis for decisions requiring large capital investments. However, with modern computational tools, even simple trend analyses may be integrated into larger multi-model assessments when applied carefully.

In addition, the methods, such as the shift-share technique, can be used to project and forecast the population of an area by its relation to available projections and forecasts for a larger area made by the cohort-survival method. This method produces only aggregate data.

Population projections and forecasts are best prepared at the regional or metropolitan planning level utilizing the cohort-survival method under the direction of an experienced demographer. At least three projections should be prepared: a higher projection that may be used in cases where overestimation rather than underestimation constitutes a conservative approach to the planning; an intermediate projection; and a lower projection that may be used in cases where underestimation constitutes a conservative approach. One of the three projections, or a modification of one of the projections, is selected as the forecast level for use in planning. The forecast level is then allocated to the constituent counties of the regional or metropolitan area, and then within each county to planning districts encompassing urban growth centers located within the counties. This last allocation should be made on the basis of land use development objectives set forth in an adopted regional or metropolitan plan that is not ideologically driven but is grounded in reality. Recent practices also incorporate resilience goals, environmental constraints, and social equity metrics into the allocation process.

Figure 6.3 illustrates an actual application of this approach. The upper left-hand graph of Figure 6.3 shows the population projections prepared for a large, seven-county, urbanizing region located in the midwestern area of the United States. The other seven graphs show the population projections prepared for each of the seven constituent counties of the region. Each of the three population growth projections for the region reflects an application of the cohort-survival methodology under a combination of assumed fertility, mortality, and inter-regional migration rates carefully selected from the range of possible values. The highest fertility rates considered applicable were assumed for the high-growth projection, and the lowest for the low-growth projections. Mortality rates, being relatively stable over time, were not changed. Migration rates for the high-growth projection were linked to independently prepared high-growth economic conditions, taking into account attendant employment and labor force participation rates. Similarly, migration rates for the low-growth projection were linked to independently prepared low-growth economic conditions. The cohort-survival methodology was then independently applied in a similar manner to each of the constituent counties considering inter-county migration rates. The results were then balanced with the regional projections to achieve consistency. The balanced county projections were then distributed to 60 planning analysis districts within the seven-county, 2700-square-mile planning area. The districts were delineated within the urbanized area as logical groupings of cities and villages, and within the rest of the metropolitan area as cities or villages together with related unincorporated areas. This final distribution was made on the basis of agreed-upon areawide land use development objectives. In contemporary practice, such district-level forecasts are increasingly refined using geospatial modeling, parcel-based land use data, and community engagement inputs.

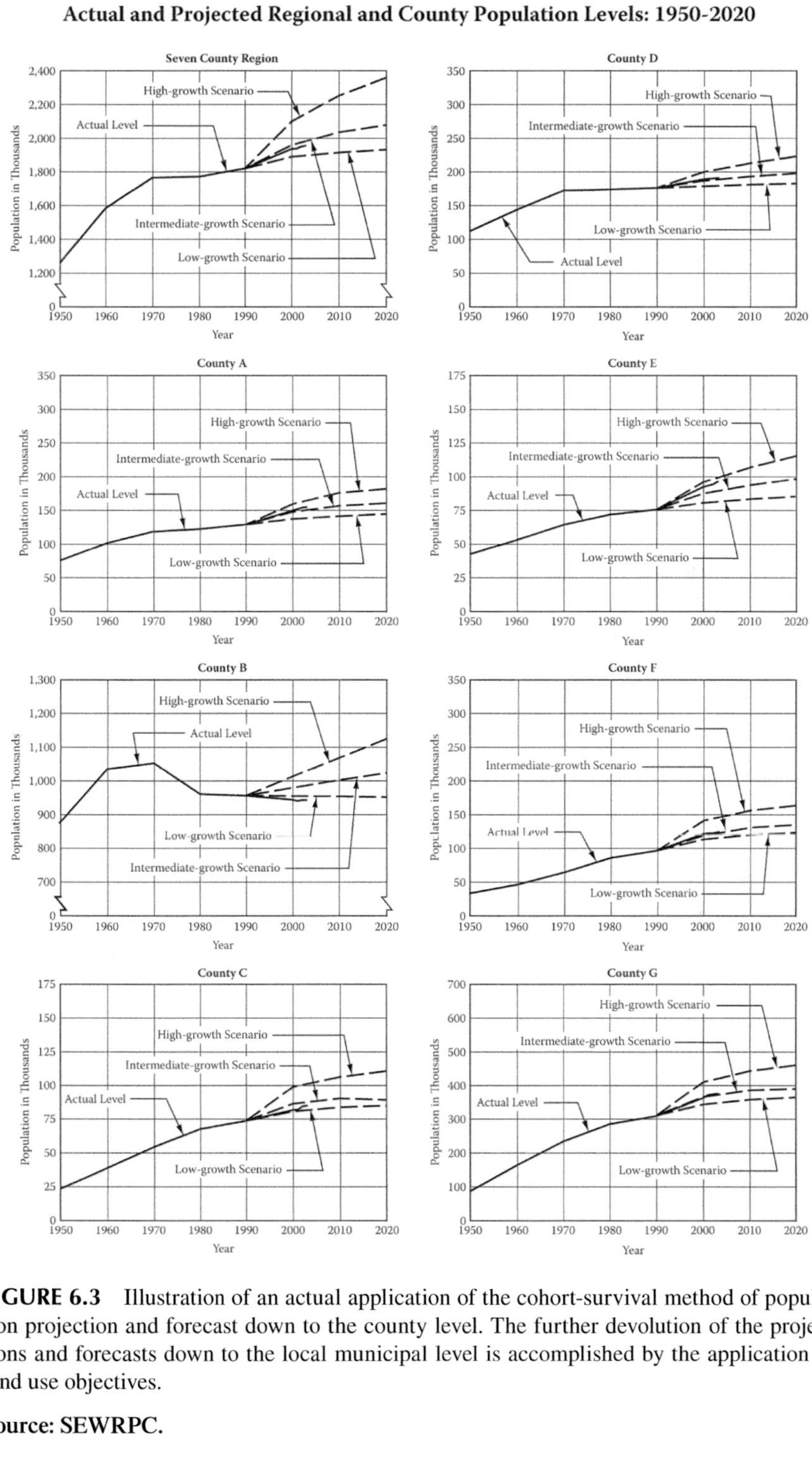

FIGURE 6.3 Illustration of an actual application of the cohort-survival method of population projection and forecast down to the county level. The further devolution of the projections and forecasts down to the local municipal level is accomplished by the application of land use objectives.

Source: SEWRPC.

OTHER POPULATION CHARACTERISTICS

As important as population size and distribution are to city planning, certain other population characteristics have implications for such planning. Some of these characteristics, such as age composition, have indirect as well as direct implications for planning, since they affect the rate of population growth and change as well as the need for housing, schools, and transportation. Other of these characteristics, such as the number, size, and composition of households, have direct implications for land use and infrastructure planning. Increasing attention is also being paid to characteristics such as digital access, language spoken at home, and disability status in inclusive planning efforts.

Age Composition

The age composition of a population is important to city planning, since age governs the time at which persons enter or leave schooling, enter the labor market, marry and form families, and retire from the labor force. Each of these events has implications for land use, housing, transportation, and community facility planning. Since each age group exerts different demands on society for facilities and services, it is important to know the number of persons in each age group now and the probable number that may be expected in the future. The needs of older groups are quite different from those of younger groups, and each group contributes differently to society.

Data on age are commonly tabulated by five-year age groups and facilitate analyses of changes through births, deaths, and migration. For fertility analyses, the total number of women aged 15–44 is significant; for school planning, the age groups 5–17 and 18–24 are significant, the latter range also being the prime military service age; for economic development analyses, the 25–64, or workforce age, group is significant; and the 65 and over, or retired age, group is significant for transportation, housing, and some types of social services planning. Sociologists have developed a number of measures of population characteristics. With respect to age composition, these include median age. For example, the median age in the United States rose from 35 in 2000 to approximately 39 in 2020, with trends indicating continued aging of the population.

Dependency Ratio

The dependency ratio is defined as the ratio of the population under 18 and over 65 to the population from 18 to 64, multiplied by 100. This ratio measures how many dependents each 100 persons in the productive years must support. Recent data indicate a rising old-age dependency ratio due to population aging, which has implications for fiscal planning and elder services.

Marital Status

Marital status is defined as single, married, widowed, or divorced. Marital status when related to other population characteristics may have implications for some types of planning. For example, the number, or percentage, of births in an area to married women as opposed to single women may have implications for educational, social, and even economic planning. In addition, trends in cohabitation and delayed marriage are influencing housing demand, particularly for smaller and multi-family units.

Family

A family is defined as two or more persons living in the same household and related by descent, marriage, or adoption. A family constitutes a household or occupied housing unit. Not all households, however, contain families, since a household may contain a group of unrelated persons or one person only. The number of married-couple families has been decreasing in the United States, while the number of families headed by single women has been increasing.

Households

The household is the basic consumer unit. Households generate the demand for urban land and for housing and are an important determinant of trip generation rates in transportation planning. Households are by definition composed of all persons who occupy a single room or group of rooms that constitute a housing unit—that is, a separate living quarter. Persons not living in households are classified as living in group quarters, such as hospitals, nursing homes, dormitories, and barracks. The number of households in an area increases with, but not proportional to, population. Household size is measured in terms of the number of persons occupying a housing unit. Household size has been decreasing for some time in the United States. As of the 2020 Census, the average household size was approximately 2.6 persons. Contributing factors to the decline in household size include smaller family sizes, increases in the number of single-person households, increases in divorce, desire of elderly persons to remain in their own households, and the desire of young persons to form their own households. Recent data indicate that family households now constitute just under 65 percent of all households in the U.S.

Other population characteristics of increasing relevance to planners include disability status, language proficiency, internet access, and citizenship status. These factors influence equitable access to housing, services, employment, and civic participation.

COMPONENTS OF POPULATION CHANGE

The crude birth rate is defined as the number of births per 1,000 members of a total population—male and female—in a fixed geographic area. Recent U.S. crude birth rates have generally declined, reflecting demographic transitions, with values around 11.0 in 2010 and below 11.0 in more recent years. The crude death rate is defined as the number of deaths per 1,000 for a population in a fixed geographic area. The crude natural increase rate is defined as the difference between the crude birth and death rates.

The age-specific fertility rate is defined as the number of births per 1,000 women of a given area and of given five-year age intervals from ages 15 to 44. The age-specific mortality rate is computed separately for males and females, and also by race, and is defined as the number of deaths in a given area by five-year age intervals.

While direct tracking of internal person movements remains limited, data from the U.S. Census Bureau's ACS, Internal Revenue Service (IRS) migration files, and change-of-address records now provide detailed insight into population mobility. The net migration rate can also be estimated by computing the net balance between total population change from one census to another and the computed intercensal natural increase, the difference being assumed to represent the net migration rate.

SPATIAL DISTRIBUTION

The spatial distribution of the population of a planning area usually changes only over relatively long periods of time; districts and neighborhoods within cities generally increase to a maximum population value then decline, and the cycle may then be repeated. The spatial distribution of the population also changes from daytime to nighttime. This type of change is not given by census data that enumerates population only by place of residence, and can be accurately calculated only from comprehensive travel—origin-destination—surveys. Planners may also now use data from mobile devices and transportation analytics platforms to supplement traditional survey data. The daytime–nighttime population levels may be important in land use, sewerage, water supply, and transportation planning.

LABOR FORCE PARTICIPATION RATES

Population and economic activity levels in a planning area are clearly related. Certain characteristics of the population, such as the number of school-age children, number of households, and size and character of the labor force, have an impact on employment patterns. The economic vitality of an area, in turn, is a major determinant of overall population levels and income patterns of an area. To relate population to employment levels, it is necessary to consider the unemployment rates, which usually range from 5 to 6 percent, but could range from 3 to 10 percent of the labor force, and multiple job holding, which may range from 5 to 10 percent of the labor force. The civilian labor force of an area is defined as that segment of the resident population who are 16 years of age or older and either employed at one or more civilian jobs or temporarily unemployed and seeking work. Thus, retired persons, homemakers, students, and institutionalized persons not seeking work are not counted in the labor force. Labor force data are often referred to as "place-of-residence" data since the labor force is enumerated by the Census on the basis of the residence of the individual members. Certain characteristics of the labor force, in addition to size, may be of interest in city planning. These include distributional, educational, and occupational characteristics.

The labor force participation rate is defined as the ratio of the number of persons in the civilian labor force to the total labor force age population, also defined as all persons 16 years of age or older. For example, in the United States those rates have recently been as follows:

	1970	1980	1990	2000
Male	0.79	0.79	0.76	0.70
Female	0.45	0.54	0.60	0.57
Total	0.62	0.66	0.68	0.63

The labor force participation rate can be used to relate the employment in an area to the resident population age 16 and older. In comparing the total population as given by cohort-survival analyses to the population derived by applying the labor force participation rate to independently prepared employment projections and forecasts, allowance must be made in the comparison for the number of individuals in the 0–16 age group.

FURTHER READING

Goodman, William I., et al. "Principles and Practice of Urban Planning," Chapter 3, *Population Studies*, International City Managers' Association, 1968.

Meyer, Zitter, U.S. Bureau of the Census. "Population Projections for Local Areas," Public Works, June 1957.

Population Reference Bureau, Inc. "Population Statistics: What Do They Mean?" *Population Profile*, Washington, D.C., March 1972.

7 Economic Data and Forecasts

If you can look into the seeds of time, And see which grains will grow and which will not, Speak then to me—

William Shakespeare
Macbeth 1606

INTRODUCTION

Current and historical data on the economy of an area are important to city planning. Such data contribute to an understanding of existing development patterns and historical trends in development and provide a framework for preparing employment projections and forecasts. Current and historical data are required on the number and type of jobs in the planning area, on the labor force of the area, and on income levels within the area. More specifically, economic studies have three purposes with respect to city planning:

1. To provide information that will assist the community in formulating economic development objectives and standards
2. To provide an independent basis for projecting and forecasting probable future population levels
3. To provide quantitative estimates of existing and probable future employment levels in the planning area for use in plan preparation

More refined employment data on occupational classifications and related data on unemployment rates, real property valuations, and tax rates may constitute additional important inputs to economic development planning as well as to comprehensive planning. Additional data—such as detailed industry sectors, educational attainment, job accessibility, broadband access, and real property values—may enhance both economic development and comprehensive planning.

Economic data provide insights into the factors that explain the origin, historical development, and potential for development of the planning area. Such data help identify the principal factors of change in the economy and of important forms of imbalance. Some of the latter may be self-adjusting, such as wage rates, housing supply, and labor force skills, while some may be non-self-adjusting, such as sharp cyclical shifts and lack of economic diversification.

Key measures of economic activity are available from the U.S. Census Bureau, Bureau of Labor Statistics (BLS), and Bureau of Economic Analysis (BEA), including labor force participation rates, job counts by sector, unemployment, income, and

DOI: 10.1201/9781003500384-7

commuting patterns. Job data are generally available at the county or metropolitan level, though distinguishing between full-time and part-time employment, as well as resident versus non-resident workers, may require supplemental sources or local surveys. Care should be taken in interpreting employment data from businesses with multiple outlets, as employment may be reported at a central location rather than distributed across the planning area. Geographic information systems (GIS) and business directories can help correct for this issue.

ECONOMIC BASE

The concept of the economic base is an old one, extending back to early economic theories such as mercantilism during the eighteenth century. The concept is still employed in studies of the urban economy. The economic base may be defined as that portion of the total urban economy that is composed of the producers of goods, services, and capital who export all, or a predominant part, of their output from the urban area, and thereby bring a flow of purchasing power into the area. For planning purposes, the export segment of the urban economy is best quantified in terms of the number of employees engaged in basic economic activities. The difference between this basic and the total employment is then defined as non-basic or local service employment. The ratio of basic to service employment approximates 1:2 to 1:3 in many developed urban areas, depending on the structure of the local economy. The ratio of total employment to population approximates 1:2.5 to 1:4, reflecting increases in labor force participation and multiple job holding. Therefore, the ratio of basic employment to population approximates 1:5 to 1:10. These ratios between basic and service employment vary from community to community and over time. Contemporary trends in remote work, gig economy participation, and automation may further affect these relationships. These variations are conceptually illustrated in Figure 7.1. Diagram A indicates that as an urban area grows, the quantitative relation of basic to service employment may be expected to reverse over time. Diagram B indicates that under a constant population level, basic employment may be expected to remain stable, but service employment may be expected to increase over time. Diagram C indicates that as the density of development increases, with the population level remaining constant, service employment may be expected to decrease. Justification for this view rests principally on accessibility. Diagram D indicates that the absence of direct competition with another service market may be expected to increase service employment over time. Diagram E indicates that as basic economic activities evolve from primary activities, such as agriculture, mining, and fishing, to secondary activities, such as manufacturing, to tertiary activities, such as financial services, service employment may be expected to increase.

In the application of the concept, the delineation of the study area concerned is of particular importance. The labor market area, as delineated by the U.S. Census Bureau, may be used for this purpose. This delineation uses a maximum commuting time limit of 90 minutes, approximating a travel distance radius of 30–40 miles from the center of the study area. Another basis that may be used for delineation of the local market area limits is the metropolitan area as delineated by the U.S. Bureau of the Census. For smaller planning areas, the convenience trade

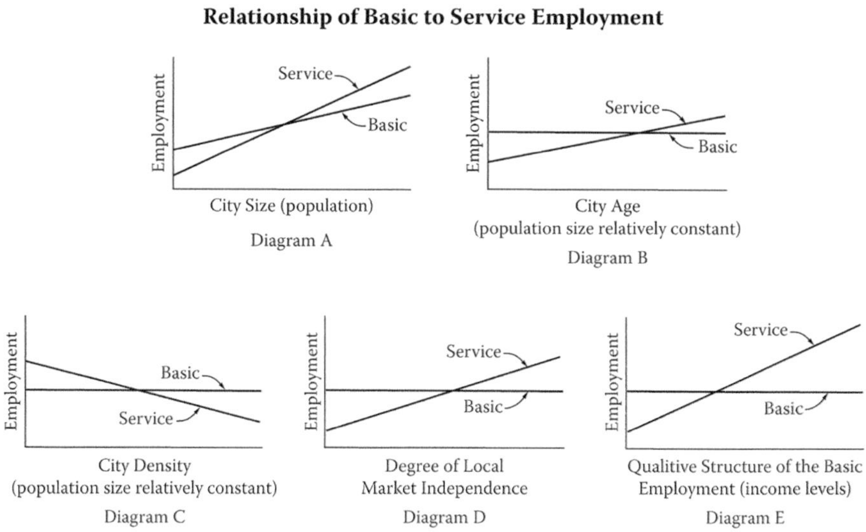

FIGURE 7.1 Illustration of the effects of specific variables on the quantitative relation of basic to service employment.

Source: International City Managers Association and SEWRPC.

area for goods and services may be used. This delineation may require locally administered surveys.

STRUCTURE OF THE URBAN ECONOMY

The U.S. Office of Management and Budget historically utilized the Standard Industrial Classification (SIC) system to collect and publish employment and other economic data. This system divided the economy into 21 major divisions (e.g., agriculture, construction, and manufacturing), further categorized into major groups and establishments, each with assigned SIC numbers.

Example:	Division:	Manufacturing Transportation Equipment Establishment	D SIC 37 GM Janesville plant

Several techniques are used to analyze an urban economy's base, including assumption, location quotient, minimum requirement, and direct investigation. The assumption technique classifies industries like manufacturing and mining as basic and sectors like education and retail as service, but it lacks empirical support.

The location quotient technique compares local employment in a sector to national averages; excess employment is assigned as basic employment. For example, if 10 percent of local employment is in food manufacturing but only 5 percent nationally, the additional 5 percent is considered basic employment.

The minimum requirement technique compares employment shares in industry sectors across similar regions, assigning the minimum observed share as service employment and the remainder as basic.

The direct investigation technique uses structured interviews with business leaders and internal data to determine export orientation, offering high accuracy but at a high cost.

In 1997, the SIC system was replaced by the North American Industry Classification System (NAICS) to better reflect the structure of the modern economy, especially in light of economic integration across North America. NAICS groups establishments with similar production processes into 20 sectors—five for goods and fifteen for services. NAICS enhances the ability to capture activity in rapidly growing sectors such as information services, technical consulting, entertainment, and health care. For instance, the manufacturing sector now includes detailed subsectors like computer and electronic products.

NAICS uses the terms "sector" and "subsector," replacing the SIC terms "division" and "major group." It also offers improved granularity with six-digit codes (compared to SIC's four), making it better suited for modern economic planning applications.

PROJECTION AND FORECAST TECHNIQUES

Economic activity projection and forecasting techniques combine expert judgment with data-driven models and are increasingly supported by computational tools and datasets. For city planning purposes, employment levels are the most useful indicators of economic activity. Four techniques are commonly used for the preparation of employment projections and forecasts: economic base analysis, dominant-subdominant industry analysis, industrial cluster analysis, and mathematical modeling using input-output and computable general equilibrium (CGE) models.

Economic Base Analysis

The economic base concept may be used to prepare employment and population estimates, projections, and forecasts. In the application of this concept, the basic and service employments are determined using one of the four techniques described in the previous section. The sum of these two employments is then taken as the total employment in the planning area.

Mathematical Model—Input-Output Method

The mathematical model–input-output method attempts to develop a quantitative description of the origins of the purchases procured and the destinations of the sales of the goods and services produced by each of the sectors of the urban economy. The technique uses a monetary measure of purchases and sales, but it is possible to convert the dollars concerned to employment by the development of ratios. The technique creates a dynamic input-output matrix that can be used to simulate the operation of the local economy and that is particularly valuable in exploring the ramifications of change within the economy. The technique also provides a "multiplier" for the study area—a factor—that can be used to assess the impact of any particular sector, and of changes in that sector, on the flow of money into the study area. The input-output technique is sound and is used in national-level economic analyses. Input-output models are widely used at local, regional, and national scales and have been extended to include environmental and energy-economic linkages. While historically expensive to implement due to data collection demands, recent advancements in automated data integration and regional modeling platforms (e.g., IMPLAN, REMI, RIMS II) have made input-output modeling more accessible to planners.

EXAMPLE APPLICATION

Figures 7.2 and 7.3 illustrate an actual application of the dominant-sub-dominant industry analysis technique to the preparation of employment projections and forecasts for a seven-county urbanizing region. The analyses identified eight dominant industries using the two-digit sector level of the SIC system, which together accounted for over 63 percent of the total employment in the region. The analyses identified five subdominant industries that together accounted for an additional 13 percent of the total employment of the region. A team of regional analysts analyzed the historical data for each dominant and subdominant industry and prepared projections and a forecast level of employment for each industry. In these analyses consideration

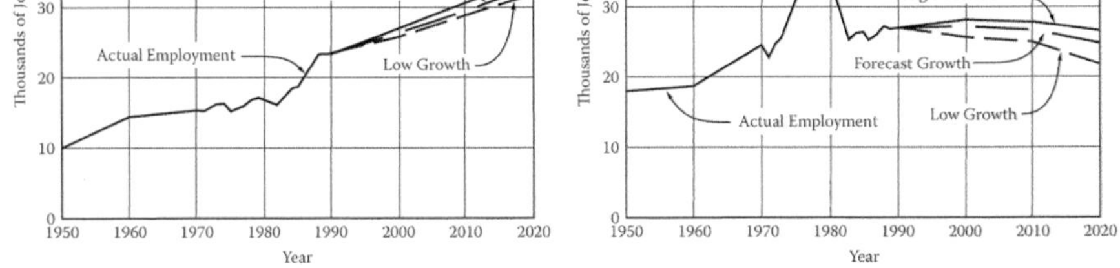

FIGURE 7.2 Illustration of the actual employment projections and forecasts made by industry under application of the dominant-subdominant industry analysis technique for the preparation of employment projections and forecasts. The year 2000 employment level forecast for the printing and publishing industry was 26,000, actual employment in 2000 was 24,500, an error of 6 percent in 10 years. The year 2000 forecast employment for the fabricated metals industry was 27,000, actual employment was 25,600, an error of 5 percent in 10 years. Similar forecasts were made for six other dominant industries in the region, and in 2000 for five subdominant industries, and an allowance was made for non-dominant, non-subdominant employment to project the total employment for the region and counties shown in Figure 7.3.

Source: U.S. Bureau of Economic Analysis and SEWPRC.

Actual and Projected Regional and County Employment Levels: 1970–2020

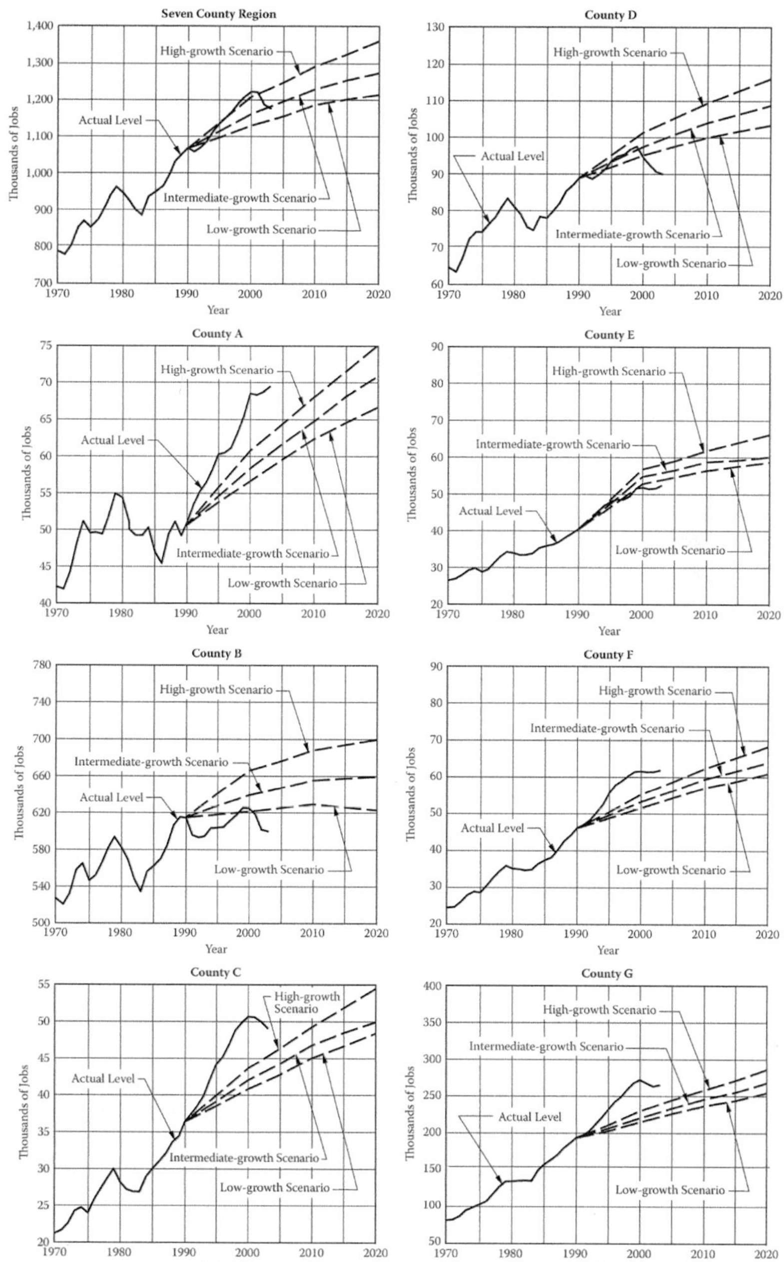

FIGURE 7.3 This figure, together with Figure 7.2, illustrates actual employment projections and forecasts made by application of the dominant-subdominant industry analysis technique. Note that while the regional forecast and actual levels of employment in 2003 were virtually identical, large errors existed in the forecasts for the individual counties.

was given to, among other factors, past industry trends, available indicators of future trends nationally, and relative industry and sector strength within the region. Particular care was given to distinguishing between what might be cyclical and what might be structural changes in the industries concerned. Actual examples of such industry-specific projections and forecasts are provided in Figure 7.2. The projections and forecasts as prepared for each industry were aggregated, and an allowance was made for other non-dominant and non-subdominant industry employment to produce the regional employment projections and forecasts shown in Figure 7.3. The regional projection was then distributed to the individual constituent counties on the basis of historical ratios to produce the county-level projections and forecasts also shown in Figure 7.3. The actual employment levels are also compared to the forecast levels to assess accuracy and bias. The regional and county projections and forecasts were further distributed to individual planning districts used in population projections, on the basis of historical experience and agreed-upon land use development objectives. The employment projections and forecasts were used directly in regional, county, and local planning efforts and were also used to assist in the preparation of the population projections and forecasts presented in the population projection and forecast example given in Chapter 6. The population and employment projections and forecasts were related by use of the labor force participation rate, the comparison being made for the 16 and over age cohorts.

PERSONAL INCOME

Another economic indicator of interest in city planning is personal income. Data of particular interest include per capita personal income and the median household income, which are reported by the U.S. Bureau of the Census for the year immediately preceding the census year. More recent estimates can be obtained from the American Community Survey (ACS), which provides annual updates at various geographic levels. Historical data should be converted to constant dollars referenced to a base year to control for inflation and enable accurate comparisons. Personal income data have particular implications for land use, housing, and transportation planning. Personal income level data are also important to private sector market-oriented planning such as for the location of new retail outlets.

PROPERTY TAX BASE

Data on the historical, current, and probable future property tax base of a planning area are important to city planning and engineering, and particularly to capital improvement programming as a plan implementation device. The data should include assessed property values, ratios between assessed and full market values (assessment ratios), property tax levies, and property tax rates. The data should be provided by major land use categories, such as residential, commercial, and industrial, and by neighborhoods and other planning analysis areas within the community. Availability of parcel-level data through GIS platforms has further improved spatial analysis of the tax base. Comparative data for other comparable communities are also useful.

IMPACT OF DECARBONIZATION ON URBAN ECONOMIES

As cities transition toward low-carbon futures, the decarbonization of urban economies is reshaping the landscape of employment, infrastructure investment, and fiscal planning. This transformation involves the gradual replacement of fossil fuel–intensive industries with clean energy sectors, electrified transportation systems, and circular economic models. These shifts influence the composition of local labor markets, often requiring workforce retraining and new economic development strategies. Decarbonization also affects municipal revenue structures—altering the property tax base and utility income—and demands long-range planning for capital improvements such as green buildings, microgrids, and carbon-neutral public assets. For planners, integrating decarbonization pathways into economic forecasts and infrastructure priorities is essential to achieving resilient, equitable, and sustainable urban growth.

CONCLUDING COMMENTS ON EMPLOYMENT PROJECTIONS AND FORECASTS

The dominant-subdominant industry analysis technique is widely considered the most practical technique to use for the preparation of employment, projections, and forecasts for city planning and engineering purposes. The data collection and analyses required for the application of this technique should be conducted under the direction of an experienced economist. Similar to population studies, at least three projections should be prepared: a higher, an intermediate, and a lower projection. One of the three projections should then be selected as the forecast level for use in planning. As with population studies, employment and related economic analyses should begin at the regional or metropolitan level and be allocated to constituent counties, then further distributed to planning districts within those counties. Using appropriate labor force participation rates, the employment projections and forecasts should then be correlated with population projections and forecasts prepared using the cohort-survival method to achieve consistency between the two sets of projections.

FURTHER READING

Goodman, William I., et al. "Principles and Practice of Urban Planning," Chapter 4, *Economic Studies*. International City Managers Association, 1968.

Isard, Walter. *Methods of Regional Analyses: An Introduction to Regional Science*. MIT Press, 1960.

8 Land Use and Supporting Infrastructure Data

> I often say that when you can measure what you are speaking about, and express it in numbers, you know something about it; but when you cannot measure it, when you cannot express it in numbers, your knowledge is of a meager and unsatisfactory kind.
>
> **William Thomson (Lord Kelvin)**
> *Lecture to the Institution of Civil Engineers*
> *3 May 1883*

INTRODUCTION

Land use studies aim to provide basic data on the various activities occupying land in the planning area. These data are used in analyzing the current pattern of urban land use and serve as the framework for formulating the long-range land use plan. The land use plan establishes the character, quality, and pattern of the physical environment to support the activities of people and organizations in the planning area. Land use planning depends on reliable population forecasts, sound economic projections, and a thorough understanding of the interrelationship of all types of urban land use, that is, those for living, livelihood, and leisure. A series of basic studies is required to furnish information on the use, nonuse, and misuse of urban land. These may include a compilation of data on the physiographic features of the planning area, a land use survey, a vacant land survey, a flood hazard survey, a structural and environmental quality survey, a cost-revenue study of land use, a land value study, a study of aesthetic and historic features of the urban area, and a survey of public attitudes and preferences regarding land use. The most basic and important of these surveys and studies is the land use survey. The information provided by the land use survey is essential to all types of physical planning. Closely related to the land use survey are the transportation and thoroughfare studies that provide concomitant data on the movement of people and goods. Also closely related to the land use survey are utility studies: sanitary sewerage, water supply, and stormwater management.

Modern planning practice increasingly integrates these studies within Geographic Information Systems (GIS), enabling real-time spatial analysis and connecting land use patterns with environmental impacts, mobility equity, and resilience planning. Traditionally, land use was implicitly defined based on surface-level urban activities aligned with economic functions such as production, distribution, consumption, and services—resulting in categories like industrial, commercial, residential, and institutional. However, modern land use frameworks recognize that land is a

DOI: 10.1201/9781003500384-8

multidimensional asset, encompassing vertical and underground uses (e.g., rooftop solar installations, underground utilities), ecosystem services, climate adaptation needs, and social equity concerns. Previously undervalued, rural and open spaces such as agricultural lands, wetlands, and forests are now acknowledged as critical components of sustainable land systems—providing food security, flood mitigation, carbon sequestration, and biodiversity conservation.

A better definition would also recognize that many land uses are, in fact, mixed uses, and this complicates any land use inventory. Mixed-use situations may be represented by different uses, such as commercial, office, and residential, on the various floors of a multi-story building occupying an ownership parcel. Alternatively, they may include mixed office, industrial, and wholesale uses occupying an ownership parcel in one or more buildings. Similarly, a parcel identified as a farm may contain residential, crop and pasture, and woodland uses, while the latter may be used for productive purposes or for resource conservation purposes. Contemporary land use classification systems (e.g., LBCS, MLUCS) accommodate this complexity using multi-attribute tagging, allowing for more nuanced representation of overlapping uses and land functions. These more complex aspects of the term land use should be kept in mind as land use classification systems and inventories are designed and carried out and as their findings are considered. Additionally, modern land use inventories are increasingly updated using remote sensing, LiDAR, mobile mapping systems, and AI-driven image classification—improving the frequency, objectivity, and detail of datasets used in planning.

MAPS FOR LAND USE STUDIES

Good maps are required for conducting land use studies. These include topographic maps showing the hypsometry and planimetry of the planning area and cadastral maps showing the location and configuration of property boundary lines. By using these maps and recent aerial photographs, if available, a series of base maps of the planning area can be prepared to collect and display definitive land use data. Contemporary land use mapping now relies heavily on digital base maps produced using Geographic Information Systems (GIS), which allow for scalable, multi-layered, and georeferenced data visualization. Instead of fixed paper map scales, users can zoom in and out dynamically on high-resolution base layers such as digital orthophotos, LiDAR-derived elevation models, and vector cadastral data. These base maps are typically compiled using national and regional geospatial data sources, such as the U.S. Geological Survey (USGS), National Map, state GIS clearinghouses, and commercial providers. Modern tools like drones, mobile mapping systems, and satellite imagery platforms (e.g., Sentinel, Landsat, or PlanetScope) provide timely updates and facilitate automated land use classification using machine learning techniques. Furthermore, mobile GIS applications enable field data collection directly into the digital environment, replacing manual annotations on paper field maps. Cloud-based collaboration tools also allow planners, engineers, and stakeholders to access, edit, and visualize land use maps simultaneously.

LAND USE SURVEY

With the widespread availability of GIS, drones, LiDAR, and AI-based image recognition, land use surveys are now more precise, scalable, and automatable than in the past. However, traditional field-based techniques remain foundational and relevant, particularly for validating remotely sensed data. Depending on the size of the area, the size of the planning staff, and the time available, the actual land use survey can be accomplished by remote sensing (e.g., satellite imagery or drone surveys), interpretation of aerial photography, on foot or by automobile, that is, by "windshield inspection," or by a combination of these methods. In addition to maps for recording data, a land use classification system must be established in advance, together with a standardized system of field notations corresponding to the land use classifications. A standardized classification system has been developed for use in transportation system planning, which is also suitable for land use and utility system planning. This classification system, together with a digital and color-coding system, is set forth in Table 8.1. The field surveys, office compilations, and analyses may be complicated by the presence of mixed uses and multi-storied buildings. Digital tools such as GIS software (e.g., ArcGIS or QGIS), spatial databases, and 3D city models can assist in managing this complexity. Measurements of the areas mapped may be made from the field inventory maps by planimeter or nomograph, or using GIS and remote measurement tools if the maps are available in digital format.

The traditional or classic land use classification system outlined in Table 8.1 has long served city planning and engineering purposes effectively, especially when integrated into transportation system planning involving the development and application of traffic simulation models that relate trip generation and travel to land use. In current practice, Table 8.1 is often implemented within GIS platforms as a base classification layer for zoning, infrastructure assessment, and scenario modeling. Its simplicity and long-standing use make it ideal for baseline comparisons and historical continuity. The development of computerized, parcel-based land information systems permits more complex land use classification systems to be developed that more fully recognize the complexity of the land use concept. One such system has been termed the land-based classification standards system, a system promulgated by the American Planning Association.

The land-based classification standards system represents a major departure from the traditional system. It proposes to provide data by parcel on activity, function, structure type, site development character, and ownership. Activity is defined as the actual use of the land as indicated by observable, physical features such as residential, commercial, and industrial. This classification apparently corresponds, more or less, to the primary categories used in the traditional classification system. The activity classification has 115 activity categories and codes. Function is defined as the type of establishment using the land. It provides an indication of the type of economic enterprise to which the land is devoted. Thus, an activity may be categorized as "office," while the function may be categorized as "finance and insurance," and, in greater detail as "bank, credit union, or savings institutions." The function classification has 367 function categories and codes. Structure type is defined as the type of building on the parcel, such as "warehouse or storage facility," and in greater

TABLE 8.1
Land Use Classification System

Land Use Code	Land Use Description	Land Use Color	Land Use Code	Land Use Description	Land Use Color
	Residential			***Government and Institutional***	
111	Single-Family	Yellow		Administrative, Safety, and Assembly	
120	Two-Family Low Rise (1–3 Stories)	Orange	611	Local	Dk. Blue
				Educational	
141	Multi-Family High Rise (4 or More Stories)	Orange	641	Local	Dk. Blue
150	Mobile Homes	Yellow	642	Regional	Dk. Blue
199	Residential Land Under Development	Yellow		Group Quarters	Dk. Blue
	Commercial		661	Local	Dk. Blue
210	Retail Sales and Service—Intensive	Red	662	Regional	Dk. Blue
220	Retail Sales and Service—Nonintensive	Red		Cemeteries	Dk. Blue
299	Retail Sales and Service Land Under Development	Red	681	Local	Dk. Blue
	Industrial		682	Regional	Dk. Blue
310	Manufacturing	Lt. Gray	699	Governmental and Institutional Land Under Development	Dk. Blue
340	Wholesaling and Storage	Lt. Gray		***Recreational***	
360	Extractive	BlackCross-Hatch		Cultural/Special Recreation Areas	
399	Industrial Land Under Development	Lt. Gray	711	Public	Dk. Green
	Transportation		712	Nonpublic	Dk. Green
	Motor-Vehicle-Related			Land-Related Recreation Areas	

(Continued)

TABLE 8.1 *(Continued)*
Land Use Classification System

Land Use Code	Land Use Description	Land Use Color	Land Use Code	Land Use Description	Land Use Color
411	Freeway	Black	731	Public	Dk. Green
414	Standard Arterial Street and Expressway	Pink	732	Nonpublic	Dk. Green
418	Local and Collector Streets	Black		Water-Related Recreation Areas	Dk. Green
425	Bus Terminal	Dk. Purple	781	Public	
426	Truck Terminal	Dk. Purple	782	Nonpublic	
	Off-Street Parking		799	Recreation Land under Development	Dk. Green
430	Multiple-Land-Use-Related	White		***Agricultural***	
431	Residential-Related	Yellow w/ Red Hatch	811	Cropland	White or Tan
432	Retail Sales and Service-Related	Red Hatch	815	Pasture and Other Agriculture	White or Tan
433	Industrial-Related	Black Hatch	816	Lowland Pasture	White or Tan
434	Transportation-Related	Dk. Purple w/Hatch	820	Orchards and Nursery	White or Tan
435	Communication- and Utilities-Related	Dk. Purple w/Hatch	841	Special Agriculture	White or Tan
436	Government- and Institutional-Related	Dk. Blue w/Hatch	871	Farm Buildings	White or Tan
437	Recreation-Related	Dk. Green w/Hatch		***Open Lands***	
	Rail-Related		910	Wetlands	Lt. Green
441	Track Right-of-Way	Dk. Purple		Unused Lands	
443	Switching Yards	Dk. Purple	921	Urban	Cream
445	Stations and Depots	Dk. Purple	922	Rural	Cream
	Air-Related		930	Landfills and Dumps	Black Cross-Hatch
463	Air Fields	Dk. Purple	940	Woodlands	Med. Green

(Continued)

TABLE 8.1 *(Continued)*
Land Use Classification System

Land Use Code	Land Use Description	Land Use Color	Land Use Code	Land Use Description	Land Use Color
465	Air Terminals and Hangers	Dk. Purple	950	Surface Water	Lt. Blue
485	Air Terminal	Dk. Purple		***Supplemental Land Use Suffix Codes***	
499	Transportation Land Under Development	Dk. Purple	X	High-Density Residential	
	Communication and Utilities		M	Medium-Density Residential	
510	Communication and Utilities	Dk. Purple	L	Low-Density Residential	
599	Communication and Utility Land Under Development	Dk. Purple	S	Suburban-Density Residential	
			F	Woodlands	
			G	Wetlands	
			H	Unused Lands	
			P	Agricultural Land Preservation Area	

detail, “refrigerated warehouse for cold storage.” The structure-type classification has 399 structure-type categories and codes. Site development character is defined as the state of development of the land, such as “developed site – no structures,” and in greater detail, “outdoor storage area.” The site development classification has 46 site categories and codes. Ownership is virtually self-defining. The ownership classification has 47 categories and codes.

The structure, site development, and ownership categories of the proposed land-based classification standards system clearly have applicability in the development of computerized, parcel-based land information systems, as does the function category. While computerized land information systems can readily accommodate complex inventory systems, such as the land-based classification standards system, that have multiple data fields for each parcel, the use of such systems has certain drawbacks. These include greater costs associated with system implementation, data maintenance, user training, and a lack of backward compatibility with legacy datasets. In small or resource-constrained jurisdictions, these barriers may limit adoption. The use of the proposed system as a substitute for the traditional land use classification system in city planning and engineering is problematic due to needless complexity, attendant costs, and a lack of comparability with historical land use data. Pending further development of the state of the art, continued use of traditional land use classification systems for conducting land use inventories in city planning and engineering is probably advisable.

Analysis and Presentation of Data

Once the measurements of the areas devoted to the various land uses have been completed, a statistical summary is made for each subunit of the planning area and for the planning area as a whole. The subunits may be census tracts, well-defined residential neighborhoods, commercial or industrial districts, or other areas. An existing land use map is then prepared for display purposes. A standard color-coding system, such as that provided in Table 8.1, should be used in the preparation of the map. The survey data represent conditions existing on a given date, and the utility of the data will decrease with time as changes in land use occur. To extend their usefulness, dynamic updating through integration with permit databases, parcel records, and real-time satellite or drone imagery is increasingly practiced. Although the data can be kept reasonably current using building and demolition permit data, new surveys should be conducted decennially to align with census data. Many planning agencies now aim for more frequent updates (e.g., annually or biennially), especially in fast-growing urban areas, using automated data pipelines and digital tools.

In addition to the application of the land use inventory data in land use and transportation planning, the land use inventory data provides information for a number of special planning applications. In his book entitled Land Uses in American Cities, Harland Bartholomew has stated the case for land use data most succinctly. The statement in that book, quoted below, serves as a capsule summary of land use operations, as an admonition to keep the data current, and as an introduction to the problems of forecasting future land requirements.

The applications of land use data for planning purposes are manifold. For example, they can be used to determine commercial markets, to locate institutions such as churches and schools, or for zoning purposes. Therefore, the type of statistical analysis in any given situation will be determined by the problems under study. In planning studies, with which we are concerned here, it is essential to know the amount of land used for various purposes. Computations of lot and parcel areas, arranged according to each major type of use, should be made for individual blocks, then summarized for permanent unit areas or neighborhoods, and finally for the entire community. The result of these summaries is generally expressed as an area in acres, in percentages of total areas, and as a ratio of land used to given units of population. Zoning that is based on the facts of actual use of land will have far greater validity than that based upon opinions unsupported by such facts.

In conclusion, the value of land use surveys tabulated in the manner described lies in comparative statistics. But, as with all comparisons of this nature there are definite limits of applicability. A community's future land use requirements cannot be projected with complete accuracy on a basis of current ratios. Likewise, a comparison of land uses between two or more communities will disclose differences due to character and physiography. However, in both cases such comparisons can be instructive.

In combinations with other basic studies, and with good judgment, current land use data offer a factual base for improved planning and zoning practices.

Example Land Use Map and Summary Table

An example of a land use inventory for a small village is provided in Figure 8.1 and Table 8.2. The existing land use map is compiled on a base map assembled from reductions of the large-scale cadastral map sheets available for the area. The various

Example of an Existing Land Use Map

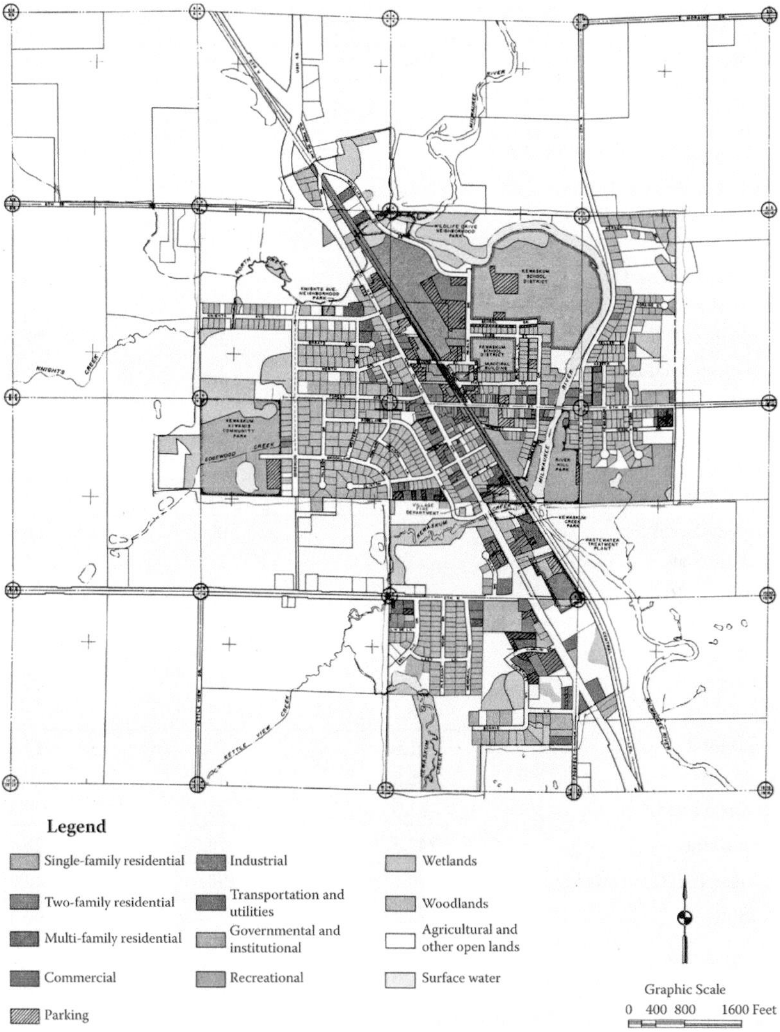

FIGURE 8.1 Illustration of an existing land use map utilizing the color code given in Table 8.1. The attendant quantified data are given in Table 8.2.

TABLE 8.2
Summary of Existing Land Use Attendant to Existing Land Use Map Shown in Figure 8.1

Land Use Category	Number of Acres	Percent of Subtotal (Urban or Nonurban)	Percent of Total
Urban[a]			
Residential			
Single-Family[b]	179.9	34.4	20.6
Two-Family	9.4	1.8	1.1
Multi-Family	20.7	4.0	2.4
Subtotal	**210.0**	**40.2**	**24.1**
Commercial	32.4	6.2	3.7
Industrial	33.6	6.4	3.8
Transportation and Utilities			
Arterial Streets and Highways	27.3	5.2	3.1
Collector and Local Streets	76.5	14.6	8.8
Railways	10.2	1.9	1.2
Communications, Utilities, and Others	5.1	1.0	0.6
Subtotal	**185.1**	**35.4**	**21.2**
Governmental and Institutional	69.7	13.3	8.0
Recreational[c]			
Public	38.2	7.3	4.4
Private	20.1	3.8	2.3
Subtotal	**128.0**	**24.4**	**14.7**
Urban Land Use Subtotal	**523.1**	**100.0**	**60.0**
Nonurban			
Natural Areas			
Water	25.9	7.4	3.0
Wetlands	49.2	14.1	5.6
Woodlands	26.5	7.6	3.0
Subtotal	**101.6**	**29.1**	**11.6**
Agricultural	166.5	47.7	19.1
Other Open Lands[d]	81.0	23.2	9.3
Subtotal	**247.5**	**70.0**	**28.4**
Nonurban Land Use Subtotal	**349.1**	**100.0**	**40.0**
Total	**872.2**	—	**100.0**

Source: SEWRPC.

[a] Includes related off-street parking areas for each urban land use category.

[b] Includes farm residences but not farm buildings included in the agricultural land use category.

[c] Includes only those areas used for intensive outdoor recreational activities.

[d] Includes unused lands.

land uses are displayed utilizing the color code given in Table 8.1. The inventory findings are summarized in Table 8.2.

Note that residential use is the singularly largest land use category, with over 40 percent of the urban land use pattern. This is true for most urban areas. The nature and extent of this important land use are major determinants of the type and location of the transportation facilities, utilities, and community facilities needed to serve local residents. The next largest land use category consists of public streets, which make up almost 20 percent of the urban land use pattern. This is also true for most urban areas. Commercial and industrial uses have surprising small percentages of the urban land use pattern, yet provide much of the economic base. It should also be noted that there was still a substantial amount of land in agricultural use within the boundaries of the village concerned, clearly a temporary use, intended to be converted to urban use as the population of the village grows. Also note that based on a resident population of 2,658 persons, the total amount of land devoted to urban use approximates 0.20 acre per capita. Based on a size of 2.72 persons per household, the amount of land devoted to residential use approximates 0.54 acres per household, or approximately 9600 square feet per household. The latter figure would approximate an average-sized residential lot in the village.

Use in Forecasting Space Requirements

One of the purposes of the land use survey is to develop data that can be used in estimating probable future land use requirements. This may be done by relating the findings of the land use survey to current population and employment data. The resulting findings may be expressed as existing ratios, such as persons per acre, or employees per acre for each of the various land use categories. These ratios can then be applied to relevant population and employment projections and forecasts to provide estimates of future land requirements. A brief review of the estimating sources and measuring units for the major types of land use may be helpful as an introduction to the process of forecasting space requirements, some of which are based on a level of activity, and others predominantly on a neighborhood- or district-level need.

While no strict universal standards exist, widely referenced planning benchmarks and guidance documents (e.g., APA, HUD, ITE) are now routinely used as starting points for assumptions. These standards are also modified by sustainability goals, zoning reforms, and technology-driven changes. Reasonable estimates can be made of the future space requirements for each class of use in a community, and these estimates can be employed in the preparation of the land use plan. The measures used to estimate space requirements are frequently based on current space use, modified by anticipated impacts of new technology and legal requirements in zoning, land subdivision control, and housing codes. Allowances for auxiliary space needs for off-street parking, off-street loading, and landscaping may be necessary. Space standards are based on a unit of measurement, such as a person, household, worker, or shopper. For this reason, population and employment data, together with land use data, are fundamental in determining guidelines for use in future space requirements.

Industrial Land

Industrial land use requirements may be expressed in terms of acres of site area per employee. Employment per net industrial acre tends to vary with the location of the establishment, the nature of the manufacturing process, and the extent of automation. Central city plants may be multi-storied to economize on land requirements. Outlying plants require considerably more land to accommodate the newer single-storied plant structures, for off-street loading and parking space, and for landscaped open space. Employment by peak shift rather than total employment may be a more realistic measure for determining density, that is, the number of employees per acre. The site area occupied by a specific industrial use may include area reserved for the normal expansion of an existing plant as well as for accommodating new industrial processes. Recent trends such as e-commerce logistics, automated warehouses, and last-mile distribution facilities increasingly influence industrial land requirements, often resulting in larger sites with fewer employees per acre. Planners also consider the integration of renewable energy infrastructure, freight hubs, and adaptive reuse of older industrial lands.

Wholesale Land

Employment has also been used as a measure for analyzing wholesale space needs. However, modern GIS-linked business registries, parcel-based databases, and NAICS-coded employment data now provide detailed information on wholesaling activity. This allows more accurate forecasting, especially when distinguishing between bulk distribution centers, decentralized last-mile facilities, and specialized wholesale clusters.

Commercial Land

Needs for commercial use spaces, which are primarily retail stores and office buildings, can be based on a number of units of measure. Future population levels are extensively used as a simple measure. However, projection of retail sales volume, based on future income estimates, has also been applied by market analysts in determining space needs of new shopping centers. This technique has also been used in the planning of commercial areas in new towns and in urban renewal areas.

To determine the amount of land to be allocated to office buildings, employment is considered a more suitable indicator than population. Today, employment data at finer geographic scales is available through federal sources (BLS, Census LEHD) and commercial datasets, allowing more precise office space projections. Remote work, coworking spaces, and hybrid office models have significantly altered demand for centralized office space, requiring planners to consider both decentralization and repurposing of surplus office buildings. Land area requirements in commercial areas that serve regions should be distinguished from those in commercial areas that serve communities and neighborhoods, and separate projections should be made for each category. Neighborhood requirements are related directly to the population to be served.

Daytime population is another unit of measure that is used for estimating nonresidential space requirements. The daytime population is particularly useful in estimating future space needs of central business districts. Today, anonymized mobile phone location data, Global Positioning System (GPS)-based travel demand models, and traffic APIs complement traditional surveys, providing real-time and high-resolution measures of daytime population.

Governmental and Institutional Land

The amount of land needed for public buildings differs in each planning area, particularly where a number of governmental units, such as city, county, state, or federal, are involved. While these space needs are related most closely to population growth and the stage of urban development, other factors may be decisive. There are no general standards, except possibly for educational, hospital, and recreation facilities. Nevertheless, present deficiencies, or surpluses, can be determined by survey, and future land requirements estimated accordingly. In all of these calculations, projections of population groups to be served are a basic requirement. In addition to traditional facilities, modern planning considers emergency management centers, integrated health campuses, and digital service hubs, some of which reduce the amount of physical space required.

Transportation, Communications, and Utilities

Pending up-to-date survey data, or specific long-range plans, space needs for this group of linked uses can be related to population growth, the status of present facilities, and expected technological advances. Particular attention should be given to probable distribution of the future population, to trends in industrial growth, to public and private policies on industrial development, and to expansion in communication channels. The importance of conducting parallel transportation studies and preparing a comprehensive arterial street plan cannot be overemphasized. Modern land use forecasting also accounts for electric vehicle (EV) charging infrastructure, 5G and broadband networks, renewable energy systems, and micromobility hubs. These elements increasingly shape spatial requirements for infrastructure planning.

Residential Land and Neighborhood Facilities

Finally, the amount of land to be devoted to residential use depends on the rate at which new families are formed, on in-migration, and on the effect of housing codes, slum clearance, and private demolition on the supply of dwelling units. Population forecasts can be converted to household forecasts—that is, to dwelling unit forecasts—by assuming an average household size for the forecast period. The actual amount of space allocated to housing will depend on the densities selected and the assumptions made regarding the percentages of single-family units, duplexes, row houses, garden apartments, and high-rise apartments to be built. Neighborhood facilities are closely related to residential development, so that schools, playgrounds, and local shopping area requirements can be computed on a per-household basis or according to the total population of the neighborhood. Modern approaches also consider accessory dwelling units (ADUs), micro-apartments, and higher-density housing made possible by zoning reforms. Demographic changes—such as smaller household sizes and aging populations—are factored into forecasting models, along with equitable access to neighborhood amenities.

UTILITIES

Urban land uses are supported by private and public utilities that provide the individual land uses with sanitary sewerage, water supply, stormwater management, electric power, natural gas, telecommunications, and solid waste collection and disposal. Urban development is highly dependent on the availability and quality of the utility services, which significantly influence the pattern of land use development. Moreover, certain utility facilities are closely linked to the surface and groundwater resources of an area and may, therefore, directly affect the quality of the resource base. This is particularly true of sanitary sewerage, water supply, and stormwater drainage facilities, which are, in a sense, modifications or extensions of the natural lake and stream system of an area and of the underlying groundwater reservoir. Knowledge of the location and extent of the service areas and of the capacities of these utilities is, therefore, essential to intelligent city planning.

Planning for the provision of electric power, natural gas, telecommunications, and solid waste collection and disposal services is now typically coordinated between private providers, regional authorities, and municipal governments, subject to regulation and oversight to ensure resiliency, equity, and sustainability. Increasingly, electric utilities also plan for renewable integration, distributed energy resources (DERs), and microgrids. Telecommunications planning now includes broadband and 5G networks, which are treated as critical infrastructure in many regions. For city planning purposes, the inventories of these facilities and services should include not only service areas and quality/cost data, but also resiliency measures, climate risk assessments, and capacity for future technological upgrades.

Planning for the provision of sanitary sewerage, water supply, and stormwater drainage is usually the primary responsibility of the city engineering function, carried out in cooperation with the city planning function. For city planning purposes, the inventory of these facilities may be limited to delineation of the existing and proposed service areas; a map of the configuration of the facilities concerned is shown in Figure 8.2.

For the sanitary sewerage system, the location, capacity, and adequacy of the treatment plant and of any pumping or lift stations and attendant force mains should be known, together with any inadequacies in the gravity flow and pressure collection and conveyance facilities that would result in overflows and bypasses and that might constrain service extensions. The disposal point of treated effluent to receiving surface waters should be known, along with the level of treatment required to meet water quality standards in the receiving water areas. For the water supply system, planners must consider not only source adequacy, treatment, pumping, storage, and distribution, but also groundwater sustainability, climate change impacts on water availability, and water conservation technologies such as reuse and recycling.

Urban stormwater management facilities may include curbs and gutters, catch basins and inlets, gravity flow sewers, storage and infiltration facilities, and, in some cases, pumping or lift stations and force mains. Suburban facilities usually include road ditches, parallel and cross culverts, and interconnected surface water courses. For stormwater management facilities, it is important to know the service areas, facility configurations, points of discharge to and identification of receiving watercourses,

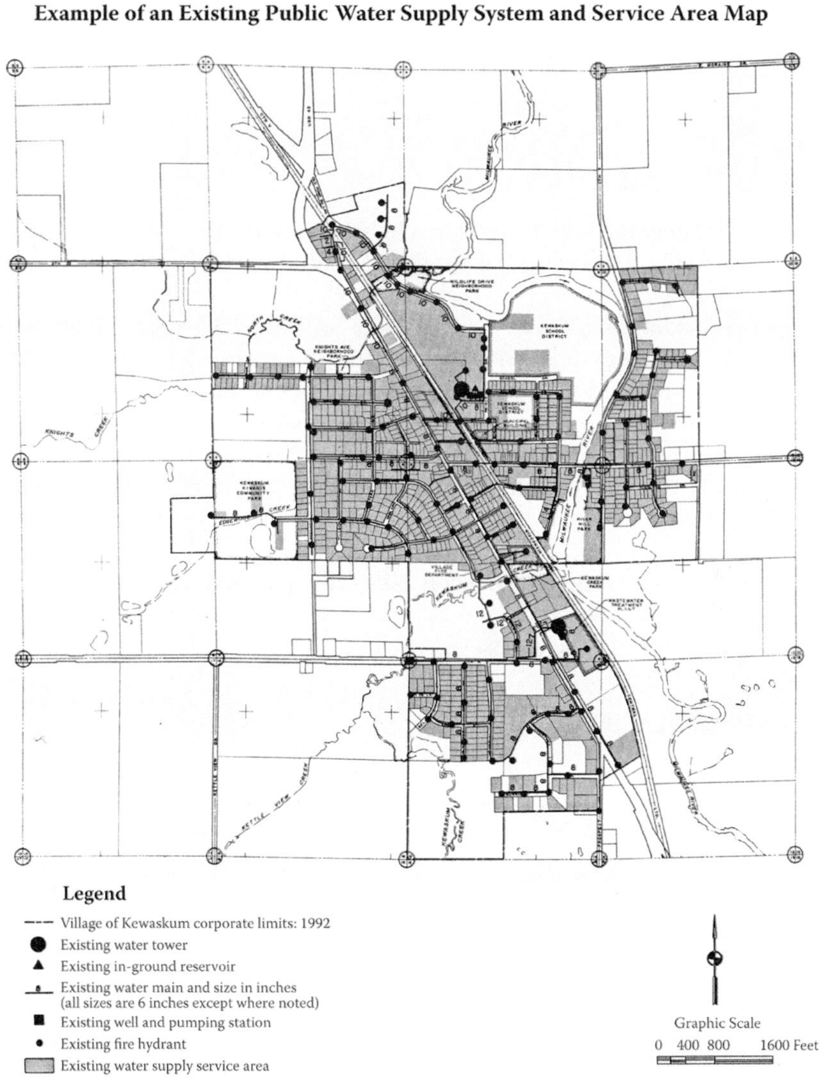

FIGURE 8.2 Illustration of a public water supply system and service area map compiled in support of land use planning operations. More detailed inventory maps are required in support of actual utility system planning operations.

and areas of ponding and flooding. Additionally, green infrastructure strategies such as permeable pavements, bioswales, rain gardens, and constructed wetlands should be considered, as they also address non-point source pollution.

The inventory requirements for city planning purposes are not intended to meet the needs of the detailed systems engineering and design that are the responsibility of the city engineering function. Indeed, such systems engineering is not addressed in this chapter. The detailed inventories required are described in engineering texts and are now best managed through GIS-based public works management systems and integrated smart-city infrastructure platforms.

COMMUNITY FACILITIES

Certain community facilities are required to help promote the public health, safety, and general welfare. These are generally, but not always, publicly provided and include primary and secondary level schools, public libraries, fire and police stations, government buildings such as city halls and post offices, child care facilities, health care facilities including clinics and hospitals, and cemeteries. These facilities should be identified in the needed land use inventories. For city planning purposes, basic information on facility locations, service areas, condition, accessibility (including ADA compliance), and capacity should be collected. Equity in spatial distribution is an increasingly emphasized consideration, especially in underserved communities. Detailed planning and operation of these facilities are typically managed by specialized agencies—such as school districts, library boards, and public safety departments—working in coordination with city planning staff. Collaboration ensures alignment with broader land use and transportation plans, climate adaptation goals, and demographic trends.

FURTHER READING

Bartholomew, Harland, et al. *Land Uses in American Cities*. Harvard University Press, 1955.
Goodman, William I., et al. "Principles and Practice of Urban Planning," Chapter 5, *Land Use Studies*. International City Managers' Association, 1968.
Silberstein, Jane & Maser, Chris. *Land-Use Planning for Sustainable Development* (2nd edition). CRC Press, 2013.

9 Natural Resource Base Inventories

> To waste, to destroy our natural resources, to skin and exhaust the land instead of using it so as to increase its usefulness, will result in undermining in the days of our children the very prosperity which we ought by right to hand down to them amplified and developed.
>
> **President Theodore Roosevelt**
> *Speech to Congress of the United States, 1908*

INTRODUCTION

City planning and engineering must recognize the existence of a limited natural resource base to which both rural and urban development must be properly adjusted in order to ensure a pleasant and habitable environment for life. Land and water resources are limited and subject to grave misuse through improper land use and public works development. The preparation of city plans must be based, in part, on careful assessment of the effects of alternative plan proposals on the underlying and supporting natural resource base.

Such assessment requires the collection and analysis of a great deal of information about the natural resource base of the planning area and its ability to sustain development. Such information should include definitive data on climate, air quality, physiography, geology, mineral resources, soils, surface water resources and associated shorelands and floodlands, groundwater resources and associated recharge and discharge areas, woodlands, wetlands, significant wildlife habitats, and areas having scenic, historic, scientific, and recreational value. Such definitive data should be in a form readily relatable to data on the cultural base of the planning area, particularly to data on land use.

CLIMATE

Climate, especially extreme variations in temperature, precipitation, and snow cover, directly affects the growth and development of an area. Climate determines the recreational pursuits that can be followed in an area. It also has economic implications affecting the kinds of crops that can be grown and the yields. It also affects the design of buildings and the cost of providing public facilities and services. Certain climatological data are, therefore, relevant to city planning and engineering. These data include definitive historical data on ambient air temperatures, including seasonal and monthly maximum and minimum average temperatures; data on precipitation, including annual and seasonal patterns and frequencies; data on frost depths;

DOI: 10.1201/9781003500384-9

data on snowfall and snowfall accumulations, including seasonal patterns and frequencies; data on droughts, including intensity, duration, and frequency; and data on prevailing air movements within the planning area. In many areas, data on the frequency, intensity, and paths of severe storms—such as hurricanes, tornadoes, and atmospheric rivers—are critical for hazard mitigation planning. Additionally, with increasing climate variability, planning now often incorporates climate change projections and vulnerability assessments based on future scenarios. Data on wind magnitude and direction, the amount of daylight and sky cover, and evaporation may be desirable, if not necessary, for some types of planning. Among the particularly important and useful data on climate are rainfall frequency–duration–intensity data, which have direct application in city engineering.

Much of the needed climatological data are readily available from the National Climate Data Center of the National Oceanic and Atmospheric Administration (NOAA). In addition, the U.S. Climate Resilience Toolkit, the Intergovernmental Panel on Climate Change (IPCC), and regional climate centers provide scenario-based climate projections, risk indices, and adaptation tools. The available data, however, must be collated and presented in a form suitable for use in city planning and engineering.

AIR QUALITY

Air quality clearly affects the growth and development of an area, and particularly so when ambient air quality levels exceed federal standards and result in the delineation of "non-attainment" areas. Certain data on ambient air quality and air pollutant emissions are therefore relevant to city planning. These include data on particulate matter, sulfur oxides, carbon monoxide, lead, hydrocarbons, and nitrogen dioxide, the latter two being precursors to the formation of ozone. Pollution sources are classified as point, line, and area. Major point source emissions include major industrial establishments and electric power generating stations. Major line sources include motor vehicles operating on the street and highway system of the planning area. Area sources include small industrial establishments, dry-cleaning operations, gasoline marketing, and incinerators, as well as residential heating. The interaction of the physiography, weather, and air quality may be important in some areas where events such as inversions may cause significant health concerns. Nowadays, urban heat islands and the proximity of vulnerable populations have also become important factors in air quality management and equity-focused planning.

The U.S. Environmental Protection Agency (EPA) has promulgated ambient air quality standards in the form of the minimum levels to be maintained to ensure the protection of human health—primary standards—and welfare—secondary standards. Individual states can adopt more stringent standards. The determination of ambient air quality levels, the inventory of pollutant sources, and the preparation of air quality management plans are complex technical tasks requiring the installation and operation of air quality monitoring stations, the conduct of massive pollutant source inventories, and the development and application of air pollutant emission and ambient air quality simulation models. These tasks are therefore best left to state regulatory and regional or metropolitan planning organizations that have the formidable funding and staff resources required. These agencies now also increasingly integrate satellite remote sensing data, low-cost sensor networks, and air quality modeling platforms such as the EPA's CMAQ (Community Multiscale Air Quality) model into planning workflows. Air quality planning efforts are closely related to federally required areawide transportation planning efforts and to the traffic demand and level of service condition data produced by the transportation planning programs. County and local municipal planners and engineers, while not usually directly engaged in air quality management planning, should, however, understand the processes involved and be able to interpret and present the inventory data and plan proposals produced by the state and areawide planning efforts for local planning purposes.

PHYSIOGRAPHY

The physiography of an area may be defined as the surface features of an area, including the landforms and surface drainage pattern. As a practical matter, the large-scale topographic maps described in Chapter 5 in effect provide most of the physiographic data necessary for city planning and engineering purposes. The rationale for this conclusion is also presented in Chapter 5.

GEOLOGY

Knowledge of the bedrock and surficial deposits overlying the bedrock is important to city planning and engineering. Bedrock and overlying surficial deposit conditions, and particularly depth to bedrock, directly affect the construction costs of such urban development projects as public utilities, particularly those requiring extensive trenching or tunneling. Moreover, the placement of urban improvements in relation to the bedrock and overlying surficial deposits may directly or indirectly affect the quality and quantity of the groundwater resources of an area. The geological bedrock formations underlying a city planning area have usually been mapped and characterized by state agencies—such as the state geologist—or by state academic institutions. Absent such available data, collation and analyses of existing well-boring logs, foundation investigation studies, and supplemental borings made under the direction of an experienced geologist would be required. In recent years, 3D subsurface models and Geographic Information System (GIS)-based geological datasets have become more widely available through state geological surveys and the U.S. Geological Survey (USGS), enhancing planners' and engineers' ability to assess geotechnical constraints and opportunities.

MINERAL RESOURCES

The most common mineral resources associated with expanding urban areas are non-metallic deposits of sand, gravel, and building stone. These may be important sources of construction materials required for continuing development. City planning should take into consideration the location of these resources, since urbanization of lands overlying the deposits may make it economically impossible to utilize the resources. This may eventually result in severe shortages of, and concomitant increases in, the costs of these essential materials, or the need to mine sensitive natural resource areas. State resource agencies and geological surveys often maintain updated maps and databases identifying active and potential mineral resource extraction zones, which can be integrated into land use planning efforts to avoid future conflicts between development and resource protection.

SOILS

The soils of an area are among the most important elements of the natural resource base, influencing both rural and urban development. The soil resource has been subject to grave abuse and misuse through poorly planned, improper land use development. Serious health, safety, and water pollution problems have been created by failure to properly consider the capabilities and limitations of soils during the planning and design stages of both rural and urban development projects. Such problems are usually very costly to correct and include malfunctioning on-site sewage treatment and disposal systems, surface and groundwater pollution, footing and foundation failures, excessive infiltration of clear water into sanitary sewerage systems, excessive operation of foundation drain sump pumps, soil erosion, and stream and lake sedimentation. Knowledge of the soil resource and its ability to sustain development not only helps avoid such problems but also can contribute to reducing urban development and maintenance costs. The placement of streets and highways on unstable soils and the excavation of basements and utility trenches in areas of high groundwater result in additional site preparation, construction, and maintenance costs. Such costs may include the costs of the removal of poor soils and their replacement with stable materials and the use of tight sheathing and dewatering systems to control groundwater seepage during construction.

To help avoid further abuse and misuse of this important element of the natural resource base, definitive data are required about the geographic location of the various kinds of soils, about the physical, chemical, and biological properties of the soils, and about the capability of the soils to support various kinds of rural and urban land uses. For city planning and engineering purposes, the necessary soil studies should permit, on a uniform areawide basis, preliminary assessment of:

1. The engineering properties of the soils as an aid in the design of desirable patterns of residential, commercial, and industrial land use development
2. The biological properties of the soils, including soil–plant and soil–wildlife relationships, as an aid in the design of desirable patterns of permanent agricultural and recreational green belts and open space
3. The suitability and limitations of soils in specific engineering applications such as on-site sewage treatment and disposal, footings and foundations for light buildings, water retention areas, and embankments as an aid to the location and design of specific development proposals—such as land subdivisions; in the location of highways, railways, airports, pipelines, and other transportation facilities; and in the application of plan implementation devices such as zoning, land subdivision control, and official mapping.

The necessary areawide soil capability study is not intended to, and does not, replace the need for subsequent on-site engineering foundation investigations and laboratory testing of soils in connection with the final design and construction of specific engineering works. Such an areawide soil study is intended to provide the means for broadly assessing the suitability of land areas for the location of various land uses and public works facilities, and thereby permit, during the planning stages of development, the adjustment of land use patterns to the natural resource base.

Standard Soil Survey

The standard soil survey as conducted by the Soil Conservation Service (SCS)—now known as the Natural Resources Conservation Service (NRCS)—of the U.S. Department of Agriculture, if accompanied by appropriate interpretations, can meet the basic soil inventory needs of a city planning and engineering program. These surveys are made by carefully examining the soil in its natural state and delineating areas of similar soils on aerial photographs. The areas so mapped are keyed to a national classification system in which all soils identified as belonging to a given series have, within defined limits, similar physical, chemical, and biological properties, these properties being determined by field and laboratory tests. This makes it possible to predict the probable behavior of the mapped soils, based on past experience with similar soils, under various proposed land uses.

The surveys have certain limitations, particularly with respect to the depth surveyed and to the possible inclusion of soils with somewhat different properties within the mapped areas because of map scale, time, and cost limitations. Nevertheless, these surveys represent the best available source of accurate, areawide soil information. The surveys are carried out by experienced soil scientists and constitute a valuable basic scientific inventory, which, when accompanied by the necessary interpretations, has multiple planning and engineering applications. More recently, digital soil survey data and web-based tools like the USDA NRCS Web Soil Survey have expanded access to and usability of this information in GIS-based planning workflows.

Mapping

Historically, the field mapping of the soils was accomplished at a scale of one inch equals 1,320 feet, the field sheets being produced from uncontrolled mosaics of U.S. Department of Agriculture aerial photography. Finished photo maps were produced in standard published soil survey reports, each report usually covering a county. Key planimetric features such as highways, railways, streams, lakes, cemeteries, and major structures were identified on the photo maps together with U.S. Public Land Survey System township, range, and section identifications. Contemporary practice uses high-resolution orthophotography and digital base maps, georeferenced to the North American Datum of 1983 (NAD-83), to produce interactive digital maps accessible through online platforms. The mapping scale now most commonly used is one inch equals 1,000 feet. Soil maps and accompanying datasets are now typically distributed as digital files, compatible with GIS and Computer-Aided Design (CAD) systems, and made available through the NRCS and local planning agencies.

Soil Data Interpretations

Beginning with surveys completed in 1966 for a 2,700-square-mile urbanizing region in southeastern Wisconsin, the properties of the mapped soils were interpreted for urban planning and engineering as well as for agricultural and forestry applications. The mapping procedures were changed to use controlled—ratioed and rectified—aerial photographs as both field sheets and published photo maps, thus facilitating the interpretation of the soil photo maps into planning data banks. Each kind of mapped soil was rated in terms of its inherent limitations for specific land uses and engineering applications. These ratings included presentation of the pertinent properties of each soil type, including, importantly, properties influencing engineering applications such as liquid limit, plastic limit, plasticity index, maximum dry density, optimal moisture content, mechanical analysis, percolation rate, bearing strength, and depth to water table. Interpretations of the properties of each soil type are provided for planning purposes, including suitability ratings for residential, commercial, industrial, recreational, and agricultural use. Importantly, the classification of the soils under the American Association of State Highway and Transportation Officials (AASHTO) and unified—U.S. Army Corps of Engineers—classification systems are provided.

The soil information may be used either graphically, for example, to show how soils having various properties are distributed relative to each other, to other elements of the resource base, and to existing and proposed land uses, or quantitatively, for example, to determine the total area covered by soils having certain properties.

The soil maps can be readily transformed into interpretive maps showing the potential use for seven land uses: agricultural, residential with conventional onsite sewage treatment and disposal, residential with mound-type onsite sewage treatment and disposal, residential with sanitary sewer service, commercial and industrial, transportational, and recreational. Each interpretive map can show up to six degrees of soil limitations: very slight, slight, moderate, severe, severe to very severe, and very severe. These degrees range from representing soils with essentially no limitations to overcome to soils with limitations very difficult and very costly to overcome. Figure 9.1 illustrates the kind of interpretive maps that can be made.

SURFACE WATER RESOURCES

The surface water resources, consisting of the lakes, streams, and associated floodlands of a planning area, constitute the singularly most important element of the natural resource base. Surface waters are focal points for water-related recreational activities, provide attractive sites for properly planned residential development, and enhance the aesthetic aspects of the environment. In some areas the surface waters may constitute parts of a major transportation system and provide settings for port- and water-transportation-dependent industries. The surface waters may also constitute a source of hydroelectric power generation and a source of cooling water for coal-fired and nuclear electric power generating stations. Surface waters may constitute the major source of water supply of an urban area. Surface waters also serve as major stormwater drainage facilities and usually receive the discharges of sewage treatment plants.

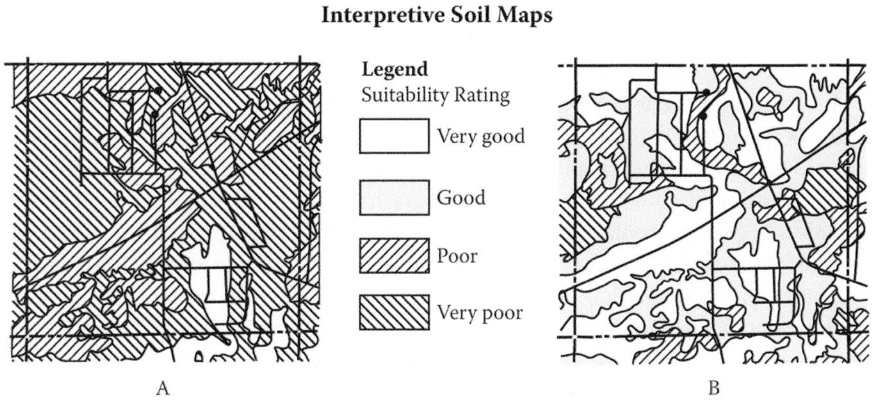

FIGURE 9.1 Illustration of the kind of interpretive maps useful in city planning and engineering that can be made from detailed operational soil survey maps. Map A shows the suitability of an area for small lot residential development utilizing onsite sewage treatment and disposal systems. Map B shows the suitability of the area for such development with public sanitary sewer service.

Source: SEWRPC.

Lakes and streams are extremely susceptible to deterioration through improper use. Water quality can deteriorate as a result of excessive nutrient loads from malfunctioning onsite sewage treatment systems, inadequate sewage treatment plants, and careless agricultural practices, and by excessive development of lacustrine and riverine areas in combination with the filling of peripheral wetlands. The surface water resources of an area must be properly managed if the economic development and quality of life of a planning area are to be sustained. The inventories, analyses, and plan design activities necessary for such management are highly complex and costly. They must, moreover, be carried out on a watershed basis—a geographic area that transcends the corporate limits of almost all municipalities, of most counties, and even of states. Watershed-based approaches are now widely adopted in integrated water resource management.

Quantitative Aspects

With respect to the quantitative aspects of the resource, the inventories required include the detailed mapping of watershed boundaries, the definitive determination of the hydrologic and hydraulic characteristics of the land and stream and watercourse systems of the watershed, and the collection of data on water distribution and use. The work requires the establishment and operation of a network of stream flow and flood crest gauges within each watershed, the analyses of stream flow regimens and lake levels, and the development, calibration, verification, and application of mathematical stream flow and stage simulation models. The models are used to calculate stream flows and stages, including regulatory flood stages—the regulatory flood stage generally being the 100-year recurrence interval flood stage—and to map the flood hazard areas along the streams and watercourses, that is, the limits of the floodways and floodplains.

Flood hazard areas should be determined and mapped under a comprehensive watershed planning program[1] that relates the calculated flood flows and stages to existing and planned land use patterns within the watershed and makes recommendations for structural and non-structural flood control measures. The studies, which should include all of the perennial streams of the planning area, may thus produce three sets of flood hazard delineations: one under existing land use and channel conditions, one under planned land use and existing channel conditions, and one under planned land use and planned channel conditions together with the attendant hydrological and hydraulic data. To be most useful in city planning and engineering, the flood hazard areas should be delineated on both the large-scale topographic and cadastral maps described in Chapter 5, and the hydraulic inventories and stage data should be based on and integrated into the control survey network used for the mapping. It is the finished flood stage data and flood hazard mapping that are most useful—and indeed required—for city planning and engineering purposes. An example of the type of map required is given in Figure 9.2. Developing such data and mapping is beyond the fiscal capability, as well as the geographic jurisdiction, of most city planning and engineering organizations. Therefore, the necessary watershed-based planning can best be accomplished by state, regional, or metropolitan-level planning agencies.

The Federal Emergency Management Agency (FEMA) conducts a national flood hazard mapping program, the maps being known as flood insurance rate maps (FIRM). While the Agency attempts to use the best base mapping available, the flood hazard delineations are usually displayed on orthophotos having a scale ranging from one inch equals 500 feet to one inch equals 2,000 feet and are based on NAD-83 and NGVD-88 horizontal and vertical survey control. The FEMA data must be used with caution, for in some cases the flood stages concerned are determined by approximate methods and may be unsuitable for use in city planning and engineering. The U.S. Army Corps of Engineers also conducts flood control studies under Congressional direction, and these studies may be the source of useful planning data. The National Oceanic and Atmospheric Administration provides definitive data on levels for the five Great Lakes.

In considering the quantitative aspects of the surface water resource, a distinction needs to be made between flooding and inadequate drainage as problems requiring resolution. Flooding may be defined as the inundation of an area by surface waters moving out of streams and watercourses under high flow conditions and occupying their natural floodplains, often with attendant damage to poorly placed land uses and risk to public health and safety. Flooding, as a problem, must be properly addressed on a watershed basis. Inadequate drainage is caused by the inability of stormwater runoff to reach natural streams and watercourses, with attendant local ponding often along

[1] A distinction is being made here between planning for the relatively small watersheds that often, if not usually, exist in urbanizing regions and that may encompass several hundred square miles in area, and the massive river basin planning efforts carried out by agencies such as the Tennessee Valley Authority and that may encompass thousands of square miles.

Typical Floodplain Map

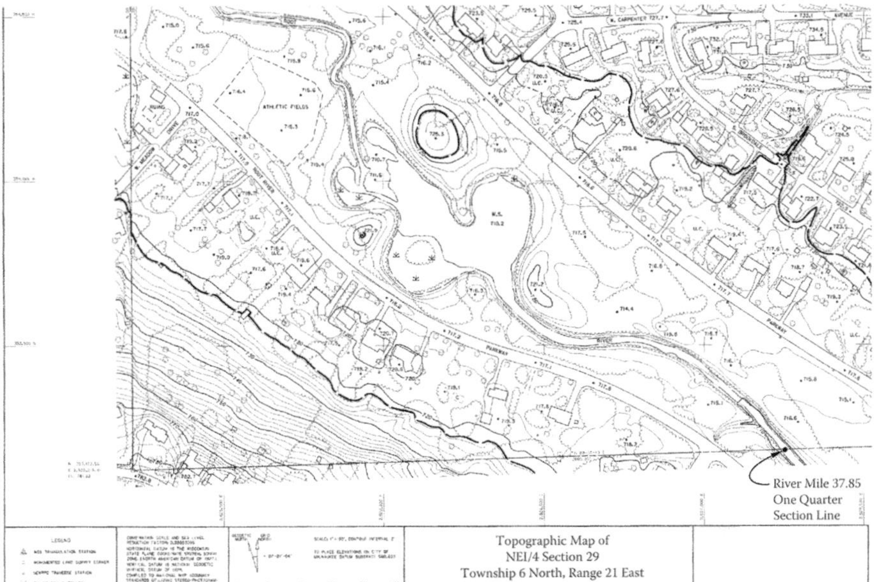

FIGURE 9.2 Illustration of a typical flood hazard map prepared for urban planning and engineering purposes. The heavy dashed lines indicate the boundaries of the 100-year recurrence interval flood hazard line—the limits of the floodplain. Note the U.S. Public Land Survey System one-quarter section lines and corners, the attendant State Plane Coordinate System values, and the map projection tick marks. The mapping utilizes the same horizontal and vertical datums as those used for the development of the local parcel-based land information system. This map was prepared from the flood stage profile shown in Figure 18.4.

Source: SEWRPC.

streets and in adjacent areas. Inadequate drainage can be addressed by local engineering studies and the improvement of local stormwater management facilities.

Qualitative Aspects

Consideration of the surface waters of a planning area as a resource must include the qualitative as well as quantitative aspects of the resource. Water quality and the practices and devices by which it is managed dramatically influence the lives of all residents of a planning area. The quality of the surface waters directly affects the public health conditions of an area. Waterborne contagious diseases such as cholera and typhoid fever were once epidemic in the United States, and serious outbreaks of Cryptosporidium and other waterborne parasitic diseases still occur. The type and cost of water treatment for supply are affected by the quality of surface water sources. Water-based recreation is another important aspect of human life directly affected by surface water quality conditions and management practices. More indirectly related are the methods and costs of wastewater treatment, stormwater management, and the aesthetic and ecological effects of changes in the natural condition of lakes and streams. The potential for commercial and industrial development is also affected by surface water quality conditions. Without proper attention, surface water quality management can become a major impediment to the continued social and economic development of an area.

Sources of surface water pollution include point sources such as separate and combined sewer overflows and sewage treatment plant and industrial waste discharges. Sources related to sanitary sewage may spread disease, increase the cost and complexity of providing water supplies, contribute to lake and stream sedimentation and fertilization, degrade or destroy the habitat of fish and other aquatic life, destroy recreational opportunities, reduce property values, and create aesthetic nuisances. Industrial pollutants may also have similar adverse effects. Non-point sources of pollution include both urban and agricultural area stormwater runoff. In addition to waterborne diseases, such problems as birth defects, decreased stability of biological populations, and both chronic and acute toxicity have become increasing concerns associated with both point and non-point sources of surface water pollution.

Ambient surface water quality conditions may be quantitatively assessed by comparison to established standards for dissolved oxygen, fecal coliform count, temperature, pH, ammonia nitrogen, and phosphorus, among others. These standards can be related to desired water uses to develop, for example, water quality indices for streams and trophic classifications for lakes. These existing conditions can then be related to needed levels of pollutant control, the pollutant contribution being measured in terms of biochemical oxygen demand, pathogenic organisms, suspended and dissolved solids, and temperature, among others.

As is the case for consideration of the quantitative aspects of the surface water resources of an area, consideration of the qualitative aspects requires highly complex and costly inventories, analyses, and plan design activities. And like the quantitative aspects, the qualitative aspects must be considered on a watershed basis. The inventories required include the establishment and operation of a network of water quality monitoring stations and the development, calibration, verification, and application of mathematical water quality simulation models. The models are used in pollution abatement plan design and evaluation. Water quality management planning, like flood management, is best accomplished by state, regional, or metropolitan-level planning agencies. These plans also align with the principles of integrated watershed management and sustainable development. Section 208 of the federal Clean Water Act (formerly the Water Pollution Control Act, Public Law 92-500) requires the preparation of regional or metropolitan water quality management plans. The plans are intended to identify the pollutant abatement measures required to attain, to the extent practicable, essentially fishable and swimmable surface waters.

The required areawide water quality planning efforts can be an invaluable source of surface water quality data for city planners and engineers. Certain findings and recommendations of the areawide plans may directly affect the city planning and engineering efforts. These findings and recommendations may relate to the location, size, and level of treatment to be provided by municipal sewage treatment plants; the areas to be served by sanitary sewage facilities and desirable levels of clear water inflow and infiltration into sanitary sewage systems; and needed stormwater management practices to abate non-point pollution. State agencies may use permitting requirements to enforce some of the plan recommendations.

GROUNDWATER RESOURCES

Groundwater resources also constitute an extremely valuable element of the natural resource base. The groundwater reservoir may not only sustain lake levels and provide the base flow of streams and watercourses but also may be a major source of water supply for domestic, industrial, and municipal water users. Groundwater is susceptible to depletion in quantity and to deterioration in quality. As urban development intensifies, protecting aquifer recharge areas and preventing contamination from land use activities becomes increasingly important. The protection of the quantity and quality of this important resource is an important consideration in city planning and engineering.

The inventories, analyses, and plan design activities necessary for the proper planning for and management of groundwater resources are, like those for surface water resources, highly complex and costly. Like the surface water resource, consideration of the groundwater resource has both quantitative and qualitative aspects. The rock units and overlying unconsolidated material that make up the groundwater reservoir of an area may differ widely in the yield of stored water. Moreover, the rock units may be part of more than one aquifer, being separated by rock formations that function as aquicludes. The inventories and analyses required include definitive data on the hydrogeology of the aquifers, as well as data on historical use of groundwater for municipal, industrial, and agricultural supply; the attendant historical progression of water levels or potentiometric surfaces, throughout the primary aquifers; the permeability, transmissibility, and specific capacities of the aquifers; the recharge areas; and the susceptibility of the aquifers to contamination. Understanding groundwater recharge, discharge, and flow paths is essential to managing both quantity and quality across multiple jurisdictions. Because lake levels, base flows of streams, and wetland complexes are often maintained by groundwater discharges, the relationships between the groundwaters and surface waters of the planning area must also be established. The inventories may build upon existing geological soil-survey data and well-boring logs to develop definitive information on factors such as depth to bedrock, depth to water table, groundwater flow patterns, potential for contamination, and sources of contamination. Modern GIS tools and 3D hydrogeologic models are increasingly used to integrate these datasets and visualize subsurface conditions. Test well borings may be required. Definitive data are also required on the quality of the groundwater resources. The data should define the physical, chemical, and bacteriological quality of the water from each of the aquifers concerned. In some instances, data on radium, heavy metals, and pesticide content may be required. Monitoring of emerging contaminants such as PFAS (per- and polyfluoroalkyl substances) has also become critical in recent groundwater quality assessments.

The necessary work should include the development, calibration, verification, and application of mathematical models that can be used to assess the performance of the aquifers concerned under varying climatic conditions, locations, alternative land use patterns, and demands of major pumping centers. The models should be specifically designed to assess how land-use activities affect groundwater resources. This includes predicting drawdown effects, identifying the need for recharge-area protection, assessing well interference, optimizing groundwater use, and determining optimal locations for new water supply wells.

Because of the complexities and costs involved, the necessary groundwater studies are best conducted at the state, regional, or metropolitan levels. This is particularly true in areas where the coordination of the utilization of the resource by multiple communities in an urbanizing region is essential. In urban regions, groundwater management should be coordinated with surface water and stormwater planning under an integrated water resources management (IWRM) framework. However, groundwater studies may be carried out by municipalities in isolated rural settings, particularly if financially supported by the water utility concerned.

WOODLANDS

Woodlands may be defined as areas of one acre or more in size having at least 50 percent canopy cover and a density of between 15 and 20 trees of four-inch diameter or larger at breast height (dbh) per acre. Lowland wooded areas such as tamarack swamps are classified as wetlands. The woodlands of an area have value beyond monetary returns for forest products. Under good management, woodlands can serve a variety of uses and provide multiple benefits. Woodlands reduce stormwater runoff, contribute to atmospheric oxygen and water content, reduce soil erosion and stream and lake sedimentation, provide wildlife habitat, provide opportunities for wholesome outdoor recreation, and provide a desirable aesthetic setting for certain types of urban development. Woodland types include northern and southern upland hardwoods, northern and southern lowland hardwoods, northern and southern upland conifers, and northern and southern lowland conifers. Woodlands are subject to deterioration through mismanagement, which may affect the age distribution and health of the stands of trees involved. Urban encroachment, invasive species, fragmentation, and climate change also pose increasing threats to the ecological integrity and sustainability of urban and peri-urban woodlands.

The woodlands of a planning area should be inventoried and mapped as a part of the land use inventory. The inventory should be conducted by field survey utilizing orthophotos as field sheets for recording the inventory findings. The orthophoto field sheets should be based on the same projection, datum, and survey control network as those adopted for planning and engineering use in the planning area. The inventory should include an assessment by an experienced forester of the quality and management needs of the mapped woodlands and should clearly identify those woodlands that should be protected and preserved for their environmental, scenic, and recreational value. Where feasible, woodland inventories should be integrated with ecological assessments and urban forestry datasets to support conservation zoning, green infrastructure planning, and biodiversity enhancement strategies.

WETLANDS

Wetlands perform an important set of natural functions, which include sup-port of a wide variety of desirable, and sometimes unique, forms of plant and animal life; stabilization of lake levels and streamflows; entrapment and storage of plant nutrients contained in stormwater runoff, thus reducing the rate of enrichment of surface waters and noxious weed and algae growth; contribution to atmospheric oxygen and water content; reduction in storm-water runoff and flood flows by providing areas for stormwater drainage and floodwater impoundment and storage; protection of lake and stream shorelines from erosion; entrapment of soil particles in runoff and reduction of stream and lake sedimentation; provision of groundwater recharge and discharge areas; and provision of opportunities for wholesome outdoor recreational activities. They also play a critical role in carbon sequestration and in mitigating the impacts of climate change.

Wetlands have severe limitations for residential, commercial, and industrial development. Generally, these limitations relate to the high compressibility, instability, and low bearing capacity of wetland soils, as well as to the high water table. In addition, the use of metal conduits in some wetland soils may be constrained because of a high corrosion potential. If ignored in land use planning and development, these limitations may result in flooding, wet basements, unstable foundations, failing walkway and driveway pavements, excessive clear water infiltration and inflow into sanitary sewerage systems, and failing sewer and water supply lines. In addition, significant onsite preparation and maintenance costs are associated with the development on wetland soils, particularly as related to roadways, building foundations, and public utilities.

Somewhat differing definitions of the term "wetlands" are used by the U.S. Environmental Protection Agency and the U.S. Department of the Army, Corps of Engineers, for the administration of the federal Clean Water Act; by the U.S. Department of Agriculture, Natural Resource Conservation Service, for administration of the federal Food Security Act; and often by state agencies for the administration of state wetland regulations. The U.S. Army Corps of Engineers and U.S. Environmental Protection Agency jointly define the term as

> those areas that are inundated or saturated by surface or groundwater at a frequency and duration sufficient to support, and that under normal circumstances do support, a prevalence of vegetation typically adopted for life in saturated soil conditions.

This definition emphasizes hydrologic and vegetative conditions and requires that wetland vegetation actually be present for an area to be identified as a wetland. The rationale for the latter requirement is based, in part, on the concept that the functional values provided by wetlands are largely provided through the vegetative cover.

The U.S. Natural Resource Conservation Service defines the term as

> areas that have a predominance of hydric soils and that are inundated or saturated by surface or groundwater at a frequency and duration sufficient to support, and under normal circumstances, do support, a prevalence of hydrophytic vegetation typically adapted for life in saturated soil conditions, except lands in Alaska identified as having a high potential for agriculture development and a predominance of permafrost soils.

This definition also emphasizes hydrologic and vegetative conditions but adds the need for the presence of hydric soils. Under this definition, some areas included under the joint U.S. Army Corps of Engineers–U.S. Environmental Protection Agency definition—such as some fresh wet meadows, some wet prairies, and some wetland forests—may be excluded from identification as wetlands.

An example of a state wetland definition, used by Wisconsin, is

> an area where water is at, near, or above the land surface long enough to be capable of supporting aquatic or hydrophytic vegetation, and which has soils indicative of wet conditions.

This definition emphasizes hydrologic and vegetative soil conditions. It provides, however, that wetland areas need only be capable of supporting aquatic or hydrophytic vegetation; that is, wetland vegetation need not be present for an area to be identified as a wetland.

The wetlands of a planning area, like the woodlands, should be inventoried and mapped as a part of the land use inventory. The inventory should be conducted by an experienced wetland ecologist, utilizing orthophotos as field sheets for recording the inventory findings. The orthophoto field sheets should be based on the same projection, datum, and survey control network as those adopted for planning and engineering use in the planning area. High-resolution aerial imagery, LiDAR, and GPS-assisted delineation methods have become standard tools in supporting more accurate mapping. An example of the kind of wetland mapping required is provided in Figure 9.3. The definition of what is to be recognized and mapped as a wetland should be given careful consideration and be based on the applicable federal and state regulatory situation. Generally, the joint U.S. Army Corps of Engineers–U.S. Environmental Protection definition should be used. The inventory should include an assessment of the quality and management needs of the mapped wetlands and clearly identify those wetlands that should be protected and preserved for their environmental value. Wetland inventory data suitable for use in city planning and engineering may be available from cognizant state regulatory agencies and from such federal agencies as the U.S. Natural Resource Conservation Service.

It should be noted that wetlands are constantly changing—expanding and contracting—in response to changes in drainage patterns and precipitation conditions. Nevertheless, if the wetland boundaries as mapped fall within a normal range, as should be the case, the maps should provide a sound basis for comprehensive city planning and engineering purposes. In view of the dynamic nature of wetlands, detailed field investigations are sometimes necessary to precisely identify wetland boundaries on individual tracts of land at a given point in time.

Areas Having Scenic, Historic, Scientific, and Recreational Value

In addition to the elements of the natural resource base heretofore described, some planning areas may contain certain elements of the natural resource base, or elements related to that base, that are unique to the area. These may include natural wildlife habitat areas, including sport and commercial fisheries, critical species habitat areas,

Example of a Wetland Inventory Map

FIGURE 9.3 Illustration of an actual wetland delineation and mapping effort. The boundary was identified and staked in the field by an experienced ecologist using an orthophotograph as a field sheet. The annotations denote various wetland types, such as emergent wet meadow, shrub, or forested, among others. The staked boundary was then accurately mapped using global positioning technology, the map projection, and the datum adopted for use for the local parcel-based land information system. The figure shows U.S. Public Land Survey System quarter-section lines and corners, along with the corresponding State Plane Coordinate System values.

Source: SEWRPC.

remnant prairies, scenic overlooks, and sites that have historic value. Consideration of these types of areas in city planning requires the conduct of special inventories specifically designed to identify and map the areas concerned. These assets contribute significantly to community identity, tourism, and quality of life, and therefore merit early and prominent consideration in land use plans.

There is a special case, consisting of a combination of natural resource base and cultural elements, that requires mention here: that of prime agricultural areas. Prime agricultural areas may be defined as areas particularly well suited for highly productive agricultural use. Such areas must have the climate, water supply, and particularly the highly productive soils characteristic of such areas—soils that meet the U.S. Natural Resource Conservation Service standards for classification as national prime farmland or farmland of statewide importance. They must also, however, contain farm units of viable size, and those units must occur in blocks large enough to make the provision of support services such as feed mills, feed-processing plants, and agricultural implement suppliers viable. In addition to performing a vital economic function, prime agricultural areas, as a land use, provide open-area settings for the development of urban centers, provide invaluable opportunities for passive recreation, and serve to protect, preserve, and enhance certain elements of the natural resource base. Planning for the protection and preservation of prime agricultural areas is best accomplished at the state, regional, or county level. City planning should, however, be coordinated within the framework plans provided by the broader levels of planning.

Environmental Corridors

Protection and preservation of the natural resource base of an area can be greatly facilitated by the identification of areas that contain a concentration of the most important elements of that natural resource base. To that end, the Southeastern Wisconsin Regional Planning Commission, building on earlier efforts of pioneering Milwaukee area park and parkway planners and on work done at the University of Wisconsin, in 1963 developed and applied the concept of the environmental corridor. Such corridors were defined as areas in the landscape that contain the best remaining elements of the natural resource base. In southeastern Wisconsin, the environmental corridors, as shown in Figure 9.4, generally lie along stream valleys, around major inland lakes, along the Lake Michigan shoreline, and in still undeveloped areas of glacial topography.

Operationally defining the corridors involved identifying and mapping all the important elements of the natural resource base: the surface waters and associated shorelands and floodlands; groundwater recharge and discharge areas; organic soils; woodlands and wetlands; wildlife habitat; remnant prairies and other natural areas; and areas of steep slope. The mapping was originally done on ratioed and rectified aerial photographs—but currently is best done on orthophotographs—utilizing a common map projection, datum, and survey control network to facilitate accurate correlation of the layers of mapping in analog and digital form. Corridors were delineated by combining contiguous overlying polygons while observing specified minimum-size criteria: length 2 miles, width 200 feet, and area 400 acres. The resulting network of corridors is shown in Figure 9.4. In southeastern Wisconsin, the delineated corridors occupy 17 percent of the 2,700-square-mile planning region but encompass all the surface waters and undeveloped shorelines, 62 percent of all the undeveloped floodlands, 62 percent of all the remaining woodlands, 77 percent of all the remaining wetlands, 56 percent of all the remaining wildlife habitat, 94 percent

Primary Environmental Corridors

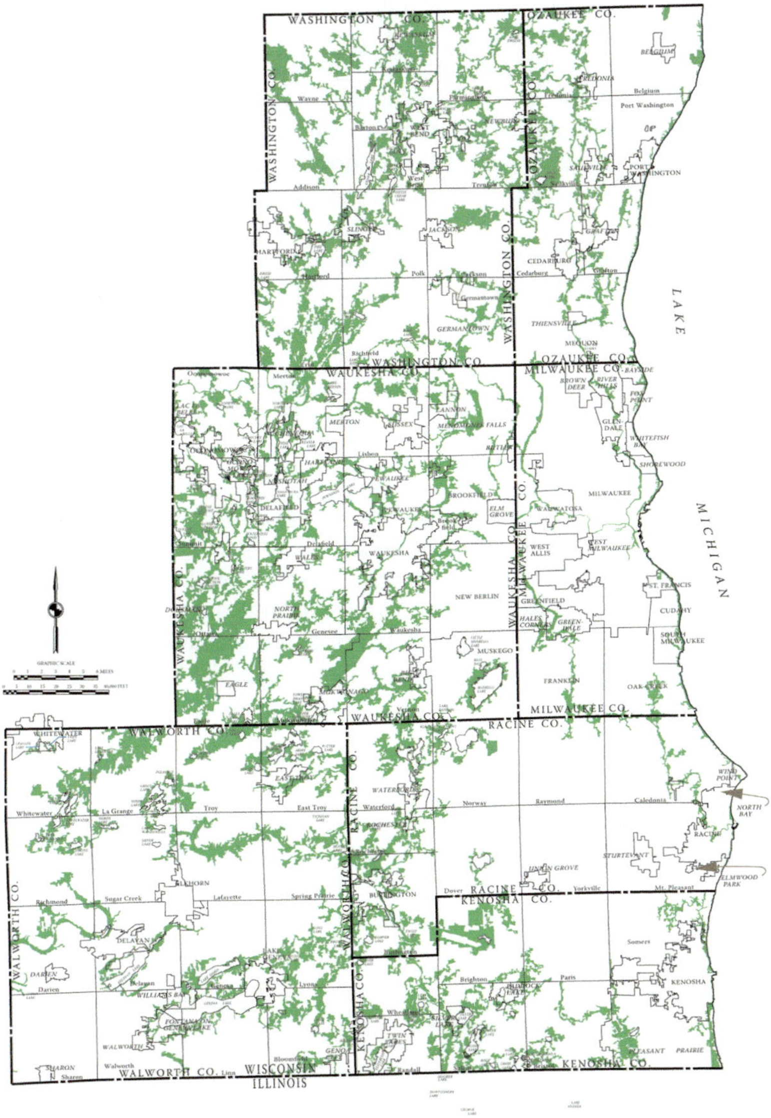

FIGURE 9.4 Illustration of an environmental corridor delineation. The corridor encompasses about 17 percent of the total seven-county region but contains almost all of the best remaining woodlands, wetlands, undeveloped floodlands, and prime wildlife habitat areas within the region. The corridors also consist largely of lands poorly suited to urban development. The corridors are intended to be incorporated in county and municipal comprehensive plans and preserved in essentially natural open uses through public purchase, zoning, and land subdivision control.

Source: SEWRPC.

of all the remaining remnant prairies and natural areas, and about 50 percent of the mapped groundwater recharge and discharge areas.

Delineation, protection, and preservation of the environmental corridors of a planning area in essentially natural open uses will do much to preserve and enhance the quality of the environment for life. Such protection and preservation will contribute to the maintenance of surface water quality and the base flows of streams, the reduction of flood flows and stages, the abatement of air and noise pollution, and the maintenance of groundwater recharge. The corridors help maintain wildlife, protecting plant and animal diversity and rare, threatened, and endangered species. The linear nature of the corridors facilitates the movement of wildlife and the dispersal of plant seeds. The corridors provide opportunities for outdoor recreational activities such as hiking, cross-country skiing, hunting, and fishing. The corridors also provide locations for high-quality natural park and parkway sites and for pedestrian trails. Importantly, the corridors lend form and structure to urban development, providing needed greenbelts to separate developing neighborhoods and communities.

Excluding urban development from the corridors will also help prevent the creation of serious and costly environmental problems. These problems may include surface and groundwater pollution, inadequate drainage and flooding, failing on-site sewage treatment and disposal systems, excessive infiltration of clear water into sanitary sewerage systems, wet basements and excessive operation of sump pumps, and settlement and structural failure of roadway, walkway, and parking area pavements and buildings. Such exclusion will also help avoid development on steep slopes.

The delineation of environmental corridors is best accomplished at the regional or metropolitan level, but may also be accomplished at the local level. Delineated corridors should, in any case, be integrated into local city planning and plan implementation activities.

FURTHER READING

Beatty, Marvin T., et al. "Planning the Uses and Management of Land," Number 21 in the *Series Agronomy*. American Society of Agronomy, 1979.

Molks, Manuel C. *Ecology, Concepts and Applications* (4th edition). McGraw-Hill, 2008.

Wang, Yeqiao (ed.). "The Handbook of Natural Resources, Second Edition," CRC Press, 2023.

10 Institutional Structure for City Planning and the Comprehensive Plan

… the preparation and maintenance of the general plan is the primary, continuing responsibility of the city planning profession.

T. J. Kent, Jr.
The Urban General Plan
1964

INTRODUCTION

In the United States, municipalities—cities, villages, and towns—are, as noted in Chapter 3, generally creatures of the states. As such, municipalities can exercise only such powers as may be delegated to them by the states. Therefore, the institutional structure—or organization—for the conduct of city planning must be related to the state law authorizing and governing such planning.

In some states, the institutional structure for the conduct of city planning may be specified in considerable detail in the state enabling legislation. In other states, municipalities may be granted authority to select an institutional structure for the conduct of city planning under broad charter or home rule provision of the state statutes. Consequently, there is no uniform institutional structure for city planning in the United States. Moreover, the issue is often complicated by a tendency to confuse the need for a structure to carry out the physical development planning function and the need for a policy planning function as an aid to the chief executive of larger cities. The latter should be a separate function and have a separate structure from the former.

As indicated in Chapter 1, this book is focused on planning for the physical development of cities. Therefore, it recommends the city plan commission as the most desirable form of institutional structure for the conduct of city planning. This is the form that emerged historically in the city beautiful and city efficient movements that marked the renaissance of public planning in the United States in the twentieth century. It is the institutional structure envisioned in the federal Model City Planning Enabling Act of 1928. It is conceptually sound, and experience has shown that it can work well over time, especially in small and medium-sized communities. The proposition that the plan commission is the most desirable institutional structure for city planning is not without its critics, but these critics have been unable, over a period of almost 70 years, to develop an alternative that has been as widely adopted.

DOI: 10.1201/9781003500384-10

The Wisconsin city planning enabling legislation is herein used as an example of enabling legislation that requires the creation of city plan commissions as the institutional structure for the conduct of city planning. The original Wisconsin planning enabling legislation closely followed the federal Model City Planning Enabling Act of 1928. The Wisconsin act was, however, amended in 1998 to broaden the scope of the required comprehensive plan and to strengthen plan implementation.

THE PLAN COMMISSION

The city planning enabling legislation in Wisconsin is contained in Section 62.23 of the Wisconsin Statutes. This statute delegates the city planning function to advisory bodies known as plan commissions. This delegation entrusts this important municipal function to a continuing body of public officials and citizens qualified for the task, free from the distraction of routine matters of administration and somewhat detached and insulated from partisan politics. The enabling legislation requires the creation of such commissions in order for municipalities to exercise the planning function and specifies the composition, functions, duties, and powers of the commissioners. The statutes make the provisions of the enabling act applicable to villages and to towns that have adopted village powers.

Plan commissions are to consist of seven members appointed by the mayor, who is to be the presiding officer of the commission. The mayor may appoint, in addition to himself, other city-elected or -appointed officials, except that the commissions are always to have at least three citizen members who are not city officials. The citizen members are to be persons of recognized experience and qualifications. Historically, but no longer, the statutes provided that the city engineer and an alderman—preferably the president of the common council—were to be members of the plan commission. Under this historical composition, the chief executive, the legislative body, and the city official most knowledgeable about the urban infrastructure concerned were represented on the commission. The members of the commission are to hold staggered, three-year terms of appointment, a length of term provided to ensure some desirable continuity in the membership of the commissions, the three-year term being generally longer than the terms of the office of the mayor and aldermen. The city plan commissions are given the authority to employ experts and a staff and to incur such other expenses as may be necessary and proper, not exceeding in all, however, the appropriation made annually for such commission by the legislative body.

City plan commissions are entirely advisory to the executive and legislative branches of the municipal government. The commissions may make reports and recommendations relating to the development of the city, not only to municipal officials and agencies but also to public utility companies, other organizations, and citizens. Importantly, plan commissions may recommend long-term capital improvement programs to the mayor and common council. However, certain matters must be referred to the city plan commission before the common council or other public body or officer having final authority acts thereon. These matters include the location and architectural design of public buildings; the location, extension, alteration, vacation, or acquisition of land for streets, alleys, or other public ways, and parks and playgrounds; the location, extension, abandonment, or authorization for any public utility whether publicly or privately owned; and all plots of land within the city and its extraterritorial and the approval of plan jurisdiction. The planning enabling act also provides for the preparation and adoption of certain plan implementation devices, such as the official map and zoning ordinance, and grants advisory powers to the plan commissions concerning the adoption and amendment of such plan implementation measures.

THE COMPREHENSIVE PLAN

Importance

The most important function of a plan commission is to make and adopt a comprehensive plan for the physical development of the city, including any areas outside of its boundaries that may bear relation to the development of the city. In Wisconsin, such areas may extend for up to three miles beyond the corporate limits of a city. This function is not permissive but is clearly made mandatory in the statutes, which state that it shall be the duty and function of the plan commission to make and adopt a comprehensive plan for the physical development of the city. The comprehensive plan is to show the commission's recommendations for the physical development of the city and its environs. The plan is intended to be made with the purpose of guiding and accomplishing a coordinated, adjusted, and harmonious development of the city that will best promote the public health, safety, and general welfare, as well as efficiency and economy in the process of development.

The concept of the comprehensive plan is one of the oldest and most important concepts underlying the practice of city planning. A comprehensive plan for the physical development of the city is essential if land use development is to be properly coordinated with the development of supporting transportation, utility, and community facility systems; if the development of each of these individual functional systems is to be coordinated with the development of the others; if serious and costly environmental and developmental problems are to be minimized or avoided; and if a more healthful, attractive, and efficient urban settlement pattern is to be evolved. The preparation, adoption, and use of the comprehensive plan should be the primary focus of the city planning process, and all planning and plan implementation techniques should be based on or related to the comprehensive plan.

A comprehensive plan is essential to the sound guidance of urban development and redevelopment. The comprehensive plan provides not only the necessary framework for coordinating and guiding urban growth and development within a community but also the best conceptual basis available for the application of system engineering techniques and skills to urban development problems. This is because systems engineering basically focuses on the design of physical systems. It seeks to achieve good design by setting good objectives, determining the ability of alternative designs to meet those objectives through quantitative analyses, cultivating interdisciplinary team activity, and considering all the relationships involved both within the system being designed and between that system and its environment.

While criticisms of comprehensive plans have periodically emerged, no alternative framework has achieved comparable adoption or effectiveness in promoting rational and coordinated urban development. The comprehensive plan has, therefore, remained a viable and valid concept, a concept essential to coping with the environmental and developmental problems generated by urban development and redevelopment in a rational and constructive manner.

Scope and Content

The scope and content of the comprehensive plan as envisioned in the original state planning enabling legislation were very general and very broadly defined, implicitly extending to all phases of city development, but without, however, specifically identifying any plan elements. This left the local plan commissions to identify the elements to be included in the comprehensive plan based on the local situation. Each individual plan element would thus be intended to deal with the specific developmental or environmental problems present in the planning area concerned. The original planning enabling legislation also implicitly recognized that practical considerations might prohibit the preparation and adoption of an entire comprehensive plan at one time and the permitted preparation and adoption of individual plan elements over time as fiscal and staff limitations might require.

The scope and content of comprehensive plans prepared under the original enabling act usually consisted of the following elements:

1. Land Use
2. Transportation
   a. Arterial streets and highways
   b. Public transit
   c. Other
3. Utility Systems
   a. Sanitary sewerage
   b. Water supply
   c. Stormwater management
   d. Other
4. Community Facilities
   a. Parks and open space
   b. Other
5. Special

The "other" category under the transportation element could include, depending upon the local situation, parking facilities, airports, seaports and marinas, and railways. The "other" category under the utility systems element could include solid waste collection and disposal, electric power generation and distribution, and digital and broadband telecommunication services. The "other" category under the community facilities element could include schools, libraries, police and fire stations, government buildings such as city halls and post offices, hospitals and clinics, and museums. The "special" element could include neighborhood unit development plans, central business district renewal plans, and industrial district development plans. Each of the individual plan elements usually contained recommendations for implementation.

The state legislature in 1998 enacted an important addition to the original state planning enabling act. Identified as Section 1001, the addition leaves the original act in place but re-emphasizes the importance of the comprehensive plan, specifies the scope and content of such a plan in much greater detail than does the original enabling act, and requires the preparation and adoption of the plan as a prerequisite to the exercise of plan implementation powers such as land subdivision control, zoning, and official mapping. The new legislation specifies that a comprehensive plan is to specifically consist of the following elements:

1. Issues and Opportunities
2. Land Use
3. Housing
4. Transportation
5. Utilities
6. Community Facilities
7. Agricultural, Natural, and Cultural Resources
8. Economic Development
9. Intergovernmental Cooperation
10. Implementation

Four of the specified plan elements directly correspond to the four principal elements traditionally included in a comprehensive plan prepared under the provisions of the original planning enabling act. Two plan elements dealing with the physical development of the community were specifically enumerated: housing, and agricultural, natural, and cultural resources. Historically, the housing element would usually have been addressed as part of the land use element, and the natural resources as part of the park and open space element, while the implementation element would have been addressed on an element-by-element basis. Agricultural resources were often neglected under historical planning efforts, agricultural areas being commonly regarded as areas into which urban areas could be expected to expand.

The called-for issues and opportunities element is not a plan element in the traditional sense, being a somewhat confused and confusing presentation of inventory data, objectives, and problems. The new legislation departs from the older legislation in that it requires the preparation of two non-physical plan elements: an economic development element and an intergovernmental cooperation element.

Importantly, the new legislation significantly increases the efficacy of the comprehensive plan. The legislation provides that beginning January 1, 2010, certain specified actions of municipal governments must be consistent with the local adopted comprehensive plan. These include, importantly, the adoption of zoning, subdivision control, and official map ordinances. Under the new legislation, the comprehensive plan must be adopted in its entirety if the municipal government concerned is to adopt plan implementation ordinances. Moreover, the plan must be adopted by the governing body of the municipality as well as by the plan commission.

Plan Report

Comprehensive plans—or their individual elements—were once presented primarily as maps with minimal supporting documentation. Good practice, however, requires that such a plan—or each of the individual elements if prepared separately—be presented in a written report that fully documents the planning process and its resulting findings and recommendations. That process is illustrated in summary form in Figure 10.1. The report should be organized into chapters, the number of which will depend upon the size and complexity of the planning area concerned.

An introductory chapter should describe the institutional structure for the conduct of the planning process. This should include a description of the plan commission and staff structure and the relationship of the commission to the governing body of the planning area, to other municipal departments, and to other planning agencies, such as metropolitan, regional, or state planning agencies. The introductory chapter should also briefly describe the planning area and its regional setting, the need for and purpose of the planning effort, and the scheme of presentation used in structuring the report.

One or more chapters should then describe the inventories conducted for the planning effort and the pertinent findings of those inventories. Depending upon the size and complexity of the planning area concerned, separate chapters may be required to present historical and current data on the demography and economy of the area, including data on the size, distribution, characteristics, and growth trends of the population; the labor force; the economic base and structure; the retail trade area; personal income levels; and the property tax base of the planning area. The chapter should include a presentation of the pertinent population, employment, and land use demand projections and forecasts prepared under the planning effort.

A chapter should provide pertinent data on the natural resource base of the planning area, including data on such factors as climate, air quality, geology, physiography, topography, soils, surface and groundwater resources and attendant flood hazard areas, wetlands, woodlands, wildlife habitat areas, and, importantly, environmental corridors. A chapter should be devoted to findings of the land use inventory, including information on historical development patterns and on the amount and areal distribution of the various land uses. Data should be provided on the housing stock, including number of units, distribution, occupancy, vacancies, structure types, and costs in terms of valuations and rents. An existing land use map should be included in the chapter.

A chapter should be devoted to the transportation facilities of the planning area. Information should be provided on the location, configuration, and capacity of the arterial street system; on the location, configuration, and level of service of the mass transit facilities; and, as appropriate, on airport, seaport, and railway facilities. For the arterial street and mass transit systems, the chapter should include data on travel characteristics, traffic volumes, service levels, and transit fares. Where available,

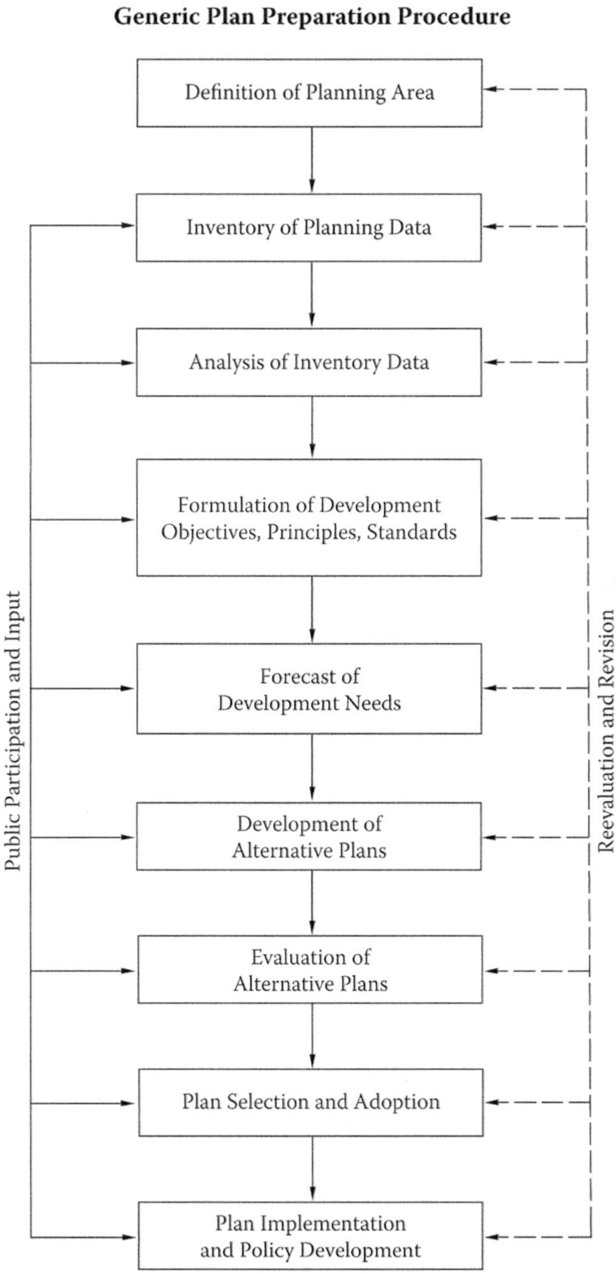

FIGURE 10.1 Illustration of the general procedure to be followed in the preparation of each of the elements of a comprehensive plan, whether the elements are individually prepared over time or whether the plan as a whole is prepared at one point in time.

Source: SEWRPC.

mobility-as-a-service (MaaS) platforms, active transportation modes, micromobility options, and emerging autonomous vehicle corridors should be documented.

A chapter should provide general information on the key utility systems of the planning area, including data on the location, service areas, configuration, and capacities of the sanitary sewerage, public water supply, stormwater management, and solid waste management facilities. Additionally, data should be provided on the service area and levels of service of electric power and telecommunication facilities. This should include fiber-optic broadband, 5G infrastructure, and renewable energy distribution networks. One or more chapters should be provided describing the significant community facilities of the planning area, including data on the location and size of existing parks and on the location of fire and police stations, administrative offices, libraries, schools, and, in some cases, public housing facilities, hospitals, and clinics.

The report should contain a chapter describing the adopted state, regional, metropolitan, and county-level plans affecting the area, including the plans of special districts such as school and metropolitan sewerage districts. Past comprehensive plans prepared for the planning area should be described, together with any facility plans prepared by the city engineering, public works, police and fire departments, library system, and school districts. The implementation devices in place should be described to include the existing zoning, land subdivision, control, and official map ordinances.

A chapter should be provided that describes the agreed-upon objectives, principles, and standards used in the preparation of the land use, transportation, utilities, and community facilities elements of the plan.

If the report is to present the entire comprehensive plan, at least four chapters should be devoted to the presentation of each of the key elements of that plan, or a chapter each for land use, transportation, utilities, and community facilities. This may be a tenable approach for presenting a plan for smaller communities. For larger and more complex communities, separate reports may have to be prepared for each of the key elements comprising the comprehensive plan. In this case, each individual report should contain the appropriate equivalent of the first three chapters of a single report as heretofore described.

In a combined report, a chapter should describe the recommended land use plan. The amount and distribution of land recommended to be used for each of the land use categories should be described together with desirable densities and intensities of the uses. A land use plan map should be included in the chapter. A chapter should present the recommended configuration and capacities of the arterial street and mass transit systems and, in some cases, address the vehicular parking, airport, and seaport facilities. One or more chapters should present the utility system plans. The chapter, or chapters, should describe the location, service areas, and capacities of the sanitary sewerage, water supply, and stormwater management systems, including the locations, configurations, capacities, and levels of treatment to be provided by the needed sewage and water treatment plants. The receiving waters for treated effluent and the sources of water supply should be identified. In some cases, the needed facilities for solid waste disposal should also be identified. One or more chapters should present the community facilities plan element. These should describe the recommended location, size, and facilities of parks and the location, size, and design of public buildings such as schools, libraries, fire and police stations, and administrative buildings. One or more chapters may be required to describe any special plan elements relating to, for example, the provision of public housing, urban redevelopment, or railway facility realignment. Modern special elements may also include climate adaptation strategies, green infrastructure systems, and equitable housing access.

The report should contain one or more chapters relating to plan implementation. These may include recommendations for the adoption of new or the amendment of existing zoning, land subdivision control, official map ordinances, and building and housing codes. The chapter, or chapters, may also present a proposed capital improvement program.

The planning report should constitute a clear graphic and written description of the inventory findings, the analyses, projections, and forecasts made, the alternative plans considered, and the recommended plan. The reasons for the selection of the recommended plan from among the alternatives considered should be clearly set forth, and the required plan implementation measures specified. Depending on the scope, a strategic environmental assessment (SEA) or environmental justice analysis may also be required. An environmental assessment of the recommended plan may have to be included in the planning report depending upon the nature of the plan recommendations. A good planning report properly describing a comprehensive plan for even a small community may constitute well over 200 pages of text, charts, graphs, and maps, and deserves careful preparation as an important educational as well as administrative tool.

STAFF ORGANIZATION

The essential functions of a city planning agency include the collection and analysis of the data required for planning; preparing necessary projections and forecasts; formulating city development objectives and standards; preparing and administering a comprehensive plan; maintaining and updating the plan regularly; fostering public understanding and acceptance of the plan; providing assistance and advice to other governmental agencies and private organizations; coordinating city development; and preparing and administering plan implementation devices, including zoning, land subdivision control, and official mapping. The proper performance of these functions requires a staff headed by a professional planner with substantive knowledge and experience in the physical development of cities. The size, composition, and organization of the staff will vary with the size and characteristics and, importantly, with the prevalent value system, political philosophy, and administrative organization of the city as a whole.

Figure 10.2 provides an example of a possible functional organization for the provision of planning services to small- to medium-sized cities. The functional organizational chart shown represents an application of the institutional concepts envisioned in the state enabling legislation. This structure represents what public administrators might call an independent commission structure. While the commission operates with some autonomy, its budgetary and personnel ties to the

FIGURE 10.2 Illustration of one possible type of functional organization for city planning. This type is in accord with the institutional structure envisioned in the federal City Planning Enabling Act of 1927 and provides a good basis for physical planning in small and medium-size communities.

broader municipal structure ensure administrative integration. It should be noted that the division of the planning department into current and long-range planning functions is particularly important. In the absence of such division, current planning activities tend to absorb ever-increasing amounts of the available staff time, with an attendant neglect of the essential long-range planning activities. The functional organizational chart shown envisions the plan commission as an objective advisory body concerned solely with the promotion of the long-term public interest. In small and medium-sized communities that value good government and insist on qualification-based staff appointments and utilize civil service-type personnel administrative procedures, this structure can provide a very high quality of planning and can effectively guide development and redevelopment in the long-term public interest, producing attractive communities that are good places in which to live and work.

Figure 10.3 provides an example of a possible functional organization for the provision of planning services to large cities that value the centralization of power in the person of the mayor. The organizational structure depicted tends to submerge, and perhaps downgrade, the importance of the physical planning functions as one of the activities carried out within the larger and more complex structure. If key staff positions are filled by political patronage instead of being qualification-based, the planning work may become politically or ideologically,

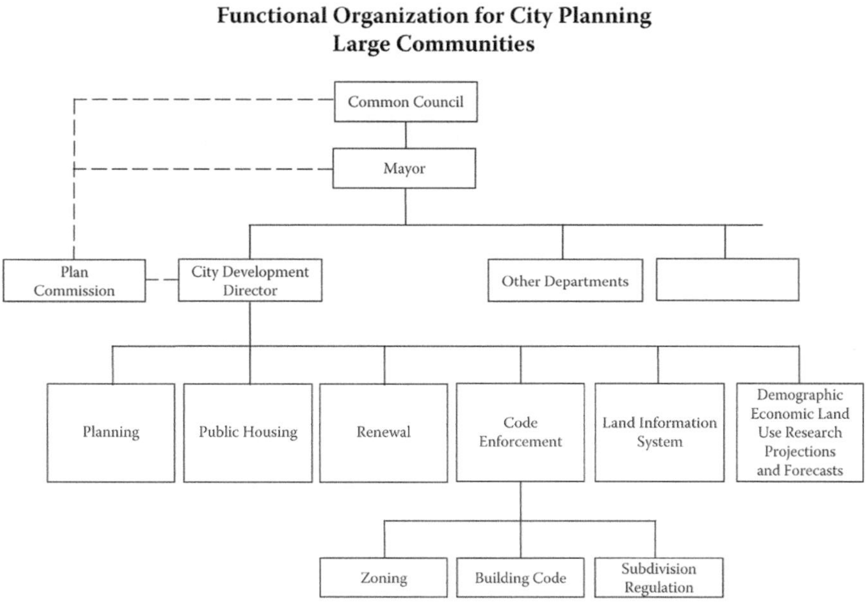

FIGURE 10.3 Illustration of one possible type of functional organization for city planning. This type is in accord with a desire for greater centralization of power in the chief executive of the community. While adapted to the more complex needs of larger communities, it tends to reduce the importance of the planning function, and if key staff positions are filled by political patronage rather than by civil service procedures, may sacrifice the continuity and objectivity needed for good long-range planning.

rather than objectively, directed. Moreover, the necessary continuity in the direction of the planning function may be lost. However, with safeguards such as professionalized staffing and protected long-range planning units, strong-mayor systems can still support high-quality planning. This type of organizational structure is often preferred by larger cities that seek a strong executive role in the management of the local government.

11 Objectives, Principles, and Standards

"Would you tell me, please, which way I ought to go from here?" asked Alice of the Cheshire Cat.

"That depends a good deal on where you want to get to," said the Cat. "I don't much care where," said Alice.

"Then it doesn't matter which way you go," said the Cat.

"So long as I get somewhere," Alice added as an explanation.

"Oh, you're sure to do that," said the Cat, "If you only walk long enough."

Lewis Carroll
Through the Looking Glass
1872

INTRODUCTION

Planning is a rational process for formulating and meeting objectives. The formulation of objectives, therefore, is an essential task that must be undertaken before plans can be prepared. The formulation of objectives for organizations whose functions are directed primarily at a single purpose or interest and, therefore, are direct and clear-cut is a relatively easy task. However, cities—even relatively small cities—are usually composed of many diverse and often divergent interests. Consequently, the formulation of objectives for the preparation of a comprehensive city plan is a difficult task.

Soundly conceived community development objectives should incorporate the combined knowledge of many people who are informed about the community and should ultimately be established by duly appointed and elected representatives legally assigned this task as represented by the plan commission and the governing body of the municipality. This consideration is important because of the value system implications inherent in any set of development objectives. The formulation of development objectives, however, is a complex task involving technical as well as value system considerations. Therefore, it is appropriate that experienced public planners and engineers initially prepare such objectives for consideration by plan commissions and governing bodies. Engaging diverse stakeholders through structured public outreach, advisory groups, and participatory tools such as workshops or online surveys is now standard practice for ensuring public involvement in objective-setting. Only by combining the accumulated knowledge and experience about the community that the advisory committee members, the plan commission members,

DOI: 10.1201/9781003500384-11

and the members of the governing body possess can a meaningful expression of the desired direction, magnitude, and quality of future development be obtained.

BASIC CONCEPTS AND DEFINITIONS

The terms objective, principle, standard, plan, policy, and program are subject to a wide range of interpretation and application. Within a planning context, these terms may be defined as follows:

1. Objective: a goal or end toward the attainment of which plans and policies are directed
2. Principle: a fundamental, primary, or generally accepted tenet used to support objectives and prepare standards and plans
3. Standard: a criterion used as a basis of comparison to determine the adequacy of plan proposals to attain objectives
4. Plan: a design that seeks to achieve agreed-upon objectives
5. Policy: a rule or course of action used to ensure plan implementation
6. Program: a coordinated series of policies and actions to carry out a plan

Some planners will distinguish between the terms *goal* and *objective*. A goal is then defined as a general description of a preferred future condition, and an objective is a statement of a more specific, measurable condition to be achieved over time by a number of measures that address a particular goal.

An understanding of the definitions of the terms *objective, principle, standard, plan, policy,* and *program,* and of the interrelationships between these terms and the concepts they represent, is essential to understanding the development objectives and standards presented in a planning report. The development objectives and standards for a comprehensive plan should address the allocation and distribution of the various land uses; the location and capacity of supporting neighborhood and community facilities, including, importantly, elementary and secondary schools; the protection of the natural resource base; the preservation of open space; the provision of recreational opportunities; the provision of safe and efficient transportation facilities; the provision of essential utilities; the provision of fire and police protection services; the provision of adequate housing and a variety of housing types; the promotion of community health and well-being through walkable neighborhoods, access to nature, and safe public spaces; the integration of digital infrastructure to support smart city functions and equitable access to information; the mitigation of and adaptation to climate change through resilient land use and infrastructure strategies; the advancement of social equity and environmental justice across all neighborhoods; and the support for a circular economy and sustainable resource management practices. Each objective, together with its supporting principles and standards, should be clearly set forth in the planning report.

OBJECTIVES

Recognizing that various public and private interest groups within a community may have varying and at times conflicting objectives, that many of these objectives are of a qualitative nature and, therefore, difficult to quantify, and that many objectives that may be held to be important by the various interest groups within the community may not be related in a demonstrable manner to physical development, two basic types of objectives may be identified. As already noted, these are general development objectives, often referred to as goals, which are by their very nature either qualitative or difficult to relate directly to physical development, and specific development objectives, which can be directly related to physical development and which can be at least crudely quantified.

General development objectives—or goals—may consist of such statements as economic growth at a rate consistent with available resources, including land, labor, and capital; primary dependence on free enterprise to provide needed employment opportunities for the labor force of the planning area; a wide range of employment opportunities through a broad, diversified economic base; an efficient and equitable allocation of fiscal resources within the public sector of the economy; development having distinctive individual character, based on physical conditions and historical factors; equitable access to essential services, infrastructure, and opportunities for all community members regardless of background; enhancing urban resilience to climate and economic disruptions through flexible infrastructure and land use policies; and integration of sustainable development principles, including renewable energy, green buildings, and efficient mobility systems.

These kinds of statements are intended as goals that public policy should promote over time. They are necessarily general but nevertheless provide a broad framework within which planning can take place and the more specific objectives can be advanced through the various functional plan elements stated and pursued. Statements of these kinds of goals are usually concerned entirely with ends and not with means, and the principal emphasis is often on those aspects of development that relate to economic development and social well-being. With respect to these kinds of goals, it is usually deemed sufficient to arrive at a consensus among the members of the plan commission and of the governing body that the physical development plan proposals considered do not conflict with the stated goals. Such a consensus represents the most practical evaluation of the ability of plan proposals to meet the goals.

Within the framework established by the general development objectives, a secondary set of more specific objectives can be postulated that is directly relatable to physical development plans and can be at least crudely quantified. The quantification is facilitated by complementing each specific objective with a set of quantifiable planning standards that are, in turn, directly relatable to a planning principle that supports the chosen objective. The planning principles thus augment each specific objective by asserting its inherent validity as an objective.

Principles and Standards

Each of the specific development objectives should be complemented by one or more planning principles and a set of planning standards. Each set of standards should be directly related to a planning principle, as well as to the objective, and should serve to facilitate quantitative application of the objectives in plan design, test, and evaluation. The planning principles, moreover, support the specific objectives by asserting their validity.

OVERRIDING CONSIDERATIONS

In applying planning objectives, principles, and standards, several overriding considerations must be recognized. First, it must be recognized that it is unlikely that any one plan proposal can meet all of the standards completely and that the extent to which each standard is met, exceeded, or violated must serve as a measure of the ability of the plan proposal to achieve the specific objectives the given standard addresses.

Second, it must be recognized that some objectives may be complementary. Thus, the achievement of one objective may support the achievement of other objectives. For example, an objective calling for the concentration of new urban residential development within neighborhood units served by public sanitary sewer, water supply, and mass transit services and facilities would be consistent with and would support an objective calling for the protection of the natural resource base. Some objectives may be conflicting, requiring reconciliation through compromise. For example, an objective calling for the preservation of agricultural lands must be reconciled with an objective calling for the allocation of land to the various urban uses.

Third, it must be recognized that the standards must be judiciously applied to areas or facilities that are already partially or fully developed, since strict application may require proposing extensive renewal or reconstruction programs. In this respect, it should be particularly noted that the land use standards concerned with natural resource protection, preservation, and wise use relate primarily to areas where the resource base has not as yet been significantly deteriorated, depleted, or destroyed. In areas where such disruption, deterioration, depletion, or destruction has already occurred, application of the standards may make it necessary to propose plans and programs that would restore the resource base to a higher level of both quality and quantity. Such restoration initiatives may also align with climate adaptation goals and green infrastructure strategies promoted in post-2010 resilience planning efforts. Such programs may indeed be specifically recommended with respect to the surface water resources of an area in watershed and areawide water quality management plans, to air resources in the areawide air quality attainment and maintenance plans, and to certain recreational resources in the areawide park and open space plans. Additional plans may also focus on reducing urban heat islands, improving soil health, or enhancing ecological connectivity through urban greenways and nature-based solutions.

An example set of objectives, principles, and standards for the land use element of a comprehensive plan is provided in Table 11.1. The set is provided solely as an example and should not be construed as a recommended set of objectives, principles, and standards. Similar sets of objectives, principles, and standards would have to be developed for each of the other elements of a comprehensive plan: the transportation, utilities, and community facilities elements. The development of such sets is a difficult intellectual task, for the objectives and standards must reflect the value system of the community concerned and the desired future conditions of the systems being planned, yet must be technically sound and defensible. These objectives must

TABLE 11.1
Example Land Use Development Objectives, Principles, and Standards

OBJECTIVE NO. 1

A balanced allocation of space to the various land use categories that meets the social, physical, and economic needs of the regional population.

PRINCIPLE

The planned supply of land set aside for any given use should approximate the known and anticipated demand for that use.

STANDARDS

1. For each additional 100 dwelling units to be accommodated within the planning area at each residential density, the following minimum amounts of residential land should be set aside:

Residential Density Category	Net Area[a] (acres per 100 dwelling units)	Gross Area[b] (acres per 100 dwelling units)
High-Density Urban	8	13
Medium-Density Urban	23	32
Low-Density Urban	83	109

2. For each additional 1,000 persons to be accommodated within the planning area, the following minimum amounts of public park and recreation land should be set aside:

Public Park and Recreation Land Category	Net Area (acres per 1,000 persons)	Gross Area[c] (acres per 1,000 persons)
Major	4	5
Other	8	9

3. For each additional 100 industrial employees to be accommodated within the planning area, the following minimum amounts of industrial land should be set aside:

Industrial Land Category	Net Area (acres per 100 employees)	Gross Area[c] (acres per 100 employees)
Major and Other	7	9

4. For each additional 100 commercial employees to be accommodated within the planning area, the following minimum amounts of commercial land should be set aside:

Commercial Land Category	Net Area (acres per 100 employees)	Gross Area[c] (acres per 100 employees)
Retail and Service		
Major	1	3
Other	2	6
Office		
Major and Other	1	2

(Continued)

TABLE 11.1 *(Continued)*
Example Land Use Development Objectives, Principles, and Standards

5. For each additional 1,000 persons to be accommodated within the planning area, the following minimum amounts of governmental and institutional land should be set aside:

Government and Institutional Land Category	Net Area (acres per 1,000 persons)	Gross Area[c] (acres per 1,000 persons)
Major and Other	9	12

OBJECTIVE NO. 2

A spatial distribution of the various land uses that will result in a compatible arrangement of land uses.

PRINCIPLE

The proper allocation of uses to land can avoid or minimize hazards and dangers to health, safety, and welfare, and maximize amenity and convenience in terms of accessibility to supporting land uses.

STANDARDS

1. Urban residential uses should be located within neighborhood units that are served with centralized public sanitary sewerage and water supply facilities and contain, within a reasonable walking distance, necessary supporting local service uses, such as neighborhood parks, local commercial, and elementary school facilities, and should have reasonable access through the appropriate component of the transportation system to employment, commercial, cultural, and governmental centers and secondary school and higher educational facilities.
2. Industrial uses should be located to have direct access to arterial street and highway facilities and reasonable access through an appropriate component of the transportation system to residential areas and to railway, seaport, and airport facilities and should not be intermixed with commercial, residential, governmental, recreational, or institutional land uses.
3. Major commercial uses should be located in centers of concentrated activity on only one side of an arterial street and should be afforded direct access to the arterial street system.

OBJECTIVE NO. 3

A spatial distribution of the various land uses that maintains biodiversity and will result in the protection and wise use of the natural resource base, including soils, inland lakes and streams, groundwater, wetlands, and woodlands.

PRINCIPLE

The proper allocation of uses to land can assist in maintaining an ecological balance between the activities of man and the natural environment that supports him.

SOILS

PRINCIPLE

The proper relation of urban and rural land use development to soil types and distribution can serve to avoid many environmental problems, aid in the establishment of better land use patterns, and promote the wise use of an irreplaceable resource.

STANDARDS

1. Sewered urban development, particularly for residential use, should not be located in areas covered by soils identified in the soil survey as having severe limitations for such development.

(Continued)

TABLE 11.1 *(Continued)*
Example Land Use Development Objectives, Principles, and Standards

INLAND LAKES AND STREAMS

PRINCIPLE

Inland lakes and streams contribute to the atmospheric water supply through evaporation; provide a suitable environment for desirable and sometimes unique plant and animal life; provide the population with opportunities for certain scientific, cultural, and educational pursuits; constitute prime recreational areas; provide a desirable aesthetic setting for certain types of land use development; serve to store and convey floodwaters; and provide certain water withdrawal requirements.

STANDARDS

1. A minimum of 25 percent of the perimeter or shoreline frontage of lakes having a surface area in excess of 50 acres should be maintained in a natural state.
2. Not more than 50 percent of the length of the shoreline of inland lakes having a surface area in excess of 50 acres should be allocated to urban development, except for park and outdoor recreational uses.
3. A minimum of 25 percent of both banks of all perennial streams should be maintained in a natural state.
4. Not more than 50 percent of the length of perennial streams should be allocated to urban development, except for park and outdoor recreational uses.
5. Floodlands[d] should not be allocated to any urban development that would cause or be subject to flood damage.
6. No unauthorized structure of fill should be allowed to encroach upon and obstruct the flow of water in the perennial stream channels[e] and floodways.[f]

WETLANDS

PRINCIPLE

Wetlands[g] support a wide variety of desirable and sometimes unique plant and animal life; assist in the stabilization of lake levels and stream flows; trap and store plant nutrients in runoff, thus reducing the rate of enrichment of surface waters and noxious weed and algae growth; contribute to the atmospheric oxygen supply; contribute to the atmospheric water supply; reduce stormwater runoff by providing areas for floodwater impoundment and storage; trap soil particles suspended in runoff and thus reduce stream sedimentation; provide opportunities for certain scientific, educational, and recreational pursuits; and may serve as groundwater recharge and discharge areas.

STANDARDS

1. All wetlands adjacent to streams or lakes, all wetlands within areas having special wildlife or other natural values, and all wetlands having an area of five acres or greater should not be allocated to urban development except for limited recreational use and should not be drained or filled.

WOODLANDS

PRINCIPLE

Woodlands[h] assist in maintaining unique natural relationships between plants and animals; reduce stormwater runoff; contribute to the atmospheric oxygen supply; contribute to the atmospheric water supply through transpiration; aid in reducing soil erosion and stream sedimentation; provide the

(Continued)

TABLE 11.1 *(Continued)*
Example Land Use Development Objectives, Principles, and Standards

resource base for the forest product industries; provide the population with opportunities for certain scientific, educational, and recreational pursuits; and provide a desirable aesthetic setting for certain types of land use development.

STANDARDS

1. A minimum of 10 percent of the land area of each watershed[i] within the planning area should be devoted to woodlands.
2. A minimum regional aggregate of five acres of woodland per 1,000 population should be maintained for recreational pursuits.

OBJECTIVE NO. 4

A spatial distribution of the various land uses that is properly related to the supporting transportation, utility, and public facility systems in order to assure the economical provision of transportation, utility, and public facility services.

PRINCIPLE

The transportation and public utility facilities and the land use pattern that these facilities serve and support are mutually interdependent in that the land use pattern determines the demand for, and loadings upon, transportation and utility facilities, and these facilities, in turn, are essential to, and form a basic framework for, land use development.

STANDARDS

1. Urban development should be located and designed so as to maximize the use of existing transportation and utility systems.
2. The transportation system should be located and designed to provide access not only to all land presently devoted to urban development but also to land proposed to be used for such urban development.
3. All land developed or proposed to be developed for urban residential use should be located in areas serviceable by an existing or proposed sanitary sewerage system and preferably within the gravity drainage area tributary to such systems.
4. All land developed or proposed to be developed for urban residential use should be located in areas serviceable by an existing or proposed public water supply system.
5. All land developed or proposed to be developed for urban residential use should be located in areas serviceable by existing or proposed mass transit facilities.
6. The transportation system should be located and designed to minimize the penetration of existing and proposed residential neighborhood units by traffic.

OBJECTIVE NO. 5

The preservation, development, and redevelopment of a variety of suitable industrial and commercial sites, both in terms of physical characteristics and location.

PRINCIPLE

The production and sale of goods and services are among the principal determinants of the level of economic vitality in any society; the important activities related to these functions require areas and locations suitable to their purposes.

(Continued)

TABLE 11.1 *(Continued)*
Example Land Use Development Objectives, Principles, and Standards

STANDARDS

1. Major industrial development[j] should be located in planned industrial districts that meet the following standards:
   a. Direct access to the arterial street and highway system and access within two miles to the freeway system.
   b. Direct access to railway facilities, if required by the industries located or proposed to be within the district.
   c. Direct access to mass transit service.
   d. Available adequate water supply.
   e. Available adequate public sanitary sewer service.
   f. Available adequate stormwater drainage facilities.
   g. Available adequate power supply.
   h. Site covered by soils identified in the soil survey as having slight or moderate limitations for industrial development.
2. Major retail development[k] should be concentrated in commercial centers that meet the following minimum standards:
   a. Direct access to the arterial street system.
   b. Direct access to mass transit service.
   c. Available adequate water supply.
   d. Available adequate public sanitary sewer service.
   e. Available adequate stormwater drainage facilities.
   f. Available adequate power supply.
   g. Site covered by soils identified in the soil survey as having slight or moderate limitations for commercial development.
3. Major office development[l] should be concentrated in commercial centers that meet the following minimum standards:
   a. Direct access to the arterial street system.
   b. Direct access to mass transit service.
   c. Available adequate water supply.
   d. Available adequate public sanitary sewer service.
   e. Available adequate stormwater drainage facilities.
   f. Available adequate power supply.
   g. Site covered by soils identified in the soil survey as having slight or moderate limitations for commercial development.

OBJECTIVE NO. 6

The preservation and provision of open space[m] to enhance the total quality of the regional environment, maximize essential natural resource availability, give form and structure to urban development, and facilitate the ultimate attainment of a balanced year-round outdoor recreational program providing a full range of facilities for all age groups.

PRINCIPLE

Open space is the fundamental element required for the preservation, wise use, and development of such natural resources as soil, water, woodlands, wetlands, native vegetation, and wildlife; it provides the opportunity to add to the physical, intellectual, and spiritual growth of the population; it enhances

(Continued)

TABLE 11.1 *(Continued)*
Example Land Use Development Objectives, Principles, and Standards

the economic and aesthetic value of certain types of development; and it is essential to outdoor recreational pursuits.

STANDARDS[n]

1. Major park and recreation sites offering opportunities for a variety of resource-oriented outdoor recreational activities should be provided within a 10-mile service radius of every dwelling unit in the planning area and should have a minimum gross site area of 250 acres.
2. Other park and recreation sites should be provided within a maximum service radius of one mile of every dwelling unit in an urban area and should have a minimum gross site area of five acres.
3. Areas having unique scientific, cultural, scenic, or educational value should not be allocated to any urban or agricultural land uses; adjacent surrounding areas should be retained in open space use, such as agriculture or limited recreation.

Source: SEWRPC.

[a] Net land use area is defined as the actual site area devoted to a given use and consists of the ground floor site area occupied by any buildings plus the required yards and open spaces.

[b] Gross residential land use area is defined as the net area devoted to this use plus the area devoted to all supporting land uses, including streets, neighborhood parks and playgrounds, elementary schools, and neighborhood institutional and commercial uses, but not including freeways and expressways and other community and areawide uses.

[c] Gross public park and recreation area is defined as the net area devoted to active or intensive recreation use plus the adjacent lands devoted to supporting land uses such as roads and parking areas. This area does not include surface water, woodlands, wetlands, or other natural resources.

[d] Floodlands are herein defined as those lands inundated by a flood having a recurrence interval of 100 years where hydrologic and hydraulic engineering data are available, and as those lands inundated by the maximum flood of record where such data are not available.

[e] A stream channel is herein defined as that area of the floodplain lying either within legally established bulkhead lines or within sharp and pronounced banks marked by an identifiable change in flora and normally occupied by the stream under average high-flow conditions.

[f] Floodway lands are herein defined as those designated portions of the floodlands that will safely convey the 100-year recurrence interval flood discharge with small, acceptable upstream and downstream stage increases.

[g] Wetlands are defined as areas in which the water table is at, near, or above the land surface and which are characterized by both hydric soils and by the growth of hydrophytes, such as sedges, cattails, willows, and tamaracks.

[h] Woodlands are defined as those upland areas having 17 or more deciduous trees per acre, each measuring at least four inches in diameter at breast height and having at least a 50 percent canopy cover. In addition, coniferous tree plantations and reforestation projects are defined as woodlands. It is also important to note that all lowland wooded areas, such as tamarack swamps, are defined as wetlands because the water table in such areas is located at, near, or above the land surface and because such areas are generally characterized by hydric soils that support hydrophytic trees and shrubs.

[i] A watershed is defined as an area 25 square miles or larger in size occupied by a surface drainage system discharging all surface water runoff to a common outlet.

[j] Major industrial development is defined as an industrial area having a minimum of 3,500 industrial employees.

(Continued)

TABLE 11.1 *(Continued)*
Example Land Use Development Objectives, Principles, and Standards

[k] Major retail development is defined as a retail area having a minimum of 2,000 retail employees.

[l] Major office development is defined as an office area having a minimum of 3,500 office- and service-related employees.

[m] Open space is defined as land or water areas that are generally undeveloped for urban residential, commercial, or industrial uses and are or can be considered relatively permanent in character. It includes areas devoted to park and recreation uses and to large land-consuming institutional uses, as well as areas devoted to agricultural use and to resource conservation, whether publicly or privately owned.

[n] It was deemed impractical to establish spatial distribution standards for open space per se. Open spaces that are not included in the spatial distribution standards are forest preserves and arboreta; major river valleys; lakes; zoological and botanical gardens; stadia; woodland, wetland, and wildlife areas; scientific areas; and agricultural lands whose location must be related to, and determined by, the natral resource base.

now also consider post-2010 planning frameworks, including environmental justice, digital inclusion, carbon neutrality, and inclusive economic development.

APPLICATION IN ALTERNATIVE PLAN EVALUATION

After alternative plans have been designed, the plans must be evaluated in order to determine the degree to which they meet the agreed-upon development objectives and standards. The techniques available for transportation and utility system plan evaluation are more highly developed than those available for land use plan evaluation. Not only have simulation models been developed for the quantitative test of the engineering feasibility of the system plans concerned, but the system development objectives and standards are more readily quantifiable than are the land use development objectives and standards. Moreover, benefit-cost and cost-effectiveness techniques for evaluating proposed investments in public works are more readily applicable to the evaluation of infrastructure facility plans than to land use plans.

Although a benefit-cost or a cost-effectiveness approach may be theoretically applicable to land use plan evaluation, these methods lose much of their effectiveness in such application because of the following limitations:

1. It is impractical to assign a monetary value to the many intangible benefits and costs that relate to the most important land use development objectives, and it is extremely difficult to assign monetary values to even the direct benefits and costs associated with a given land use plan.
2. Because of the relatively greater uncertainty often associated with land use plan implementation than with infrastructure system plan implementation, there can be no assurance that the potential benefits will even be realized, even though many of the costs associated with the development of a given land use plan may, nevertheless, be incurred through public facility and utility construction.

3. Finally, a complete benefit-cost analysis and a complete cost-effectiveness analysis of a land use plan would require the development of benefits and costs associated with the construction of the complete utility and community facility systems associated with the given land use plan, a task beyond the budgetary limitations and capabilities of public planning operations prior to the widespread availability of integrated GIS and data analytics tools.

Today, however, emerging tools in digital twin simulation, spatial data infrastructure, and AI-assisted scenario analysis can partially bridge this gap by enabling better quantification of plan outcomes, though challenges with monetizing intangible impacts remain.

In order to provide a method for quantitatively evaluating the ability of land use plans to achieve stated development objectives, the alternative plans may be scaled against the standards supporting each land use development objective, and the result evaluated by the plan commission and by advisory committees to the commission. In addition, such review may be supplemented by application of a method of plan evaluation that seeks to assign a numeric value to each alternative plan considered. The method overcomes, to a considerable extent, the difficulties inherent in the application of system integration and benefit-cost and cost-effectiveness analyses to land use plan evaluation. It is an adaptation of the rank-based expected value method used in corporate and military decision-making. It avoids the difficulties associated with the assignment of monetary values to the benefits and costs associated with alternative plans by ranking each alternative plan considered under each of the stated development objectives. It is usually much easier to rank the effectiveness of a given plan in achieving a given development objective than it is to attempt to assign a monetary value to the benefits accruing to the attainment of the same objective. The method is as applicable to transportation, utility system, and community facilities planning as it is to land use planning.

The difficult problems associated with uncertainty of plan implementation are also recognized in the rank-based expected value method of plan evaluation through the medium of probability estimation. Some alternative plans, while theoretically more desirable on the basis of their ability to attain development objectives, may have a low probability of implementation; and, in the application of the method, such plans are assigned a lower value for probability of implementation. Other plans, while theoretically less desirable on the basis of their ability to attain development objectives, may have a higher actual value because of a greater likelihood of implementation. This concept of considering the uncertainty of plan implementation in plan evaluation is particularly important in relation to land use plans prepared as a basis for the planning and design of public works facilities. Construction of the latter may require a large investment of public funds, and such an investment cannot be made on the basis of a land use plan that cannot be practically implemented.

In plan evaluation, then, the application of the rank-based expected value method involves the following sequence of activities:

1. All specific development objectives, n in number, are ranked in order of importance to the general development objectives and assigned values of n, n minus 1, n minus 2 … to n minus (n–1) in descending rank order.

2. The alternative plans, m in number, are ranked under each of the specific land use development objectives and assigned a value of m, m minus 1, m minus 2 … to m minus (m–1) in descending rank order.
3. A probability, p, or implementation is assigned to each of the plans being ranked.
4. The value, V, for each alternative plan is then determined by summing the products of n times m times p for each of the specific development objectives.

$$V = p\Sigma\left(n_1m_1 + n_2m_2 + \ldots + n_nm_n\right)$$

Table 11.2 illustrates a simple hypothetical application of the method using three specific development objectives. In the hypothetical plan evaluation shown in the table, Plan No. 3 would be selected as the plan that best meets the development objectives. This method is also increasingly supported by digital planning dashboards, which automate ranking inputs, generate visualizations, and streamline stakeholder engagement in selecting preferred alternatives.

TABLE 11.2
Example of Application of Rank-Based Expected Value Method of Alternative Plan Evaluation

		Balanced Allocation of Land to Uses	Natural Resource Conservation	Proper Spatial Distribution of Land Uses	
		Rank Order Value of Objective n = 1	Rank Order Value of Objective n = 3	Rank Order Value of Objective n = 2	Plan Value, V
Plan	Specified Development Objective	Rank Order Value of Plan, m	Rank Order Value of Plan, m	Rank Order Value of Plan, m	V = p Σ (n1 m1 + n2 m2 + n3 m3)
1	Probability of Implementation p = 0.6	3	1	3	0.6 [(1 × 3) + (3 × 1) + (2 × 3)] = 7.2
2	Probability of Implementation p = 0.5	2	2	1	0.5 [(1 × 2) + (3 × 2) + (2 × 1)] = 5.0
3	Probability of Implementation p = 0.9	1	3	2	0.9 [(1 × 1) + (3 × 3) + (2 × 2)] = 12.6

Source: SEWRPC.

As herein recommended, the specific development objectives should be expanded into a set of supporting standards that may be used to evaluate the ability of an alternative plan to achieve a given specific development objective. Any ranking of an alternative plan for a given specific development objective must, therefore, be consistent with the ability of the plan to achieve the standards formulated for that objective. To achieve this consistency, it is first necessary to compute a value for each of the alternative plans according to the standards formulated for each specific development objective before arriving at an overall value for each plan in relation to the development objectives. This subsidiary evaluation can utilize a series of tables similar to that given in the preceding example, except that the development standards replace the development objectives in the tables and that it is usually not necessary to assign a probability estimate for the standard evaluation.

FURTHER READING

Ansoff, C. H. *Corporate Strategy*. McGraw-Hill, 1965.
Schlager, K. J. "The Rank Based Expected Value Method of Plan Evaluation," *Proceedings of the Highway Research Board*. 1968.

12 Land Use Planning

> We already have land use planning of an ad hoc, accidental sort that, while it has made some people rich, has made most of us far poorer in terms of the kinds of choices we have and the quality of life we are offered ... whether we are trying to stimulate growth or stop it, we simply cannot expect to resolve the problems associated with growth on a case by case ... basis. The patterns of development that result from this approach must inevitably be both socially unfair and environmentally unsound.
>
> **The Milwaukee Journal**
> *September 30, 1974*

INTRODUCTION

Land use planning is that part of the process of city planning concerned with the type, location, intensity, and amount of land development required for the various space-using functions of urban life. Land use planning is concerned, therefore, with determining the proper relationships of various land uses to one another and to existing and proposed transportation facilities, utilities, and community facilities. Today, it also incorporates sustainability, resilience, and equity considerations, particularly in the context of climate change, housing access, and environmental justice. In this respect, it should be noted that land use planning is a process requiring expert skills that may be quite different from those required for transportation, utility, or community facilities planning.

The land use plan, quite simply and fundamentally, is a proposal, or recommendation, as to how land should be used as future community expansion or renewal occurs. The land use plan is usually expressed in a colored plan map, supported by pertinent text, tables, and graphs setting forth recommended standards for population density, commercial and industrial employment intensity, and building intensity. Modern land use plans increasingly rely on geographic information system (GIS)-based scenario models and interactive dashboards that enhance transparency and stakeholder engagement. The land use plan should be distinguished from the land use and zoning district maps. The land use map is a map showing how land and structures in a planning area are actually used at a given point in time—past or present; that is, it is a factual representation of the existing land use pattern. The zoning district map is a map that divides the community into districts for the purpose of regulating the use, density of population, and intensity of building coverage. The zoning district map and attendant regulations are intended to be a means of implementing the land use plan. Unfortunately, in some communities, the zoning district map has been confused with, and even substituted for, the land use plan.

DOI: 10.1201/9781003500384-12

This confusion can result in reactive, parcel-based decision-making that undermines long-term planning goals. It is increasingly important for communities to ensure that zoning codes are aligned with comprehensive land use plans through regular updates and integrated planning workflows.

DETERMINANTS OF THE LAND USE PATTERN

If the land use planner is to do his work well, he should have an understanding of the forces that shape the urban land use pattern. The term "land use pattern" refers to the spatial distribution of the various land uses within an urban area: the residential, industrial, commercial, institutional, and recreational uses. The land use pattern reflects the four general functions of all cities: production, distribution, consumption, and amenity service. Land uses, thus, as noted in Chapter 8, represent activities that take place on land.

It is a readily observable fact that urban communities that are allowed to grow with very little or no public regulation of the location of the various land uses—as was historically true in many North American cities—nevertheless tend to develop, as it were, "natural" groupings of related land uses. Sometimes these natural patterns show marked similarities between various communities, such as the central business districts of medium-sized cities and the main streets of smaller cities and villages. In contemporary cities, newer patterns are emerging due to e-commerce logistics, remote work, and mixed-use development trends. Examples now include lifestyle centers, innovation districts, and hybrid retail-industrial nodes. Sometimes the pattern may be unique, such as the striking corner tavern, grocery, drug, butcher, and baker shop patterns of old Milwaukee.

How these geographical patterns develop and how they change over time has been the subject of much study and research. Depending on the background of the student concerned, differing emphasis may be placed upon the importance of the possible determinants of the land use pattern. In this respect, the "four cities" concept set forth in Chapter 2 should be recalled. The major determinants of the urban land use pattern may be summarized as economic, social, and physical. In addition, the public interest should be a fourth determinant. Technological innovation, environmental constraints (such as climate adaptation and hazard risk), and land value dynamics are now also key drivers. Today, public interest increasingly encompasses sustainability, equity, and resilience in addition to traditional concerns such as efficiency and livability.

Economic Determinants

The concept of the economic determinant of the urban land use pattern is the basic tenet of the discipline of urban land economics. Scholars at the University of Wisconsin contributed significantly to the development of this discipline, and Professor Richard U. Ratcliff wrote one of the best texts in this area, a text cited herein for further reading. Recent advancements continue to build on this foundation, integrating behavioral economics, spatial analytics, and real-time data.

The hypothesis underlying the concept of the economic determination of the land use pattern holds that land will be put to that use that can obtain the highest return from it; thus, competition among bidders in the urban land market will determine the land use pattern. The urban land market is very complex, involving financial and institutional factors as well as the focus of supply and demand. The existing land use pattern is, therefore, the aggregate result of many individual decisions by landowners and developers with respect to the best economic use of many individual parcels of land. The economic base and structure are given as the explanation for the total size of the urban area and determine the intensity of bidding for sites. Thus, land values operate to determine the type and intensity of land use. Land values are, in turn, influenced by the existing land use pattern and, importantly, by the configuration and capacity of the transportation and utility facilities required to support urban land uses. Today, additional influences such as telecommuting, digital commerce, environmental constraints, and government incentives also shape demand for different types of land use.

Land economists and geographers have advanced at least four conceptual—not mathematical—models to explain the land use patterns formed by operation of the urban land market: the monocentric or concentric ring model, the sector model, the external expansion model, and the polycentric model. These models represent not so much competing views as views representing different eras in city development.

Monocentric Model

The monocentric model, which is illustrated in Figure 12.1, was proposed by Ernest Burgess of the University of Chicago in 1925. It attempts to explain why urban land markets tend to segregate land use activities and why land use intensity varies within an urban area. The model distinguishes four basic land uses: residential, manufacturing, service, and retail. Each use is assumed to face different transportation costs entailed by workers commuting and manufacturing firms receiving and shipping freight to and from a central node, services interacting with clients, and shoppers traveling to and from retail stores. The model assumes an idealized topography and a radial transportation system. Land uses in an effort to minimize transportation costs are assumed to segregate themselves in concentric rings around a core of service uses—legal, marketing, financial—located at a central import-export node. These uses can command the lowest transportation costs and the highest land values and can function in high-rise buildings. Light and heavy manufacturing uses also desire a central location, but are outbid for sites by the service uses. Major retail uses also locate centrally to maximize the number of residences within a given travel time or distance. This model conforms roughly to the land use pattern found in most North American cities before 1950. Land uses are segregated and centralized; the value of land affects density; accessibility shapes the pattern; and the price of accessibility is higher land costs.

The monocentric city model holds that cities develop around a single center or core because during the early stages of the Industrial Revolution, railway technology required the movement of freight to take place to and from a central import-export node. Manufacturing firms were located close to this center. Office firms, however, were located closest to the center because the value of accessibility to these firms exceeded the value placed on accessibility by manufacturers. Residential areas formed outside the manufacturing rings because workers preferred not to live at high densities, yet sought ready access to job locations. Expansion under this model occurs by a disruptive succession of land uses and densities or intensities of land use.

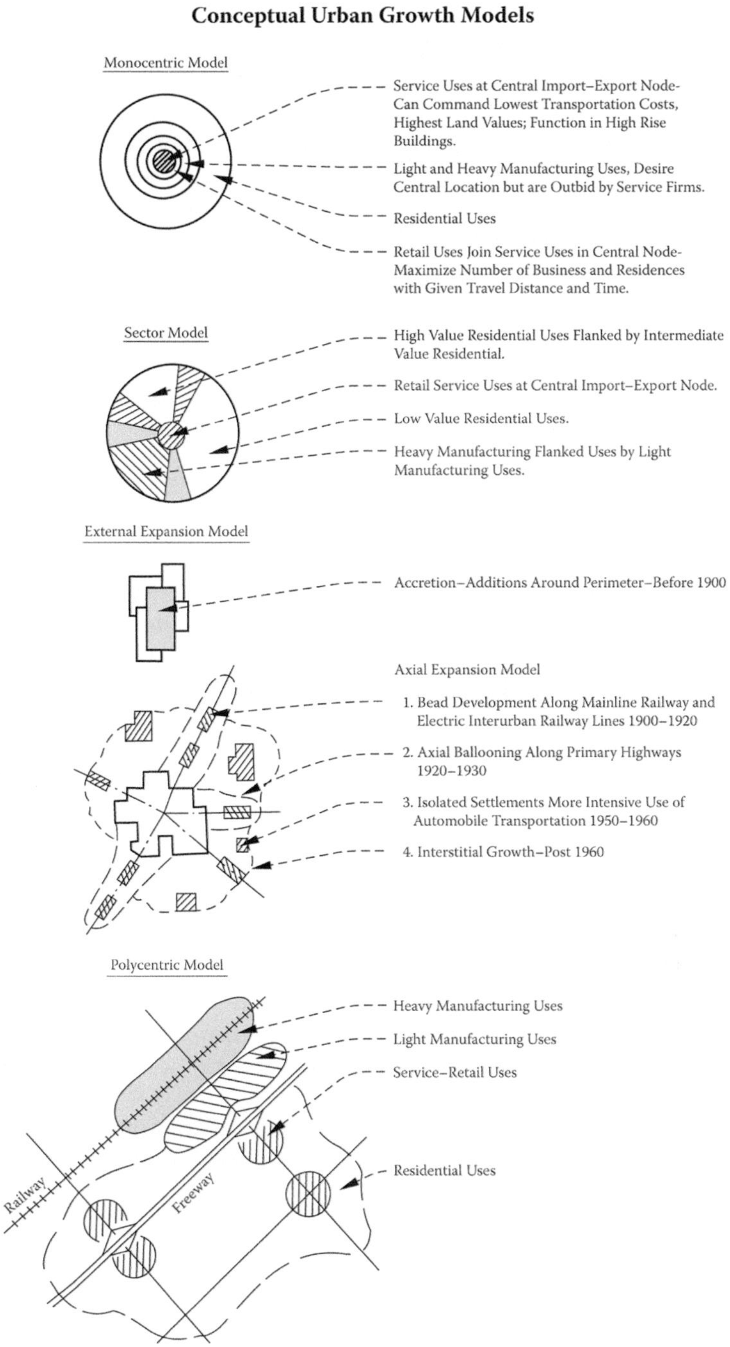

FIGURE 12.1 Illustration of the four conceptual models advanced by geographers and urban land economists to explain the urban land use patterns formed by the operation of the urban land market. The polycentric model is intended to represent past 1960 development patterns.

Sector Model

The sector model, illustrated in Figure 12.1, assumes a central core area of retail and service uses around which urban development takes place, as does the monocentric model, but postulates other uses will arrange themselves in sectors rather than in rings. This model was proposed by Homer Hoyt of the Federal Housing Administration in 1939. The sectors may comprise high-value residential uses, intermediate-value residential uses, low-value residential uses, and light and heavy manufacturing uses. Under this model, blighted areas, manufacturing districts, and high-value areas are said to "cast shadows" outward. Urban expansion under this model does not require disruption of the existing land use pattern, as does the monocentric ring model. It should be noted that the Burgess and the Hoyt models may be conceptually combined by assuming that an urban area may have a concentric ring pattern of age superimposed on an underlying sector pattern of uses.

External Expansion Model

In this model, also illustrated in Figure 12.1, changes in the technology of transportation are assumed to explain changes in city size, exemplified by the successive effects of the pedestrian, horsecar, electric streetcar, and elevated, subway, and commuter railway transportation modes on city size. These changes in technology served to increase the feasible radius of urban areas. When travel times are not correlated with distance, the shape of an urban area may be distorted into corridors. Under this model, urban expansion is assumed to have occurred successively by:

1. Accretion—that is, by additions around the perimeter of the existing urban area. This was the pattern of expansion of North American cities before 1900.
2. Areal expansion—such expansion was viewed as taking the form of bead development from 1900 to 1920, the beads forming along suburban steam railway and electric interurban railway lines, and taking the form of areal ballooning from 1920 to 1930, the ballooning occurring around the axes of primary highways.
3. Isolated settlements—the general pattern from 1930 to 1960 was marked by the intensified use of the automobile for urban transportation.
4. Interstitial growth—the most recent form of accretion, often of low density.

Today, this is supplemented by edge cities, leapfrog development, and dispersed low-density growth tied to regional highway networks and digital infrastructure.

Polycentric Model

Decreases in the cost of moving freight between cities and of moving people within cities as transportation technology developed shaped the growth of urban areas into different patterns over time. The importance of proximity to a central import-export node initially resulted in intense commercial and industrial development of central core areas. The importance of these core areas caused the development of radial transportation systems—primarily railway—which in turn reinforced the importance of the central area as an employment and service center.

With the development of the automobile and motor truck and the attendant decreasing cost of moving freight both within and between cities and increasing flexibility in choosing destinations and routes, the pattern of urban development changed significantly. This is reflected in the polycentric model also illustrated in Figure 12.1. Under this model the importance of central locations diminished dramatically, and cities became polycentric. The monocentric and sector models, however, may still be relevant with respect to the land use pattern around each center. The polycentric form of development was accentuated by the development of urban freeway systems. The automobile and motor truck allowed manufacturing to move to locations on the urban fringe to access lower cost land for new, single-story plant layouts, extended the effective labor pool commuting area, made large areas of low-priced land available for residential development, and permitted retail and service uses to decentralize following their customer base. This has been the pattern of development followed by most North American cities since 1960. Since the 1990s, this pattern has further evolved into networks of edge cities, logistics hubs, and mixed-use nodes shaped by data flows, airport access, and regional transit investments.

Concluding Comments on Conceptual Models

All of the conceptual economic-centered models of urban development patterns advanced by geographers and urban land economists recognize the importance of transportation in shaping urban development patterns. Arguably, the most significant changes in land use patterns over the last century have resulted from major changes in the characteristics of transportation. Land cannot be developed without access. The nature of the access determines the characteristics of development given the technologies involved in the production of goods and services. Historically, rail transit systems shaped the urban land use patterns before highway transportation did. However, as accessibility within an area becomes uniformly high, the relative importance of transportation as a determinant of the urban land use pattern decreases, and the importance of other influences increases, particularly the availability of sanitary sewerage and public water supply services. Nevertheless, the urban land market remains the major determinant of land use today given the characteristics of the transportation and utility systems. The private market, in contrast, provides and bears the costs of only a small proportion of the transportation and utility infrastructure systems.

Social Determinants

Social determinants of the urban land use pattern are less well understood, and the effects less evident, than the economic determinants. Sociologists seek to explain the urban land use pattern in terms of urban ecology, the processes by which humans adapt themselves to the urban environment. The processes hypothesized include the following:

1. Dominance: a process under which one area of the city bears a controlling social or economic position over others; historically, this has often been the central business district.
2. Gradient: a process marked by a receding degree of dominance from the center outward to more distant locations.

3. Segregation: a process by which like units—households, businesses, industries, and institutions—tend to cluster and form homogeneous areas; the basis for the segregation may be economic or may be founded in personal or community value systems. Examples of such land use segregations include medical centers, residential gold coasts, ethnic, religious, and racial ghettos, and dormitory suburbs.
4. Centralization–decentralization or concentration–dispersion: centralization referring to the congregation of people and urban functions in an urban center in pursuit of economic, cultural, or social satisfactions, decentralization referring to the breakdown of the urban center with the movement of people and urban functions to fringe areas or to new satellite centers.
5. Invasion and succession: a process marked by the penetration of one population group or land use by another in which the new displaces the old over time. An example of this process is the "gentrification" of blighted or slum areas.

All of these processes depend upon social and spatial mobility, the implication being that a shift in social status usually involves a shift in location. All of these processes represent observations of the composite effect of mass behavior and are value-driven—values being assumed to motivate individual and group behavior—resulting in organized action. Mass values are the consensus of values held by a majority of groups in a community and have community-wide significance. Values generally have significance for localized areas. However, where a power structure coalesces groups, the value systems concerned may take on broader significance. Planners must recognize that successful implementation depends on how well plan proposals align with both dominant societal values and emerging norms, including inclusivity, equity, and cultural identity. Social media, demographic shifts, and generational attitudes now exert strong influence on land use preferences and acceptance.

Physical Determinants

Physical determinants of the urban land use pattern include topography and drainage patterns; existing culture, including real property boundaries, transportation facilities, and land uses; availability of utilities: sewer, water, electric power, and telecommunications; soils, particularly the engineering properties of soils; wetlands and woodlands; and special hazards, such as flooding. In recent years, climate change, sea level rise, extreme heat, wildfire risk, and storm intensity have emerged as critical physical factors shaping where and how development can occur. Additionally, physical determinants are increasingly mapped with geospatial technologies and integrated into scenario-based modeling tools used in resilience planning and infrastructure investments.

The Public Interest as a Land Use Determinant

The promotion of the public interest as a determinant of the urban land use pattern is the fundamental reason for public planning. Although economic and social forces and physical conditions do indeed shape the urban land use pattern, governmental regulations in the interest of the public health, safety, and general welfare also should and do shape that pattern. Elements of the public interest include the following:

1. Health and safety: protection against contagion, accidents, hazards, excessive noise, and atmospheric pollution, and provision for adequate light and air
2. Economy: minimization of the costs of constructing, operating, and maintaining transportation, utility, and community facilities, and of providing essential services such as police and fire protection
3. Well-being: promotion of safe, inclusive, and resilient environments that enhance mental and physical health, social cohesion, and quality of life
4. Stability: for public and private investment

Urban land markets frequently fail to allocate resources equitably and efficiently, particularly when prices do not reflect full social and environmental costs or when decision-makers and the public lack adequate information about long-term impacts. Some of the costs of private development—known as "spillover" costs—may be borne by public agencies. Examples would include the costs of providing additional transportation system capacity needed to carry the traffic generated by a new development, the costs of providing additional sanitary sewerage and water supply storage and transmission capacity, and the costs of providing additional schools and police and fire protection. Spillover costs may also include the costs associated with environmental degradation that may be caused by a new development, including increased stormwater runoff, non-point source water pollution, and loss of environmentally important wetlands and woodlands. The urban land market thus operates as a mixed and imperfect system, requiring the guidance of planning, zoning, and public policy to ensure that private development serves the broader community interest. Contemporary planning practice expands the concept of the public interest to include equity, climate resilience, sustainability, and intergenerational responsibility—ensuring that land use decisions promote not only efficiency and safety, but also environmental justice and long-term community well-being.

Technological Innovation, Environmental Constraints, and Land Value Dynamics

In addition to economic, social, and physical determinants and the public interest, three increasingly influential forces shaping urban land use patterns are technological innovation, environmental constraints, and land value dynamics. These interrelated factors are redefining how cities grow, adapt, and sustain themselves in the face of rapid change.

1. Technological innovation: Advancements in transportation, communication, construction, and energy technologies continue to alter urban form and function. The rise of electric and autonomous vehicles, smart infrastructure, remote work, and digital service delivery has decoupled many land uses from strict locational dependencies. For example, the spatial demand for central office space is shifting, while warehousing, data centers,

and distributed energy systems are emerging as new spatial demands. Planning frameworks must increasingly anticipate these shifts and incorporate adaptive land use strategies.
2. Environmental constraints: Climate change has elevated environmental factors from background considerations to central planning imperatives. Constraints such as floodplains, wildfire zones, drought-prone areas, and heat islands must now be explicitly integrated into land use decisions. Regulatory pressures—such as carbon reduction goals, water conservation mandates, and habitat preservation—are also redefining the feasible and desirable uses of land. In this context, land use planning is becoming deeply intertwined with sustainability and resilience goals.
3. Land value dynamics: Land value is both a driver and a product of land use patterns. It reflects market forces, public investment, regulatory frameworks, and perceived amenity. As land becomes more scarce, volatile, or speculative in high-demand areas, market dynamics can both reinforce and undermine planned land use goals. Gentrification, displacement, and exclusionary patterns often arise where land value pressures exceed affordability thresholds. Conversely, declining land values in disinvested areas may inhibit needed development. Thus, understanding and influencing land value dynamics is a crucial task of contemporary planning.

Together, these forces demand more integrated, anticipatory, and flexible planning approaches—ones that can align long-term goals with emerging realities across infrastructure, environment, and markets.

STEPS IN LAND USE PLAN PREPARATION

Delineate Planning Area

The first step in land use plan preparation is to delineate the planning area to be considered. Such delineation requires careful study. The planning area should include the existing corporate boundaries of the municipality concerned and the currently urbanized area, and such adjacent open areas as future growth potential and projected development trajectories may suggest. Factors to be considered should include natural watershed boundaries, special topographic features such as large wetlands, woodlands, and open space reservations, environmental corridors, and prime agricultural lands. Considerations should also include adopted state, regional, metropolitan, and county plans, other local plans, and extraterritorial planning, zoning, and plat approval jurisdictions. All of these considerations, except the existing corporate boundaries and currently urbanized area, would apply to new town sites as well as to existing cities.

Assemble Needed Basic Data, Including

1. Adequate maps
2. Estimates of historical current and forecasts of probable future resident population and employment levels

3. Cultural data including existing land use data; data on the location and configuration of real property boundary lines; data on transportation facilities—type, location, configuration, capacities, traffic volumes, and levels of service; and data on utilities, including the location, configuration, capacities, loadings, and levels of service, and existing and proposed service areas of sanitary sewerage, public water supply, and other utility systems
4. Physiographic data, including drainage patterns, geology, soils, and special hazard areas, such as flood hazard areas
5. Natural resource base data, including data on climate, air quality, surface and groundwater resources, woodlands, wetlands, and environmental corridors
6. Environmental justice and equity data, including demographic and income characteristics, exposure to environmental hazards, and access to public services
7. Climate vulnerability and hazard mitigation data, including flood risk maps, urban heat island data, and carbon emission profiles
8. Land use suitability analysis and scenario-based growth projections enabled by geospatial and digital planning tools

Delineate Existing and Potential Planning Districts

Planning districts may be defined as natural or artificial districts within the planning area that are useful for analytical and plan design purposes. Generally, such districts will include the central business district, other major commercial districts, major industrial districts, and residential neighborhood units. The districts fall into two categories: those that relate to the entire urban area and those that relate to neighborhoods. The former include the central business district, major commercial and industrial districts, certain institutional districts, including universities and technical colleges, and large recreational areas. Newer planning efforts may also include innovation districts, transit-oriented development (TOD) zones, and resilience zones within this framework.

Estimate Future Land Requirements

A number of techniques are available for estimating future land requirements. These include intuitive judgment, land use accounting application of adopted standards, and land use simulation and design models.

Intuitive Judgment

The estimation of future land requirements by intuitive judgment is an approach without recourse to quantitative analysis. Under the intuitive judgment approach to estimating future land requirements, the land use planner becomes intimately acquainted with the planning area and its districts: the street pattern, topography, soil conditions, existing land uses, structural conditions, utilities, and zoning. The planner then estimates what might be developed on any remaining large undeveloped tracts within each district and in any areas subject to renewal. Thus, the anticipated land uses, densities, and amounts of land to be absorbed are subjectively determined. While largely qualitative, this method may serve as a preliminary check or be used in smaller-scale planning areas when detailed data is unavailable.

Land Use Accounting

The land use accounting method of estimating future land requirements is based on an assumed stability in the proportions of land devoted to each type of land use within an urban area. Urban areas develop as integral parts of a complex social and economic system. The amounts of land devoted to the various uses reflect the requirements of this system—particularly as evidenced by the operation of the urban land market. The land use accounting method seeks to calculate these amounts. Its application requires analyses—either taken from the literature or developed locally—of existing land use patterns.

The planning and engineering firm established by Harland Bartholomew prepared plans for a large number of cities throughout the United States, and thereby assembled existing land use data for these cities in a uniform, comparable format. Bartholomew analyzed this large database, attempting to develop stable relationships between the resident population and the amounts of area devoted to the various land uses. He reported the findings of his studies in the text *Land Use in American Cities*.

The study found no fixed relationship between the total areas encompassed within urban corporate limits. This was to be expected since such limits are delineated arbitrarily through the political process and, therefore, encompass varying amounts of "vacant" land. The study did, however, find stable relationships between developed areas and population. These relationships are summarized in Figure 12.2 and Table 12.1, adapted from the Bartholomew text. The tabulated ratios given in Table 12.1 represent average values for 53 central cities and 33 satellite cities based on land use surveys conducted by Bartholomew between 1935 and 1952. Urban development patterns change over time, and this gives rise to a need to, from time to time, review and update the data presented in the Bartholomew text. The American Planning Association conducted such an update in 1992, using data collected from 34 cities of under 100,000 population and from 32 cities of over 100,000 population. The results indicated that for the smaller cities the ratio of single-family residential use to total developed area changed from 36 percent in 1952 to 41 percent in 1992; the ratio of total residential use to total developed area changed from 42 percent in 1952 to 52 percent in 1992; the ratio of commercial use to total developed area changed from 2.5 to 10 percent; and the ratio of industrial use to total developed area changed from 6 to 10 percent. The results further indicated that for the larger cities the ratio of single-family residential use to total developed area changed from 32 to 38 percent; the ratio of total residential use to total developed area changed from 40 to 48 percent; the ratio of commercial use to total developed area changed from 3 to 10 percent; and the ratio of industrial use to total developed area changed from 6 to 10 percent. The changes in residential use ratios may be attributed to the use of lower densities in residential development over the 40-year period, although a shift back toward the 1952 ratio may be expected for a number of reasons, including changes in the availability

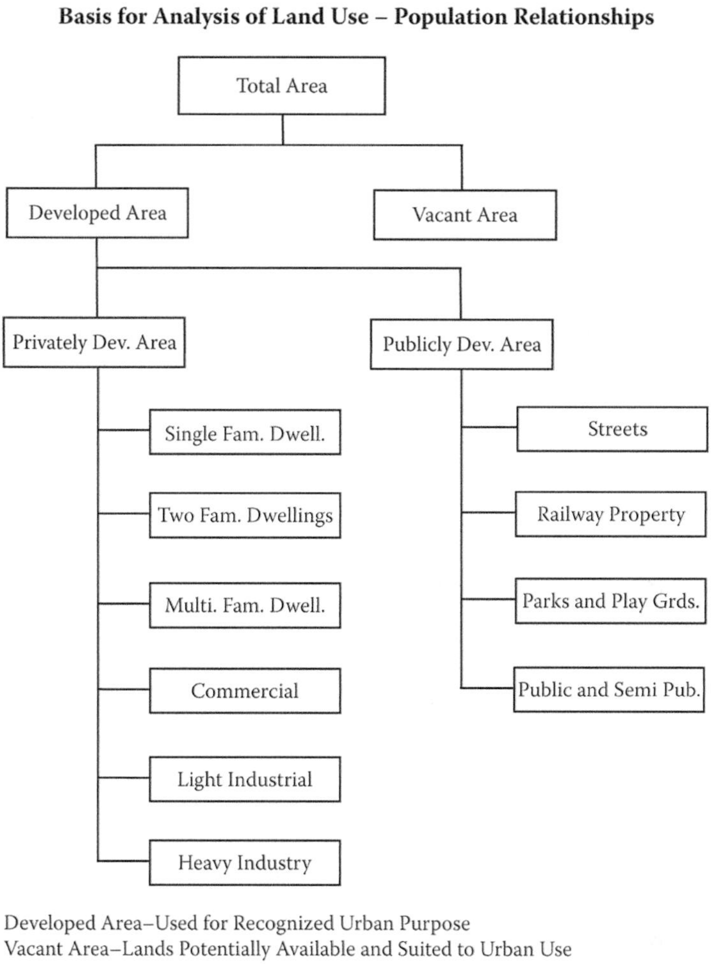

FIGURE 12.2 Illustration of the categorization of land uses adopted by Bartholomew in his analysis of land use patterns.

Source: Adapted from Land Use in American Cities, Harland Bartholomew, 1955.

and cost of petroleum-based motor fuels, declines in real personal income levels, and the promulgation of neotraditional development patterns by some planners. The changes in the ratios for commercial and industrial development may be attributed to automobile parking and truck loading area needs and to the shift by industries to large one-story plant layouts located in campus-like settings. More recent trends indicate further shifts in land use proportions due to emerging priorities such as walkability, climate adaptation, and compact mixed-use development. Accordingly, any historical data must be calibrated using up-to-date local land use inventories and regional

TABLE 12.1
Land Use Ratios

Use	Percent of Total Developed Area: Central Cities	Percent of Total Developed Area: Satellite Cities	Acres of Developed Area per 100 Persons: Central Cities	Acres of Developed Area per 100 Persons: Satellite Cities
Single-Family Residential	31.8	36.2	2.5	3.1
Two-Family Residential	4.8	3.3	0.3	0.3
Multi-Family Residential	3.0	2.5	0.2	0.2
Subtotal	**39.6**	**42.0**	**2.7**	**3.6**
Commercial	3.3	2.5	0.2	0.2
Industrial and Railway	11.3	12.5	0.8	1.1
Streets	28.2	27.7	1.9	2.4
Parks	6.7	4.4	0.5	0.4
Public and Semi-Public	10.9	10.9	0.8	1.0
Total Developed Area	**100.0**	**100.0**	**6.9**	**8.7**

Source: Adapted from Bartholomew (1955).

The tabulated ratios represent average values for 53 central cities and 33 satellite cities based on land use surveys conducted between 1935 and 1952.

market dynamics. Use of generalized land use ratios should always be supplemented with local data and verified against real estate trends, demographic shifts, and transportation demand projections.

Among the land use categories, there are some whose average values Bartholomew considered "norms," that is, uses that exhibited relatively stable ratios regardless of the size or location of the city. These are the residential, commercial, and street use categories. The studies indicated that other types of land uses did not exhibit such consistent patterns between cities. These uses included particularly the heavy industrial use category. It should also be noted that the studies consistently found that more area is devoted to residential use than to any other use, with streets constituting the second largest category of land use. The study also found a correlation between total developed area and population, the development area varying inversely with population size.

Example Application of Land Use Accounting Method

Application of the land use accounting method of estimating future land requirements in land use plan preparation may be illustrated by the following example.

Assume that a land use plan is to be prepared for a new industrial town. The town is to serve as the site for a new mineral extraction and processing plant that is to employ 3,300 persons. Estimation of the total developed area of the town is required together with estimation of the land requirements for each of the major land uses. The calculations required are set forth in Figure 12.3.

Example Land Use Accounting Computation

1. Estimate probable resident population:

   Assume ratio of basic to service employment of 1:1; therefore total employment will approximate 6,600 persons; assume ratio of total employment to resident population of 1:3; therefore design resident population will approximate 20,000 persons.

2. Estimate total developed area required:

   Use "norm" for total developed area per 100 persons for satellite cities from Table 12-1 of 8.7 acres.

   Total developed area = 8.7 × 200 = 1,740 acres, or 2.7 square miles.

3. Compute land requirements for major land uses:

Column No. 1 Land Use	Column No. 2 Percent of Developed Area	Column No. 3 Net Acreage	Column No. 4 Street Area	Column No. 5 Gross Acreage
Residential	42.0	731	280	1,011
Commerical	2.5	43	17	60
Industrial and Railroad	12.5	217	83	300
Parks	4.4	77	29	106
Public	109	190	73	263
Subtotal	72.3	1,258	482	1,740
Streets	27.7	482	--	--
Total	100.0	1,740	--	--

4. The percentage values listed in Column No. 2 for the various land uses are taken from Table 12-1 for Satellite Cities; the values for Central Cities could have been selected for use. The net acreage values listed in Column No. 3 for the various land uses are computed by multiplying the percentage values listed in Column No. 2 by the total developed area of 1,740 acres; thus the acreage to be allotted to residential use is computed as 1,740 × 0.42 = 731. The street area values listed in Column No. 4 are computed by prorating the total street area to each of the land uses. Thus for residential use the proration is given by:

$$\frac{\text{Total Street Area}}{\text{Net Area all Other Uses}} = \frac{\text{Street Allocation}}{\text{Net Residential Area}}; \frac{482}{1{,}258} = \frac{x}{731}; \; x = 280 \text{ acres}$$

5. The values in Column No. 5 are computed by adding for each of the various land uses the prorated street acreages to the net acreages to obtain the gross acreages.

FIGURE 12.3 The values for the commercial area should be reviewed in relation to off-street parking needs. The calculated values, however, indicate sound ratios based on economic need. The values for the industrial area should also be reviewed in relation to the increased site areas needed for modern single-story plant layouts and for off-street parking and truck-loading areas. Any adjustments to net use areas must be reflected in the associated infrastructure allocations, especially streets and public services. The areas to be devoted to park and to public uses—unlike the areas to be devoted to residential, commercial, and industrial uses, which should be reasonably related to the demands of the urban land market—are determined by public policies. Therefore, the areas for these uses should be adjusted on the basis of accepted planning standards.

Although this approach provides a coherent structure, reliance on mid-twentieth-century land use patterns can inadvertently replicate outdated forms. Thus, critical review and adjustment of calculated values is necessary to reflect contemporary policy goals such as sustainability, equity, climate resilience, and mobility.

The values for the residential area should be reviewed in relation to desired densities and the desired ratio of single-family dwelling units to multi-family dwelling units. If a typical post–World War II midwestern layout were contemplated for the new town, emphasis would be on the construction of single-family dwelling units on relatively large lots, resulting in an increase in the area to be devoted to residential use. If the construction of a significant number of garden apartments were also contemplated, this might offset such an increase.

Modern household sizes and housing preferences have evolved considerably, and current planning must reflect variable lot sizes, accessory dwelling units (ADUs), and a broader mix of housing types.

Application of Adopted Standards

Future land requirements can also be estimated—and, indeed, are usually estimated—by application of adopted objectives and supporting standards such as those set forth in Table 11.1 of Chapter 11. The development of the standards, however, ultimately requires recourse to some kind of land use accounting procedure.

Design Land Use Plan

The final step in land use plan preparation involves the spatial distribution of the land areas required to meet the estimated design-year land use needs as those needs are determined by the population and employment forecasts and the method used to convert those forecasts into land requirements. The American Society of Civil Engineers, in its Manual of Practice No. 16, defines the term "design" as the skillful arrangement or organization of various parts into an integrated whole that will best meet the requirements of some definite purpose. The land use design process requires great skill and experience. It is a multidisciplinary process that blends engineering, planning, environmental science, and urban design to allocate land uses in a way that aligns with community values, projected demand, infrastructure capacity, and climate resilience. This requires identification and quantification of "developable" as opposed to "undevelopable" land areas, such identification being based on the findings of the inventories conducted as a basis for the planning effort. Undevelopable areas may be represented by floodlands, wetlands, woodlands, wildlife habitat areas, areas covered by unsuitable soils, and areas of steep slope, among others. Geospatial datasets, including topographic layers, soil classification, hydrology, and biodiversity assessments, now support this step through modern GIS-based suitability modeling.

The spatial allocation may be made in the following order:

1. Protection of environmentally sensitive and hazard-prone areas (as non-developable)
2. Allocation of infrastructure-dependent land uses (e.g., industrial, commercial)
3. Integration of residential types based on proximity to jobs, services, and transportation
4. Provision of public services and recreational open space
5. Ensuring connectivity, access, and equitable distribution of land use benefits

Simulation and Design Models

Land use simulation and design tools are increasingly supported by spatial modeling platforms such as UrbanFootprint, Envision Tomorrow, and CommunityViz. These tools allow planners to test different scenarios for growth, redevelopment, and conservation. Simulation models typically explore multiple possible land use futures based on trends or policy choices. In contrast, normative design models evaluate alignment with adopted goals such as equity, climate adaptation, or mobility. While early models were highly technical and limited in accessibility, the development of computerized GIS- and parcel-based land information systems has made land use plan design using other techniques both more effective and efficient. Current tools are more user-friendly, interactive, and integrated into public engagement processes.

EXAMPLE OF LAND USE PLAN

Figure 12.4 illustrates an actual land use plan map. This plan is for the same community for which the existing land use map is given in Figure 8.1 and attendant land use areas are given in Table 8.2 of Chapter 8. The data on existing and planned land uses attendant to the land use plan map are summarized in Table 12.2. It should be noted that the existing land use areas given in Table 8.2 are net acreages, while those given in Table 12.2 are gross acreages, the differences being in the prorated street areas. The plan is intended to accommodate a resident population of about 5,000, double the base year population of 2,500, and a total employment of 2,300 jobs, an increase of about 50 percent over the base year level. Of this increase, about 12 percent are estimated to be in the service sector and 88 percent in the basic sector. The large amounts of area allocated to outdoor recreational and environmental corridor uses may be attributed to the location of the community in the Kettle-Moraine State Forest area of Wisconsin. The plan proposes the new residential development to be accommodated in seven neighborhood units, each about one square mile in area.

Example of Land Use Plan Map

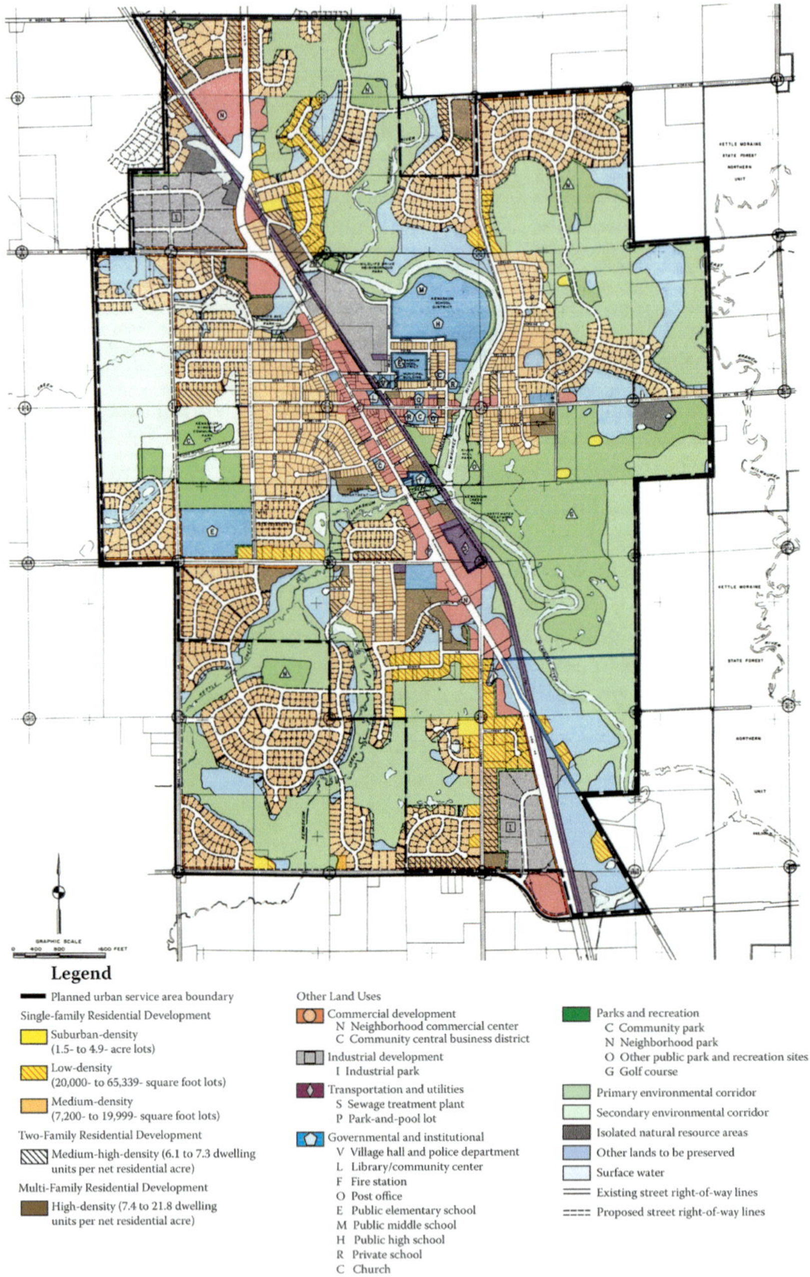

FIGURE 12.4 Illustration of an actual land use plan map. Attendant quantitative land use data are provided in Table 12.2.

Source: SEWRPC.

TABLE 12.2
Summary of Existing and Planned Land Uses Attendant to Plan Map Shown in Figure 12.4

	Existing		Planned Development Conditions	
Land Use Category[a]	Gross Area Acres	Percent of Subtotal (Urban or Nonurban)	Gross Area Acres	Percent of Subtotal (Urban or Nonurban)
Urban				
Residential				
Single-Family	224.8	43.0	950.5	53.7
Two-Family	11.8	2.2	82.7	4.7
Multi-Family	25.9	5.0	72.0	4.1
Subtotal	**262.5**	**50.2**	**1,105.2**	**62.5**
Commercial	40.5	7.7	135.6	7.7
Industrial	42.0	8.0	133.2	7.5
Transportation and Utilities[b]	19.1	16.7	41.6	2.4
Governmental and Institutional	87.1	16.7	98.8	5.6
Recreational[c]	71.9	13.7	254.0	14.3
Urban Subtotal	**523.1**	**100.0**	**1,768.4**	**100.0**
Nonurban				
Environmental Corridor	748.0	34.9	848.3	94.3
Agricultural and Other Open Lands	1,397.1	65.1	51.5	5.7
Nonurban Subtotal	**2,145.1**	**100.0**	**899.8**	**100.0**
Total	**2,668.2**	**100.0**	**2,668.2**	**100.0**

Source: SEWRPC.

[a] Street rights-of-way and off-street parking areas are included in the associated land use category.
[b] Includes only the railroad right-of-way and utility properties.
[c] Includes associated surface water areas.

FURTHER READING

Bartholomew, Harland. *Land Use in American Cities.* Harvard University Press, 1955.
Chapin, F. Stuart Jr., et al. *Urban Land Use Planning.* University of Illinois Press, 1995.

13 Neighborhood Unit Concept

... everything assumes a different shape as we pass from the abstract world to that of reality.

Karl von Clausewitz
Prussian General
"Principles of War" 1833

INTRODUCTION

The neighborhood unit concept is one of the most important concepts in modern city planning. This concept was originally developed in 1929 by Clarence Arthur Perry, a colleague of Sir Thomas Adams, as a part of the preparation of the regional plan for New York. The essence of the concept may be summarized as follows: for stability, for preservation of amenities, and for the creation of socially desirable sense of community, a city should develop its residential areas in a number of individual cellular units. These units should provide all of the facilities required by the family within the immediate vicinity of its dwelling. Each unit should have a distinctive, unifying character of its own and be well insulated from surrounding units. Figure 13.1 graphically illustrates the concept as originally envisioned by Perry.

ESSENTIAL FEATURES

The concept envisions that a neighborhood should incorporate five design features, or characteristics:

1. Size

   A neighborhood unit should provide housing for that population for which one elementary school is required. The actual area of the neighborhood will, therefore, vary with the ratio of school age population to total population; the density of development; the size of the elementary school; and the desirable walking distance to elementary school. A relationship thus exists between density—as determined by lot size—and the cost of elementary schooling including the cost of school bus transportation.
2. Isolating or Insulating Boundaries

   A neighborhood unit should be clearly delineated, the boundaries being formed by such man-made features as major streets, major parks and parkways, major institutional lands, or natural features such as streams or lake

DOI: 10.1201/9781003500384-13

Neighborhood Unit Plan

FIGURE 13.1 This figure is adapted from the original figure used by Clarence Arthur Perry to illustrate the concept of the neighborhood unit. The neighborhood unit still represents one of the most important concepts underlying the practice of city planning.

shore lines. These boundaries should serve to give a cellular structure to the unit, insulating it from other similar units and other types of land use areas.

3. Internal Street Pattern

A neighborhood unit should have an internal street pattern designed to facilitate vehicular and pedestrian circulation within the unit, but to discourage penetration of the unit by through traffic. The streets should be functionally designed and developed as local access and collector streets, with clear hierarchy and safe intersections for pedestrians and cyclists.

4. Self-Contained

   A neighborhood unit should be self-contained with respect to the day-to-day needs of the family. The unit should thus contain:

   a. A centrally located elementary school that may also serve as a community hub or shared public facility. The school site should have an area of 5–10 acres, consistent with modern educational facility standards.
   b. A centrally located neighborhood park and playground sized at approximately three acres per 1,000 resident population, or about 10–15 acres. The park and playground should be located adjacent to the school, share a common service area, and provide a proper setting for the school. The school building should be designed to serve as a shelter house and locker-room building for the park and playground. Play equipment should not have to be duplicated.
   c. A neighborhood should have a convenience shopping area. The site area should approximate one acre per 1,000 resident population, or approximately 5–10 acres. The desirability and practicality of a central location as opposed to a peripheral location is problematic. Convenience favors the former, modern marketing practices and traffic concentration the latter.
   d. Provision should be made for civic or institutional uses, such as places of worship, libraries, or community centers, with flexibility to reflect cultural diversity and changing community needs.

5. Achievement of Unity in the Design

   Unit of design is a difficult quality to define and achieve. Techniques used include providing one central feature or focal point, such as the school or park around which the design is organized so that a person entering the unit will realize at once that he or she is entering an integrated environment; use of common street "furniture": street lighting, street signs, street cross sections, hydrants; and use of common architectural design for buildings. Such unity of design is not only aesthetically pleasing, but helps ensure a stable community, for each such integrated area tends to be self-protecting against adverse outside influences. This quality may be impossible to achieve in areas where much of the land subdivision and development is done on a piecemeal basis by many different developers of small tracts, unless the community has prepared precise neighborhood unit development plans.

SIZE AND DENSITY CONSIDERATIONS

In recent years, the concept of the neighborhood unit has been taken up and, in a sense, revitalized by neo-traditionalist planners such as Andre Duany and Thomas Calthorpe. The revitalized concept, illustrated in Figure 13.2, essentially reaffirms Perry's original concept. A neighborhood unit, as envisioned both by Perry and Duany, would have an area of about 160–200 acres. At a population density of about

FIGURE 13.2 This figure reproduces figures prepared by Andres Duany, a leading proponent of the "new-urbanism." The figures illustrate the continued relevance of Clarence Arthur Perry's neighborhood unit concept to good city planning. The resemblance of the figure for an urban neighborhood to Perry's original figure reproduced in Figure 13.1 should be noted.

Source: Andres Duany et al.

30 persons per gross acre, the neighborhood unit would accommodate a resident population of about 4,500–6,000. The attendant net density might approximate 45 persons per acre, depending upon the average household size, or about 15 dwelling units per acre. This would provide an average site area of about 2,900–3,000 square feet per dwelling unit. Such densities would typically involve a mix of multifamily and compact single-family housing. Assuming that about 15 percent of the population would be of elementary school age (K-8), the potential total school loading would approximate 675–720 pupils. Further assuming that about 80–90 percent of the pupils would attend public as opposed to parochial school or be home schooled, a neighborhood school could expect to have a peak enrollment of about 550–650 pupils. It should be noted that the percentage of pupils attending parochial school tends to vary widely within the United States, from, for example, an average of about 10 percent nationally in 2000, to about 15 percent in Wisconsin, to about 20 percent in southeastern Wisconsin. A three-section elementary school could accommodate this enrollment with an average class size of 22–24 pupils. If the school were centrally located in the neighborhood unit, the maximum walking distance from home to school would approximate 0.25 mile, an acceptable distance. The extent to which the relatively high population density of such a neighborhood would be accepted by the urban land market is subject to market variation, with greater acceptance in urbanized or transit-oriented regions.

At a possibly more widely acceptable lower population density of about 8 persons per gross acre, the size of the neighborhood unit would have to be increased to about 600–700 acres to provide a resident population of about 5,000–5,500 persons. The attendant net density might approximate 10–12 persons per acre, or, depending upon the average household size, about 3–5 dwelling units per acre. This would permit development of the unit with primarily single-family structures. A three-section elementary school could accommodate the potential maximum enrollment with an average class size of 22–24 pupils. If the school were centrally located in the neighborhood unit, the maximum walking distance from home to school would approximate 0.5 mile, a not unreasonable distance.

The foregoing analyses represent crude approximations intended to illustrate the complex relationships existing between variables such as the areal size of a neighborhood unit, the density of development, the size of the desired neighborhood elementary school, and acceptable walking distances from home to school. The analyses also indicate that the concept of the neighborhood unit is applicable over a wide range of population densities and housing structure types.

In considering this concept, it should also be recognized that the total resident population and elementary school age population will vary over time. The unit may first be occupied by young families with children, and this may be expected to maximize the total resident population and elementary school age population of the unit. As these original occupant families age, the elementary school age population may decline significantly. Eventually the original occupant families will be replaced and the cycle will be repeated. This dynamic underscores the importance of flexible planning for public facilities and land use mixes over a neighborhood's lifecycle.

COMMENTS ON THE CONCEPT

A point on which most city planners and engineers will agree is that an urban area should be developed in recognizable cellular subareas instead of as a single formless mass. This is partly an aesthetic principle, partly a matter of efficiency in organizing and supplying public services, and partly a matter of convenience in living within the city. And partly it is also a matter of "humanizing" the city—of bringing the size of an area where a person lives into scale with the human individual who may feel at home in, and be a part of, a community of 5,000 or so persons, but may feel lost in a metropolis of a million.

The major advantage that a municipal engineer might advance for the neighborhood unit concept is that of a desirably defined municipal service area. Neighborhood units provide a convenient and sound basis for the design of street, utility, and stormwater drainage systems and for the effective and efficient provision of public works services such as solid waste collection, street sweeping, and snow plowing. They also align well with contemporary infrastructure planning tools such as geographic information system (GIS)-based service modeling and district-level sustainability systems.

The principal problem with the concept as originally proposed relates to the provision of centrally located neighborhood shopping facilities. Changes in retail patterns—especially the rise of e-commerce and destination-oriented retail centers—have reduced the feasibility of traditional central neighborhood shopping centers. Such a center should contain, at a minimum, essential goods and services (e.g., grocery, pharmacy, cafés, and personal care). A modern supermarket of a 25,000–60,000-square-foot floor area needs to draw patrons from a community of about 30,000–50,000 people—known as a "trade population." Large supermarkets today may stock over 30,000 products, while convenience stores providing about one-quarter of the number of products as a conventional supermarket may be viable to service a trade population of about 5,000–8,000 people. The population density required to place a large supermarket within a quarter-mile walking radius would be unrealistically high in most U.S. contexts (over 150,000 per square mile). By contrast, providing for smaller-scale convenience retail is more achievable at densities of roughly 20,000–25,000 persons per square mile. Furthermore, shopping behavior is diversified across multiple physical and digital venues, further diminishing reliance on a single local store. Table 13.1 provides selected typical characteristics of modern convenience shopping stores. As Table 13.1 suggests, even modest neighborhood shopping centers are most viable at arterial intersections, serving several neighborhoods rather than one, consistent with modern mixed-use and corridor-based planning approaches.

The concept as originally developed by Perry was, for good reasons, elementary school centered. In the United States school districts are often independent special-purpose units of government that engage in their own planning efforts. Where local education agencies do not adopt neighborhood-centered school planning, this organizing principle may be substituted by other civic or social anchors—such as community centers, shared parks, or mixed-use main streets—to preserve the neighborhood's functional and social cohesion.

TABLE 13.1
An Example for Selected Typical Characteristics of Convenience Shopping Retail Stores

Type of Store	Number of Products	Commercial Floor Space in Square Feet	Trade Area Population	Trade Area Radius
Limited Assortment Store	900	10,000	65,000	4.0 miles
Convenience Store	3,500	2,600	6,000	0.5 mile
Conventional Supermarket	12,000	25,000	40,000	2.0 miles
Superstore	17,500	40,000	55,000	2.5 miles
Food–Drug Store Combination	25,000	50,000	65,000	2.5 miles
Warehouse Store	10,500	30,000	75,000	5–10 miles
Super Warehouse Store	16,000	60,000	125,000	5–10 miles

CONNECTED COMMUNITY

A vital evolution of the neighborhood unit concept is the principle of the connected community, which emphasizes the physical, digital, and social interconnectivity among residents, places, and services. In addition to internal cohesion, modern neighborhood planning now stresses external integration with surrounding units and broader urban systems. This includes walkable and bikeable linkages to adjacent neighborhoods, access to regional transit networks, and the provision of digital infrastructure that supports civic engagement, teleworking, and remote access to education and healthcare. A connected community encourages active lifestyles, reduces automobile dependence, and fosters inclusiveness by ensuring that residents of all ages and abilities can participate in daily life. Socially, it nurtures informal interaction through shared public spaces, neighborhood-scale events, and community-driven initiatives that strengthen the local identity. The connected community expands upon Perry's original vision by adapting it to twenty-first-century needs for accessibility, sustainability, and resilience.

EXAMPLE OF AN ACTUAL NEIGHBORHOOD UNIT DEVELOPMENT PLAN

The design of neighborhood unit plans for use in the development of new towns, or of large residential or mixed-use developments on sites controlled by a single developer, can be accomplished in a manner providing all, or almost all, of the essential features of such a unit as envisioned in the original or revitalized concepts. Such contexts allow for master planning and coordinated implementation, which enable the incorporation of cohesive urban design, infrastructure efficiency, and environmental sustainability. Numerous case studies now document neo-traditional neighborhood developments across North America and globally, characterized by walkable layouts, mixed-use cores, and diverse housing options. Such design for use in the development of peripheral areas of existing cities is, however, far more difficult and almost inevitably involves compromising one or more of the desirable features of a well-planned neighborhood unit. Planners must often accommodate existing infrastructure, fragmented property ownership, and legacy zoning constraints.

Figure 13.3 illustrates an actual neighborhood unit development plan for a peripheral area of a medium-sized, developing city. The unit encompasses an area of 632 acres. As planned, the unit would have a resident population of about 4,200 persons

Example Actual Neighborhood Unit Development Plan

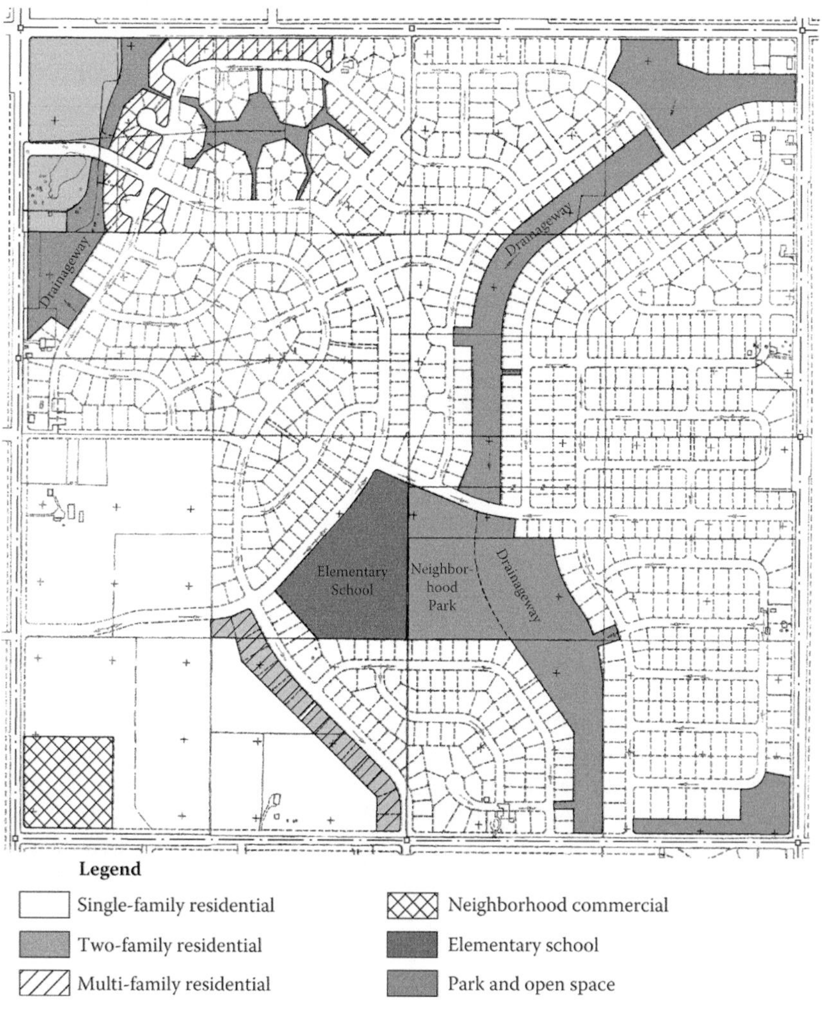

FIGURE 13.3 Illustration of an actual neighborhood unit development plan prepared for a medium-sized development city located within a large urbanized area. The example is intended to illustrate how the concept of the neighborhood unit must be adapted to the reality of each particular situation.

occupying about 1,540 dwelling units at an average household size of 2.70 persons. The land uses would be distributed as follows:

Land Use	Net Acreage	Percent of Total Developed Area
Single-Family Residential	331	53
Two-Family Residential	15	2
Multi-Family Residential	59	9
Subtotal	**405**	**64**
Commercial	8	1
Government and Institutional	15	2
Park and Open Space	70	11
Streets	134	22
Total	**632**	**100**

As planned, the unit would have a net residential density of about 10.4 persons per acre and a gross residential density of about 6.7 persons per acre—relatively low. Single-family lot sizes would approximate 10,000–12,000 square feet. The elementary school age enrollment would approximate 600 pupils, of which approximately 500 would be expected to attend public school. This level of enrollment could be accommodated in a three-section kindergarten through eighth-grade elementary school with an average class size of 18 pupils. The school is centrally located to serve the majority of the neighborhood, though the maximum walking distance slightly exceeds contemporary best practices of a quarter- to half-mile radius.

The unit would have isolating boundaries consisting of major arterial streets; the internal street pattern would facilitate access to building sites while discouraging the penetration of the unit by through traffic; and the unit would be relatively self-contained with its own park and school sites. A drainageway traverses the neighborhood unit and is to be accommodated in a greenway connected to the centrally located school and neighborhood park sites. The proposed convenience shopping area would be located on the periphery of the unit, probably too small to be successful, and in any case, automobile dependent for access. The configuration of the existing real property boundary lines may be expected to result in development over time by a number of individual developers. This may make the achievement of unit of design unlikely. To address this, municipalities increasingly rely on form-based codes or neighborhood-specific design guidelines to promote visual and functional coherence. Implementation of the plan should nevertheless achieve at least an integrated street and land use pattern and provide a sound basis for the review of individual land subdivision proposals as such proposals are submitted for approval to the municipality, ensuring that street locations and grades will facilitate the efficient provision of sanitary sewerage, water supply, and stormwater drainage facilities.

14 Principles of Good Land Subdivision Design

Land subdivision is truly the first step in the process of community building. Once land has been cut up into streets, blocks and lots, and publicly recorded, the die is cast and the pattern is difficult to change. For generations the people who occupy such land will be influenced by the character of its design.

American Society of Civil Engineers
Manual of Engineering Practice No. 16
"Land Subdivision," 1939

INTRODUCTION

Like urban planning, good land division design is both an art and a science requiring a high degree of technical skill and full realization of the importance of the design to the various interests involved and affected. Good land division design requires imagination and creativity, as well as adherence to sound principles of land planning and engineering practice and to sound development standards. For these reasons, public regulation alone is no guarantee of good land subdivision design. Collaboration between planners, engineers, developers, and the community is key.

Subdivisions should create resilient, flexible, and future-ready building sites that support evolving household needs, promote sustainable development, and enhance long-term community value. Building sites should be so arranged in relation to the rest of the community and to the natural resource base in order to provide a good environment for living, working, and recreating. The role of government should be to create conditions amenable to good land subdivision design, provide a framework for such design, and encourage and support endeavors to achieve such design. County and local governments have not only the right to regulate land division, but also an obligation to ensure that the public interest is served and that the greatest possible good is to result from a proposed land subdivision. The land subdivision design and development process requires that local governments and developers work together to bring about development that meets both the objectives of the community master, or comprehensive, plan and of its component parts, and the objectives of the developer.

THE CONTEXT OF LAND SUBDIVISION DESIGN

Land subdivision design is a process requiring professional knowledge and experience. In undertaking a subdivision design, the designer may face one of

DOI: 10.1201/9781003500384-14

three situations that will determine the manner in which the design must be approached:

1. The land proposed to be subdivided is located within a community that has not adopted a comprehensive plan.
2. The land proposed to be subdivided is located within a community that has adopted a comprehensive plan, but that plan does not include as a component detailed neighborhood unit development plans or platting layouts.
3. The land proposed to be subdivided is located within a community that has adopted a comprehensive plan, and that plan includes detailed neighborhood development plans or platting layouts.

In the case where a community has not adopted a comprehensive plan, the subdivision designer has no planning framework within which to work. Accordingly, adherence to all five principles of good design enunciated in the following section of this chapter is particularly important. Such adherence, moreover, will require substantially greater effort than would be the case if sound local planning has provided a framework for the design process. The designer will have to consider the potential need for adaptation of the design to potential external features of community-wide concern, may have to research framework plans prepared by county, regional, and state agencies and by private utilities, and may have to conduct inventories and analyses relating to onsite soil and bedrock conditions, to potential flood hazards along streams and watercourses, and to the presence of wetlands, woodlands, and wildlife habitat areas. Large-scale topographic maps meeting National Map Accuracy Standards—always essential to good design—become particularly important in this case. Although the lack of a planning context may be regarded by some developers and designers as fortuitous, such a lack of framework plans will make good design and the attendant creation of long-term stability in the development difficult to achieve. A good local comprehensive plan should recognize and incorporate market research concerning the demand for building sites to accommodate various types of housing styles and costs. However, market research by itself cannot substitute for the framework provided by a good local comprehensive plan.

In the case where the land proposed to be subdivided is located within a community that has adopted a comprehensive plan but has not adopted detailed neighborhood development plans or platting layouts as an element of that plan, the adopted community plan will provide broad but important guidance to the designer. This is particularly true with respect to the first three of the five design principles herein enunciated, calling for adapting the design to external features of community-wide concern, including the need for external and internal public infrastructure, for effecting a proper relationship between the proposed subdivision and existing and proposed surrounding land uses, and for effecting a proper relationship between the proposed subdivision and the natural resource base. The comprehensive plan should contain specific recommendations relative to such matters important to good subdivision design as the location and width of arterial streets, the generalized location of needed school and park sites, and the generalized location and configuration of needed drainageways and stormwater retention areas. The plan should also provide much of the information essential to good design, including aerial photographs, large-scale topographic and cadastral maps, detailed soil maps with attendant use in interpretations, and definitive information on the location and configuration of flood hazard areas, and of such elements of the natural resource base as wetlands, woodlands, and wildlife habitat areas. Importantly, the comprehensive plan should specify the type and intensity of land uses to be accommodated within specified subareas of the community—the intensity for residential uses usually being expressed as the number of dwelling units per gross acre. In this case, the designer can focus on the creation of a sound and attractive street, block, and lot layout and on the creation of a unified design for the proposed land subdivision.

In the case where the land proposed to be subdivided lies within a community that has adopted a comprehensive plan that includes detailed neighborhood unit development plans or platting layouts, the local government concerned will have provided not only data concerning the types and intensities of land uses to be accommodated on the land concerned, but also proposed specific locations for schools, parks, and drainageways and a specific layout for streets, blocks, and lots within the proposed subdivision. In this case, the designer will be responsible for reviewing and refining the design provided in the comprehensive plan and for ensuring a unity of design within the subdivision itself.

The adopted neighborhood unit development plan or platting layout should be viewed as a point of departure for the design of subdivisions of individual tracts of land included within the neighborhood unit. The adopted neighborhood unit plan or platting layout should not unduly constrain a designer who may find a more creative or cost-effective approach to the design of a particular subdivision within the neighborhood unit. A proposed alternative design, however, should not be permitted by the local planning agency unless it conforms to the recommendations of the adopted community comprehensive plan with respect to the types and intensities of land uses concerned, to the provision for arterial street and highway system development, and to the protection of environmentally sensitive areas. In addition, recommendations contained in the adopted neighborhood unit development plan relative to the connectivity of collector and land access streets, the provision of school and park sites, and the location of drainageways should be reflected in any alternative land subdivision design considered.

It is important to note that an adopted comprehensive plan, and particularly the detailed neighborhood unit design or platting layout element of such a plan, may not always adequately reflect current land market conditions. It should be expected that proposed land subdivision designs will propose minor changes in adopted detailed neighborhood unit design plans relating particularly to the precise details of the block and lot layouts, and such minor changes should be accommodated without resort to a plan amendment. However, when a land developer or subdivision designer proposes major departures from the adopted plan—such as changes in the type and intensity of land use, the location and configuration of proposed school sites, park sites, and drainageways, or the locations of streets providing connections into adjacent undeveloped parcels—the proposed departures should not be approved until the comprehensive plan or plan component concerned has been formally amended to reflect the proposed departures. The amendment process in such cases should include necessary public hearings and action by the local plan commission and governing body.

Recent best practices encourage the use of "living" online comprehensive plans supported by geographic information system (GIS)-based suitability analysis, LiDAR-derived elevation data, and statewide parcel and environmental databases. These interactive plans can be visualized and updated through map dashboards and cloud-based review platforms, enabling more adaptive, data-rich, and transparent decision-making while maintaining consistency with long-term planning goals.

PRINCIPLES OF GOOD DESIGN

Good land division design can be achieved through the effective application of five basic design principles. These principles are easy to enumerate but quite difficult to apply:

1. Proper provision for external features of community-wide concern
2. Proper relationship to existing and proposed surrounding land uses
3. Proper relationship to the natural resource base
4. Proper design of internal features and details
5. Creation of an integrated design

Proper Provision for External Features of Community-Wide Concern

The first principle of good land division design is that the design must properly provide for certain external features of community and areawide concern that affect the proposed land division. Land division design must provide for the proper extension of arterial streets; for the location of sanitary trunk sewers, water transmission mains, and modern utilities such as broadband internet and renewable energy systems; for the preservation of major drainageways; for adequate management of the quantity and quality of stormwater runoff; for needed school and park sites; and for convenient access to public transit facilities. Consideration should also be given to the land division's connectivity with regional commercial centers, employment hubs, higher education institutions, and multimodal transportation networks.

The proper incorporation of these external public infrastructure needs and services in the design is greatly enhanced if the community has adopted a comprehensive or master plan, and lack of such a plan can be a severe handicap to good land division design. The practice of arbitrarily requiring the dedication of a set percentage of land for school or park purposes in the absence of a comprehensive plan is particularly poor planning practice, may adversely influence land division design, and may be illegal.

Proper Relationship to the Existing and Proposed Surrounding Land Uses

The second principle of good land division design is that the design must be properly related to existing and proposed surrounding land uses. Moreover, adjacent land uses must be carefully considered in the design. Parks, parkways, and certain types of institutional uses are a definite asset and can increase the value of a land division. Others, such as railways, freeways, electric power transmission lines and other utility easements or rights-of-way, poorly subdivided and developed areas, and unsightly strip commercial and open industrial development, may be detriments, particularly to residential land divisions, and require special consideration in the design to minimize adverse impacts. Modern planning tools such as GIS-based compatibility analysis, context-sensitive design, and form-based codes can support more responsive layouts.

Proper Relationship to the Natural Resource Base

The third principle of good land division design is that the design must be properly related to the natural resource base. Woodlands, wetlands, wildlife habitat areas, areas covered by unstable soils, areas of shallow bedrock or bedrock outcrop, areas of steep slopes, and areas subject to special hazards, such as flooding and stream bank, bluff, ravine, and lakeshore erosion, must receive careful consideration in the design. Good subdivision design increasingly presents opportunities to restore and connect ecological corridors and enhance environmental quality. Such areas should be preserved for open space and nature-based solutions, such as green infrastructure, which contribute to stormwater management, habitat preservation, and climate resilience.

Proper Design of Internal Features and Details

The fourth principle of good land division design is proper attention to internal detailing. This includes careful attention to the proper layout of streets, blocks, lots, and open space areas, the organization of larger subdivisions into smaller sections, and careful adjustment of the design to the topography and the natural and cultural resource characteristics of the site.

Street System

The street system is one of the most important elements of land division design, as it is indeed one of the most important elements of the community comprehensive plan. The street pattern forms the framework for community development and to a considerable extent determines the efficiency of the other functional parts of the community. In the individual land division, the street pattern determines the shape, size, and orientation of each building site and, to a considerable extent, influences the character and attractiveness of the land division.

The street system must serve a number of sometimes conflicting purposes. The street system must provide for the efficient movement of vehicular and pedestrian traffic throughout the community while providing for convenient access to individual building sites, including access by emergency vehicles such as police, fire fighting equipment, and ambulances, and by service vehicles such as snow plows and solid waste and recycling collection trucks. Designs should also support the safe use of bicycles, electric scooters, and emerging micromobility options. The street system must also serve as an integral part of the community drainage system and must provide the location for such utilities as public sanitary and storm sewers, water distribution mains, gas mains, and electric power, telephone, broadband fiber optics, and smart city sensors. The efficiency and cost of these facilities and services will be determined to a considerable extent by the layout of the street system and, where appropriate, alley system. While historically providing light, air, and firebreaks, modern urban street systems also increasingly support sustainable infrastructure, including bioswales, permeable pavements, and green buffer strips. The importance of the street system is accentuated by the permanence of street dedications and the difficulties inherent in changing the street pattern once established by land division.

The land division layout should carefully adjust the collector and land access streets to the topography in order to minimize drainage problems and rough grading and to permit the most economical provision of sanitary sewer, water supply, and stormwater drainage facilities. Generally, collector streets should follow valley lines, while land access streets should cross contour lines at right angles. Side hill street locations should generally be avoided. Skillful adjustment of the street pattern to the topography can do much to preserve the natural beauty of the site and to lend charm and beauty to a residential area. Rigid rectangular street patterns, commonly found in older subdivisions, are often poorly suited to topographic variation and contribute to stormwater runoff issues and monotony. Context-sensitive, curvilinear, or modified grid designs are more adaptable and desirable.

The street pattern for any land division should be functional, providing as may be necessary for at least three principal types of streets: arterial streets, collector streets, and land access streets, as illustrated by Figure 14.1.

Arterial Streets

Arterial streets are intended to interconnect the various areas of the community and form its major circulation system. Arterial streets should be designed to move traffic efficiently and safely, should be of generous width as specified in the community comprehensive plan, and should be of proper grade, alignment, and continuity. Direct access to these streets from adjacent property should not be permitted. Such access should be prohibited in residential areas through good design. Early approaches to such access control involved the provision of frontage roads paralleling the arterial facility, as shown in Figure 14.2. The frontage roads of necessity were one-way travel facilities, were costly since they served only one tier of abutting lots, and presented traffic engineering problems at intersecting streets. Although no longer favored for use in residential areas, frontage roads still have application in commercial areas and along freeway facilities. The approaches illustrated in Figures 14.1 and 14.2 that utilize double frontage lots as cul-de-sacs or loop streets to eliminate direct access are preferable. In some cases an alley may be used, as also illustrated in Figure 14.2.

Intersections of other streets with arterials should be held to a minimum. Right-of-way and pavement widths for arterial streets should be determined on the basis of existing and probable future traffic demands, as determined through the preparation of the street and highway system element of the city plan. A desirable right-of-way width for urban surface arterial streets is 120 feet, which can accommodate divided pavements. Planning should also account for potential future uses such as Bus Rapid Transit (BRT), autonomous vehicle lanes, or protected cycling infrastructure.

The highest level of arterial service is provided by freeways. The function of freeways is to accommodate high volumes of intra- and inter-regional travel at high speeds. There is no access to adjacent parcels, and all intersections are grade-separated with access limited to periodic, properly spaced interchanges with surface

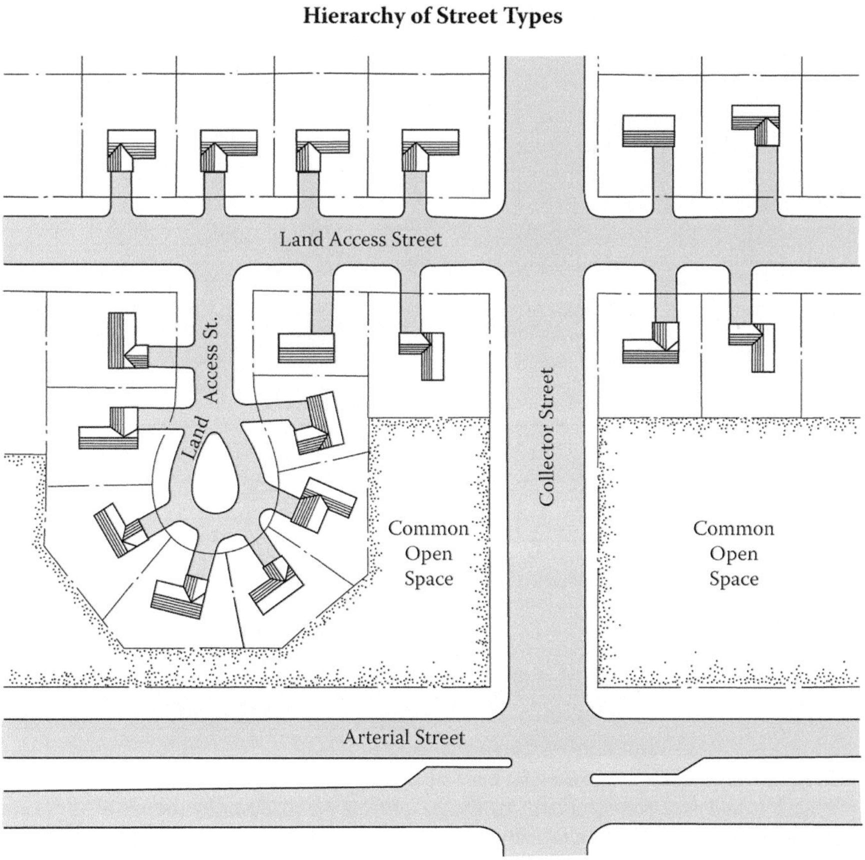

FIGURE 14.1 Streets may be classified on the basis of function into three principal types: arterial streets, collector streets, and land access streets. These three types are illustrated in this figure. Arterial streets are intended to interconnect the various areas of the community and form its major circulation system. Collector streets are intended to collect traffic from, and distribute traffic to, the land access streets, and to provide the necessary connections to the arterial street system. Collector streets are often the principal entrance streets to residential land divisions. Land access streets are intended to serve primarily as a means of access to, and egress from, abutting property.

arterials. Careful attention should be given in land subdivision design to the location of street intersections along arterials connecting to freeway on- and off-ramps. Proper spacing relative to such ramps is important in achieving both traffic safety and good block and lot arrangements.

Collector Streets

Collector streets are intended to collect traffic from, and distribute traffic to, the land access streets and to provide the necessary connections to the arterial street system. Collector streets often function as the principal entrance streets to residential areas.

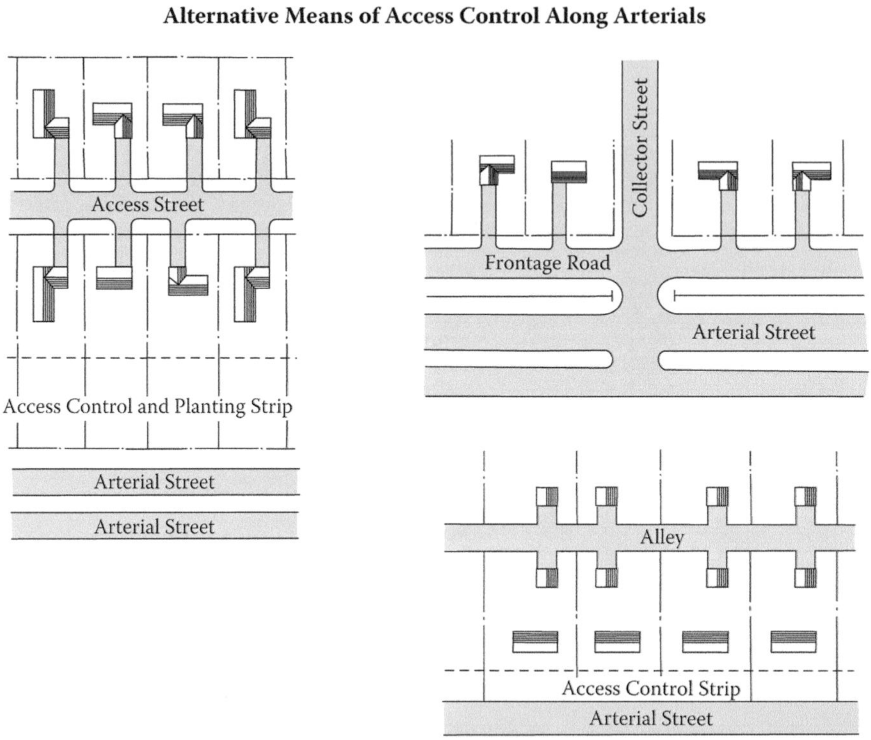

FIGURE 14.2 Access control along arterial streets can be effected by the use of double frontage lots and access control strip, by the use of frontage roads, or by the use of alleys and access control strip. The first alternative is much to be preferred. In any case, access control serves to protect the capacity and enhance the safety of arterial streets and highways.

Collector streets should be designed to carry moderate traffic volumes at moderate speeds, and may in some instances carry bus routes. A desirable right-of-way width for a collector street is 80 feet. Incorporating traffic calming measures such as bulb-outs, raised crosswalks, or landscaped medians can improve safety and aesthetics on collector streets.

Land Access Streets

Land access streets are intended to serve only as a means of access to, and egress from, abutting property. Land access streets should be designed to discourage use by through traffic and, therefore, should be discontinuous with relatively narrow pavement widths. In residential areas, the land access streets should be designed to promote quiet and attractive neighborhoods. Land access street rights-of-way may range in width from 50 feet to 66 feet and are often designed as loop or cul-de-sac streets. Contemporary designs favor short, connected loops or courts that retain local character while improving emergency access and walkability. Figure 14.3 illustrates three special types of minor land access street arrangements.

Special Types of Minor Land Access Streets

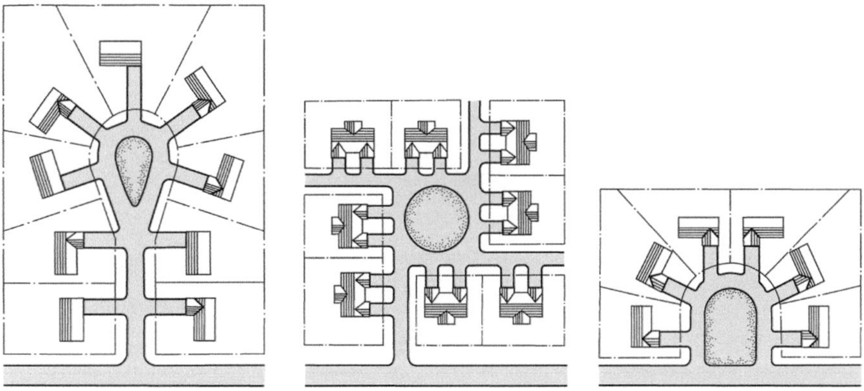

FIGURE 14.3 In residential areas, the land access streets should be designed to promote quiet and attractive neighborhoods. Land access streets should be designed to discourage use by through traffic and, therefore, should be discontinuous, with relatively narrow pavement widths. Land access streets are often designed as loop or cul-de-sac streets. Three different types of minor land access streets, each designed to eliminate or limit through traffic, are illustrated in this figure. Loop and cul-de-sac streets can be helpful tools in developing otherwise awkward remnant areas in a subdivision layout.

Stormwater Management

Particularly careful attention must be given in the land subdivision design to stormwater management. The location of each stream or watercourse channel to be preserved should be carefully determined, as should the boundaries of the attendant 100-year recurrence interval flood hazard area and the boundaries of any proposed stormwater detention or retention basins. Each open channel right-of-way to be preserved in the design should be of adequate width to hydraulically accommodate peak runoff storage and conveyance requirements, to provide space along the channel for the access and proper operation of construction and maintenance equipment, and to adequately protect any related elements of the natural resource base, such as wetlands. Designs should incorporate green infrastructure practices such as vegetated buffers, constructed wetlands, and bioswales where feasible. Proper land division design may require some relocation or reconfiguration of natural watercourses and the integration of the smaller uppermost reaches of some channels and swales into a planned system of storm sewers or open drainage channels. Particular care must be taken that the proposed development will not obstruct drainage or cause flooding. Stream or watercourse channels and detention and retention basins should generally be included within dedicated parks, parkways, or other public or common open space lands. Figure 14.4. illustrates two ways in which small streams and watercourses may be incorporated into a subdivision design.

An urban stormwater management system should comprise a major and a minor component. The minor component should be designed to avoid nuisance-level

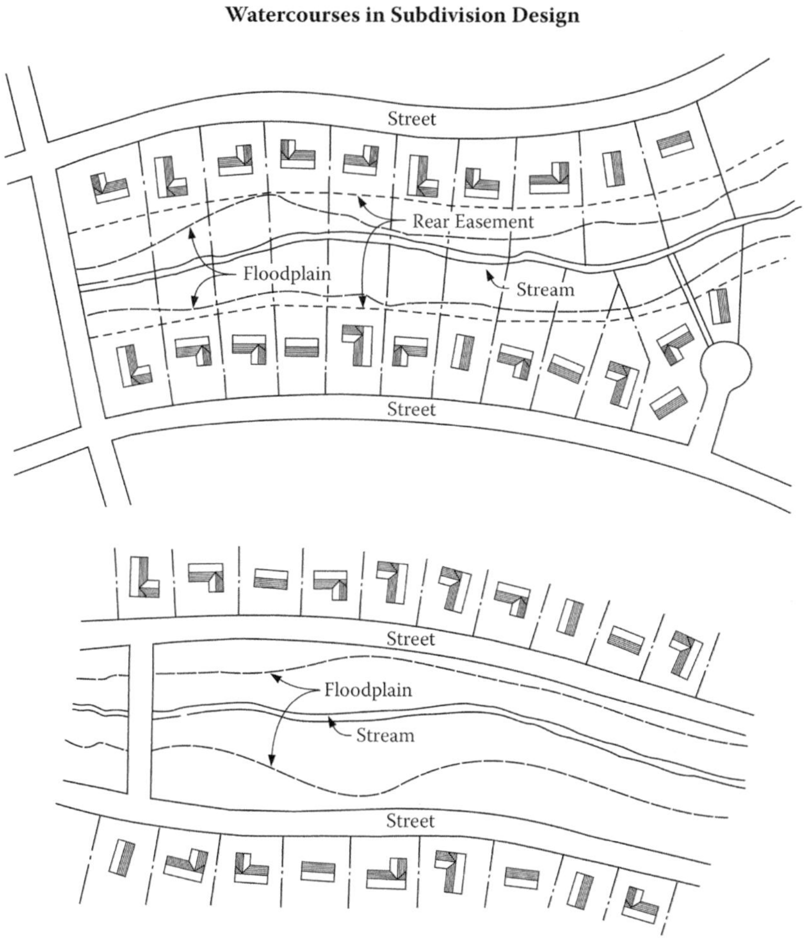

FIGURE 14.4 Illustration of alternative ways of treating small streams and watercourses in subdivision design. The intent should be to maintain the stream or watercourse as an attractive community asset, and to avoid placing the stream or watercourse in a large and costly storm sewer.

flooding of building sites, streets, and street intersections, avoid undue disruption of traffic, and facilitate access to various land uses by emergency vehicles. The minor component should also be designed to abate the surcharging of sanitary sewers due to excessive clearwater infiltration and inflow. The minor component, although functioning during all rainfall and snowmelt events, should be designed to accommodate runoff from rainfall events up to and including the 10-year recurrence interval event. The minor component will generally consist of piped storm sewers and appurtenant inlets and catch basins. Smart stormwater sensors and real-time monitoring systems may also be integrated to track flows and predict maintenance needs.

The major component of the stormwater management system consists of the minor component plus street cross-sections and associated surface swales and waterflow paths to receiving streams and watercourses. Portions of the street system, therefore, serve as parts of both the minor and major components of the urban stormwater management system. When providing conveyance of overland runoff to piped storm sewers, the streets function as part of the minor component. When utilized to convey excess flow from surcharged piped storm sewers and culverts and overflowing roadside swales and ditches, the street system functions as part of the major component. This major component should be designed to avoid the flooding of basements and first floors of buildings, with the attendant monetary damages and hazards to public health and safety. The major component is intended to function only when the capacity of the minor component is exceeded, and should be designed to accommodate runoff from rainfall events greater than the 10-year and up to and including the 100-year recurrence interval rainfall event. The proper design of the major component requires careful attention to the establishment of street grades, the delineation of attendant surface water flow paths, and, particularly, the avoidance of midblock sags in street profiles. If such midblock sags are to be permitted, then adequate overland drainage from the sag should be provided to accommodate the runoff attendant to a 100-year recurrence interval rainfall event without flooding the basements or first floors of buildings. Use of 2D hydrologic modeling software and LiDAR-based elevation data can enhance the precision of these designs.

Both the minor and major components of a stormwater management system may, under certain conditions, utilize both constructed and natural stormwater retention or detention basins as well as constructed conveyance facilities for temporary storage of stormwater runoff. When such constructed facilities that include storm sewer and dry detention basins are intended solely to control the rates of stormwater flow, the facilities are generally designed to drain completely between rainfall events, providing little or no reduction in nonpoint source pollution loadings to receiving watercourses. When such facilities are integrated with nonpoint source pollution abatement facilities, such as wet detention basins, constructed wetlands, rain gardens, and bioretention cells, a significant reduction in nonpoint source water pollution loadings may be achieved. In any case, such facilities must be designed to function as integral parts of both the minor and major components of the stormwater management system. Practices that promote infiltration and pollutant removal in situ are especially encouraged under modern municipal stormwater permits (e.g., MS4 regulations).

Block and lot grading constitute important elements of stormwater management. The simplest grading layout provides a ridge along the rear lot lines with each lot graded to drain directly to the street, as shown in Figure 14.5. An alternative grading layout provides a cross slope across part of the block width, and yet another alternative provides a cross slope across all of the block width, as shown in Figure 14.5. A fourth alternative grading layout provides for a drainageway along the rear lot lines. This is probably the least desirable of the four alternatives in that it requires that the drainageway be maintained over time in an unobstructed manner by all of the lot owners concerned, and it requires special outlet provision at the block end, such as an inlet placed behind the public walk with a connection to a storm sewer. Proper grading is

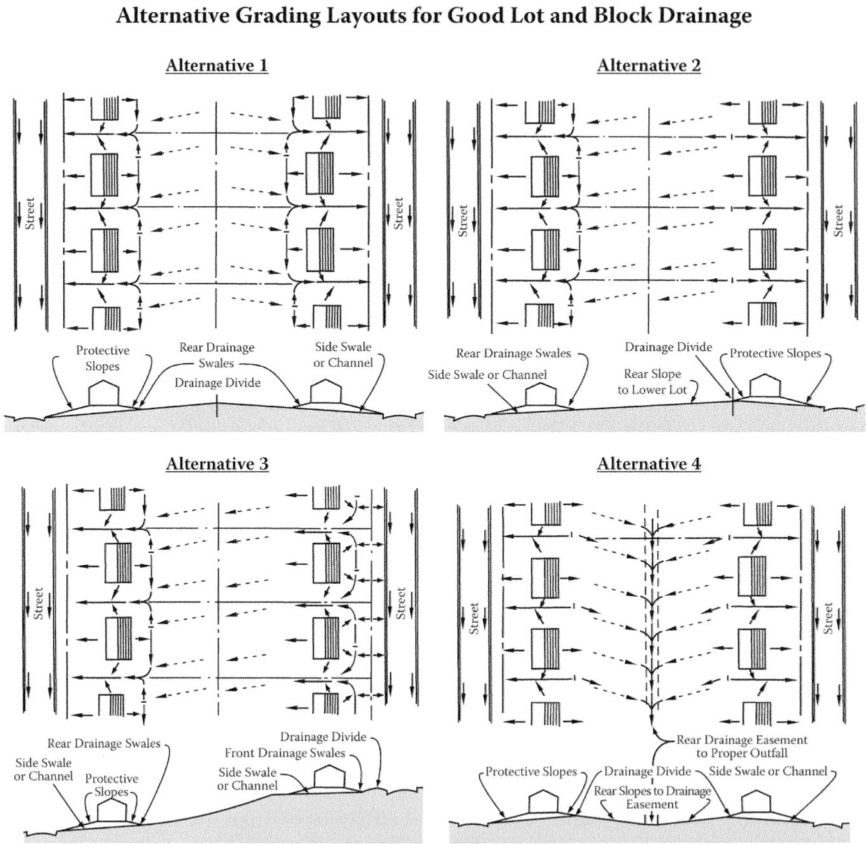

FIGURE 14.5 Illustration of four alternative grading layouts that facilitate good drainage and avoid basement flooding. The fourth alternative requires provision of an adequate outlet at the low end of the mid-block drainage swale.

essential to preventing erosion and basement flooding. It is also essential for ensuring ADA-compliant pathways and avoiding ponding on sidewalks or driveways.

Small lot developments require particularly careful attention to stormwater management because such developments have a greater proportion in impervious areas and therefore have higher rates and amounts of stormwater runoff. The smaller the lot sizes, the more complete must be the drainage facilities. In contemporary practice, even low-density subdivisions are often required to install curb and gutter systems with storm sewer infrastructure to comply with water quality and flow control standards.

Lot Layout

The primary purpose of land subdivision is the production of building sites; ideally every lot in a subdivision should provide a good building site. In every subdivision, however, there will be areas where the lots will be comparatively more valuable than in other areas due to the proximity to such features as wooded areas, parks and parkways, stream and lake shorelines, and commanding views, and to such characteristics as the directional orientation of the lot and the potential for the use of exposed basements. Great skill is required to produce comparable value in the less attractive areas of the same subdivision.

Factors affecting the selection of the lot size to be used in a particular development include zoning; topography; availability of sanitary sewer service; soils; structure type to be accommodated; location within the development, for example, along an arterial street, at street intersections, and adjacent to land uses such as parks and cemeteries; and importantly the economics concerned involving land and improvement costs and market forces. Generally, all lots within a given subdivision should have approximately the same lot area. The minimum lot area and width will usually be specified by the governing zoning ordinance. Lots must provide an appropriate site for the type and size of housing to be constructed. Lot lines should be configured in a geometric manner so as to be more or less rectangular in shape. Side lot lines should be perpendicular or radial to street right-of-way lines. Corner lots should be somewhat wider than interior lots in order to accommodate adequate building setbacks for both street yards. The ratio of depth to width of lots should generally approximate two to one. Irregularly shaped lots may be permitted where necessary to accommodate topographic constraints, preserve environmental features, or support diverse housing types.

The placement of a house on a lot, in effect, divides the lot into several exterior areas. These consist of the "public" front yard setback area, private outdoor "living" areas including patios and flower gardens, service areas including vegetable gardens and laundry drying areas, and driveway areas for access, deliveries, and trash removal. The interior area arrangement—foyer, living rooms, dining areas, kitchen and laundry, porches, and basement and storage areas—should be properly related to the various outdoor areas. Figure 14.6 illustrates the placement of a single-family

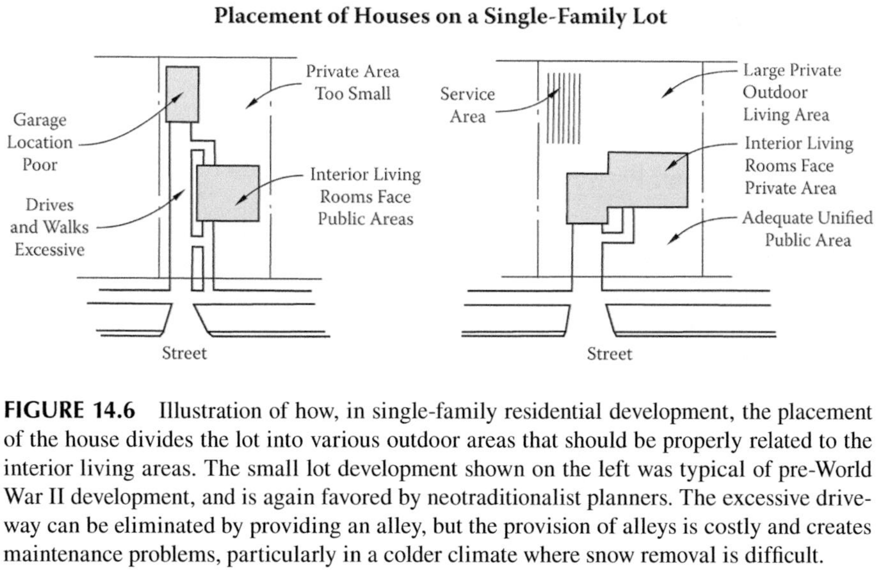

FIGURE 14.6 Illustration of how, in single-family residential development, the placement of the house divides the lot into various outdoor areas that should be properly related to the interior living areas. The small lot development shown on the left was typical of pre-World War II development, and is again favored by neotraditionalist planners. The excessive driveway can be eliminated by providing an alley, but the provision of alleys is costly and creates maintenance problems, particularly in a colder climate where snow removal is difficult.

house and the attendant creation of the public and private outdoor spaces under typical high-density, small-lot development, and under typical low-density, large-lot development. The latter readily accommodates single-story, "ranch" style houses with attached garages, while the former generally requires use of two-story structures with separate garages. The former is representative of pre-World War II, single-family residential development, the latter of post-war development. The higher-density development is favored by neotraditionalist planners, and in this respect, the new urbanism is really a return to the old urbanism.

Building setback requirements should be modest to avoid unnecessarily costly driveways and large front yard areas that reduce the size of more private outdoor living areas in rear yards. In determining building setback distances, consideration should be given to off-street parking needs. Lots should be well shaped, and the creation of narrow unusable slivers should be avoided unless necessitated by the need to provide access while protecting and preserving the natural resource base. Additionally, flexibility in setbacks can support the inclusion of accessory dwelling units (ADUs), shared open space, and passive solar orientation where applicable.

Creation of an Integrated Design

The fifth principle is the achievement of integrity in the design of the land division. Design elements that may contribute to an integrated design include focal points such as historic buildings or structures; specimen trees; schools, parks, or other public buildings; thematic architectural design of homes or other buildings; landscaping; street cross-sections and improvements; street trees; and street furniture such as streetlights and signposts. Incorporating green infrastructure, such as bioswales, permeable pavements, and rain gardens, also contributes to cohesion while supporting environmental performance. Smart technologies—like sensor-enabled lighting or connected signage—may enhance integration while improving efficiency and safety. An integrated design should also promote multimodal accessibility, resilience to climate change, and long-term adaptability for diverse community needs. Cohesiveness in visual aesthetics, functional layout, and infrastructure systems enhances both the character and livability of a subdivision.

Subdivision Design Patterns

There are three basic subdivision designs in general use today: the curvilinear pattern, the cluster pattern, and the grid pattern, the latter often as modified by neotraditional design concepts. The curvilinear pattern is typified by a predominance of curved streets, the locations of which have been carefully fitted to the terrain, and by the use of a variety of block and lot sizes and shapes, with the lots often fronting on loop and cul-de-sac streets. The cluster pattern, also known as the conservation pattern, is typified by lots tightly grouped or clustered around loop and cul-de-sac streets with each group of clustered lots separated from other similar groups by open space areas owned in common by residents of the subdivision. Figure 14.7 illustrates an urban cluster subdivision design. The grid pattern is typified by a prominence of straight streets intersecting at approximately right angles with the streets being

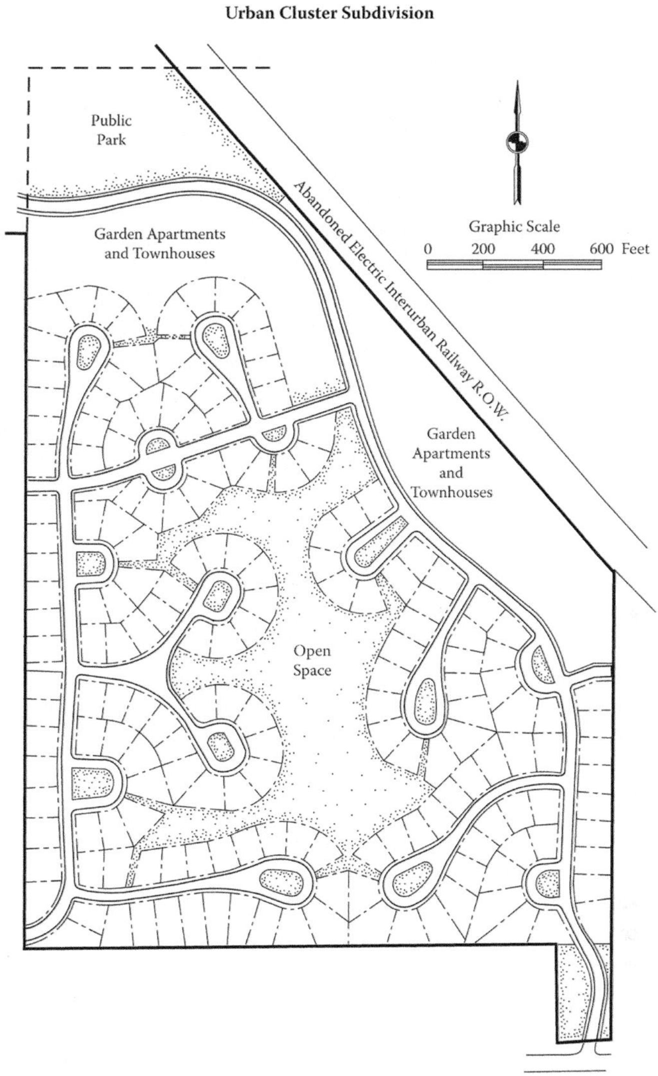

FIGURE 14.7 Illustration of an urban cluster subdivision design. Platted in 1968, the subdivision includes a mix of single, two, and multi-family dwellings and a public park. The subdivision also includes private recreational facilities, open space, and a trail network.

Source: SEWRPC.

generally laid out in the cardinal directions, and by the use of fairly uniform rectangular lots fronting on the grid streets, sometimes with alleys provided as a secondary means of access to the lots.

The subdivision design types are categorized on the basis of the basic street patterns used in the designs. It should be noted in this respect, however, that modifications or combinations of the street pattern types listed are often used in subdivision design, particularly when the subdivision involves a large tract of land. Thus, a subdivision design may utilize a grid pattern oriented in the cardinal directions in one area of the plat, combined with a grid pattern oriented in other directions in another part of the plat; or a subdivision design may combine a grid pattern in one part of the subdivision with a curvilinear pattern in another. The cluster-pattern-type subdivision, however, cannot be readily combined with the other types due to the layout and ownership structure of open space and infrastructure.

Historically, the grid pattern was the most common pattern used in subdivision design in the United States but, beginning in the 1950s, it was largely replaced by the curvilinear pattern. The grid subdivision pattern has been the subject of renewed interest in recent years, being applied in neotraditional developments. Such developments are intended to make communities more compact and livable and envision the creation of neighborhoods containing a mix of residential and commercial uses that are within walking distance of each other. In such developments, traffic circulation patterns are on a grid or modified grid pattern of streets rather than the more commonly used curvilinear pattern. This approach aligns with contemporary smart growth principles and supports walkability, transit access, and mixed-use zoning. More recently, there has been a growing interest on the part of local governments, developers, and citizens in encouraging cluster subdivisions, particularly in rural and environmentally sensitive areas, as an effective means of protecting the natural resource base. Cluster subdivision designs are quite different in urban and in rural applications.

As already noted, an example of an urban cluster subdivision design is provided in Figure 14.7. An example of a rural cluster subdivision design is provided in Figure 14.8.

The curvilinear design type is intended to maximize the use of developable land for lots while limiting open space to environmentally sensitive areas that must be protected based on recommendations set forth in the comprehensive plan or restrictions contained in the zoning ordinance. The design process focuses on the street layout, considering desired block lengths and widths and attendant lot depths. The latter require particularly careful attention along the exterior boundaries of the subdivision and around environmentally sensitive areas. The street network is designed in a curvilinear pattern that uses topographic features to create as much interest as possible while minimizing earthwork and ensuring good drainage along streets. A functional classification of streets is used, clearly distinguishing between arterial, collector, and land access streets. Cul-de-sac and loop streets are utilized as necessary or desirable. Lot lines are then added based on consideration of minimum frontage and lot size requirements. Adjustments are then made to the street layout in order to maximize use of developable land. Grading for curvilinear subdivisions is typically minimal, and this design type tends to preserve existing topography and vegetation. The large lots typically used with the curvilinear design type provide more on-site area for private use of individual homeowners, and can accommodate larger, single-story homes with attached garages. The principal advantage of the curvilinear subdivision is its adaptability to the topography of a site, minimizing grading, ensuring good relationship of lot elevations to street grades, and ensuring good drainage. However,

Rural Cluster Subdivision

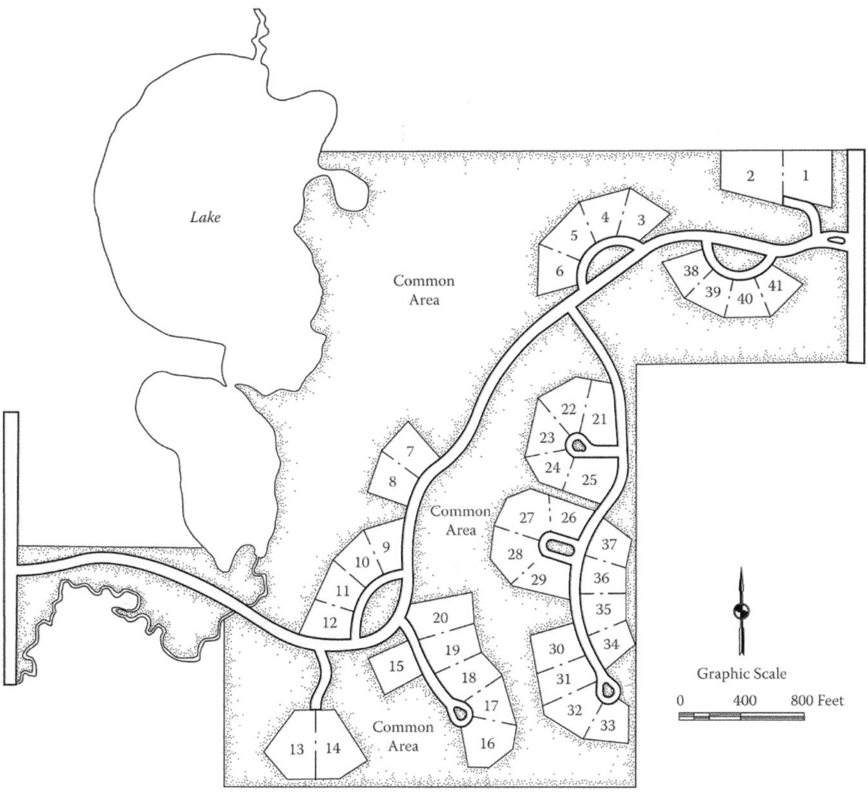

FIGURE 14.8 Illustration of a rural cluster subdivision design. The subdivision was developed in 1994 and includes 41 single-family residential lots. About 66 percent of the site has been preserved as open space, with an average density of one dwelling unit per 6.6 acres. The subdivision design preserves mature hardwood forest, wetlands, and floodplains, and provides lake and trail access for all subdivision residents.

Source: SEWRPC.

curvilinear layouts may reduce connectivity and increase travel distances, requiring careful integration with pedestrian and bicycle networks.

The cluster subdivision design type is intended to maximize the provision of common open space by minimizing individual lot sizes while maintaining the overall desired density of development. The cluster design can most effectively protect environmentally sensitive areas by maintaining such areas in open space, while concentrating lots into small groups or clusters. The design process begins with identification of a typical, model cluster that, while varying in the number of lots, maintains a similar lot and street layout. Areas of the tract to be subdivided that should be maintained in permanent open space are then identified, followed by identification of areas appropriate for development. Consideration is given to good utilization of the topography of the site, drainage patterns, and prominent overlooks or views. Once cluster locations with the attendant land access street pattern are identified, a collector street pattern is laid out that will link the clusters grouped around minor land access streets. Final layout of the lots completes the design process.

The cluster design provides small lots fronting on public streets, with all lots being located adjacent to common open space. Although substantial street length may be required to provide access throughout the site, cluster subdivision designs typically result in the lowest proportion of land devoted to open space. The locations and orientations of the lot clusters must be carefully identified to avoid the need to terrace the small lots concerned in order to adapt to the topography. Grading requirements may be significant, particularly if water features are included as an amenity.

The advantages of the cluster design include protection of natural resource features and the opportunity to maintain a significant amount of site area in common open space. This can provide attractive recreational opportunities for residents of the subdivision, maintain scenic beauty and biodiversity, and decrease the amount of impervious surface area and stormwater runoff. Cluster designs may also improve long-term environmental resilience by preserving green corridors and integrating stormwater features like bioswales and rain gardens. In situations where the design can concentrate the lot clusters on a limited portion of the site, the cost of providing infrastructure may be less than required for comparable curvilinear or neotraditional grid designs. Disadvantages of the cluster design type include the smaller lots, which require careful design of the lot arrangements and building placement to avoid the need for terracing and to provide privacy for the individual residences. The design type is not as conducive to use on sites with significant topographic variation and steeper slopes as is the curvilinear design type because of the potential need to terrace the small lots in order to provide good building pads. Architectural design controls may be needed in cluster subdivisions due to the small lots and proximity of the clustered homes. Such controls may also be used to develop an appealing neighborhood character—that is, to achieve the ever-illusive but highly desired unity of design.

Importantly, a homeowners association must be provided with cluster subdivisions to own and maintain the private common open space and related facilities. In the alternative, the local municipal government concerned would have to be willing to accept public dedication and responsibility for maintenance of the open space and related facilities, an undesirable arrangement from both the perspectives of the municipality and the homeowners. Proposed homeowners association documents must be carefully reviewed and approved by the municipality concerned to ensure that the common open areas and facilities are properly maintained. The documents must provide that each lot in the subdivision has an undivided interest in the ownership of the common open space. Provision must be made for public assumption of the maintenance if the homeowners association fails to meet its responsibilities, with the costs assessed back to the lots in the subdivision. Digital Homeowners Association (HOA) platforms may now assist in ensuring long-term compliance, coordination, and transparent budgeting.

The neotraditional grid design type is based on the urban development patterns of the past, with consideration being given, however, to open space concerns of the present. Neotraditional designs often attempt to provide a central public common that is surrounded by residential lots, or, in some larger developments, may provide for a neighborhood business or civic center. Residential lots may be double-fronted, with one face to the street and one to an alley running behind the lots, to which free-standing garages have access. This design type is better suited to use with relatively level sites where the desired higher densities of development can be more readily achieved without substantial grading. The design process begins with the identification of natural resource features that may become part of common open space. The common open space may then be defined by parallel, linear streets. The lots are smaller than typical in a curvilinear design, and homes are intended to be located closer to the front lot line. The hierarchy, or functional classification, of streets may be less distinct than in the curvilinear or cluster design types. A rectilinear grid, or modified grid, street pattern may be used. Such patterns are now often combined with complete streets policies to accommodate multimodal transportation.

The advantages of the neotraditional design type are to some extent problematic, being in part related to the smaller lots and higher densities used. The grid or modified grid street pattern may provide more paths connecting to other areas of the subdivision and result in shorter walking and bicycling trip distances. The closely spaced lots and buildings may lead to an increased sense of neighborhood for subdivision residents. The disadvantages associated with the neotraditional design type include smaller on-lot areas for private outdoor use, due both to the smaller lot sizes and to the location of the detached garages adjacent to the alley. The grid, or modified grid, street pattern may encourage the use of residential land access streets by relatively heavy volumes of faster through traffic, reducing desirable residential quiet, privacy, seclusion, and safety. Moreover, the grid street pattern cannot be readily adjusted to the topography of the site and may require excessive grading for development. The higher density of development will result in a higher proportion of impervious area and in higher rates and amounts of stormwater runoff. The smaller lots and closer buildings may require control of architectural design and home placement to ensure an attractive development.

Selection of the subdivision design type that should be used for any given parcel of land is dependent upon many factors, including market identification and conditions, surrounding land uses, and site conditions. Site conditions are a particularly important consideration in this respect as some sites are more suited to a certain design type due to the topography and the natural resource features present. For example, the new urbanism subdivision design type is more conducive for use in sites that are relatively level, while the curvilinear types can be more readily adaptable to varied terrain. A cluster subdivision is a good alternative design type to use on sites with environmentally sensitive areas that should not be developed, or natural resource features that must be protected, because lots can be clustered away from such areas without changing the overall development density.

Comparative analyses of the three alternative design types should consider such factors as average lot size, the percentage of the site occupied by lots, the amount of earthwork required for development, and drainage requirements. In evaluating the three alternatives and identifying the advantages and disadvantages of each, it is evident that there are certain elements of such an evaluation that are more subjective than objective. For example, the importance of architectural control may seem like a disadvantage to individuals interested in building a house in a style, material, and color of their own choosing. On the other hand, a very real sense of place can be achieved by building a neighborhood of homes of the same style but with reasonable options for specific designs and colors.

SITE ANALYSIS

A site analysis should be conducted before a subdivision design type is related and any subdivision layout created. The analysis should include a topographic analysis that identifies areas of steep slopes and the locations of hilltops, ridges, and scenic overviews; an analysis of drainage patterns, including stormwater flow paths and low-impact development (LID) opportunities; a vegetation analysis with particular attention to stands of fine trees and native vegetation communities; a delineation of soil types and characteristics, including soil stability and infiltration capacity; an identification of water bodies, wet-lands, woodlands, flood hazard areas and potential stormwater detention or retention areas, with consideration for green infrastructure practices such as bioretention basins and constructed wetlands; an identification of the boundaries and characteristics of environmental corridors, and connectivity to regional ecological networks; and identification of structures having historic or other cultural value recognized by local or national registries. The site analysis should also identify the classification of existing streets and highways adjacent to the development parcel and the location of desirable and undesirable points of entrance to and exit from the site; the location and characteristics of available sanitary sewerage, water supply, stormwater infrastructure, and broadband internet system connections; any proposed streets within or adjacent to the tract to be subdivided shown on the local comprehensive plan or official map; and pertinent existing physical conditions surrounding the parcel and recommendations in municipal or areawide plans affecting the parcel. Certain areas within a proposed subdivision should be considered for maintenance in permanent open space, including lands within floodplains, wetlands, and slopes of 12 percent or greater. Updated sustainability guidelines may recommend protection of areas with slopes greater than 8 percent and critical habitat zones. Primary environmental corridor lands should be preserved in natural, open uses. Conservation subdivisions and performance zoning practices are often used to align design outcomes with these preservation goals.

UTILITY SERVICE

The type of subdivision design will affect the location and configuration of the utilities serving the site. In each of the three design types described, sanitary sewers, water mains, and storm sewers will generally be located in the street rights-of-way. In a curvilinear subdivision, electric power and communication utilities are likely to be located within easements alongside and rear lot lines. In a neotraditional subdivision, the electric power and communication utilities will likely be located within the rights-of-way of streets and alleys. The location of electric power and communication facilities in an urban cluster subdivision requires particularly careful consideration on a case-by-case basis, particularly when integrating solar access easements, underground infrastructure, and fiber optic networks. Where narrower street rights-of-way are used in any subdivision design type, the utilities can be located only within the street rights-of-way to the extent that required horizontal separation distances can be maintained within the narrower streets. The use of multi-utility corridors and joint trenching techniques can facilitate this while reducing construction impact and costs.

ELECTRIFICATION AND DECARBONIZATION

Modern land subdivision design should support community-wide efforts to reduce greenhouse gas emissions by planning for the transition to all-electric buildings and transportation systems. Subdivision layouts should include infrastructure that enables electrification, such as adequately sized electrical service corridors, transformer pads, and underground conduit networks capable of supporting heat pumps, electric vehicle (EV) charging, and rooftop solar power systems. Developers should also consider integrating community-scale renewable energy systems, battery storage, and microgrids. Site design should anticipate future grid loads and incorporate building orientation and lot layouts that optimize solar access. By proactively planning for electrification and decarbonization, land subdivision can significantly contribute to municipal climate action goals while ensuring long-term energy resilience and affordability.

HISTORICAL PATTERNS OF DEVELOPMENT AND LOT YIELD EFFICIENCIES

One of the factors affecting the cost of improved building sites is the efficiency of the land subdivision design, that is, the yield in terms of the number of lots per acre that can be obtained from a particular tract of land. This yield is affected by many factors, some direct, such as lot size, block length, and street width, and some indirect, such as street pattern, topography, size and shape of the parcel to be subdivided, park, school site, and drainageway requirements, and the skill of the designer. The effect of the direct factors on lot yields can be directly, that is, geometrically, analyzed. The effect of the latter on lot yield can be determined only indirectly by an analysis of completed subdivision designs. Several historical inventories conducted in the twentieth century, including those by regional planning commissions, provide valuable baseline data on land division practices that continue to influence development today.

For the purposes of the inventory, three residential land division patterns were identified on the basis of the predominant street layout used in the land division: the grid pattern, the curvilinear pattern, and the cluster pattern. The inventories found that the most prevalent land division pattern within the region had been the grid pattern, which accounted for about 75 percent of the total plats recorded from 1920 through 1969, and by about 60 percent of the total acreage platted as of 1970. The curvilinear pattern accounted for about 24 percent of the total plats recorded, and about 40 percent of the total acreage platted. The cluster pattern of development, which was a recent innovation within the region at the time of the conduct of the inventory, accounted for less than one percent of either the plats recorded or the total area platted. Over the 50-year inventory period, the average size of the grid pattern land division was about 15 acres, the average size of the curvilinear pattern land divisions about 31 acres, and the average size of cluster pattern land divisions about 104 acres.

Public streets and alleys were found to make up about 23 percent of the gross area of grid pattern residential land divisions, about 20 percent of the gross area of curvilinear land divisions, and about 16 percent of the gross area of cluster land divisions. Areas dedicated for purposes other than streets and alleys were found to average less than 2 percent of the gross area platted, including areas dedicated for parks and schools. Other non-lotted areas within land divisions were found to make up about 3 percent of the area platted.

Design efficiency factors are computed by comparing the actual lot yield provided by a given design for a given tract to the maximum possible yield from that tract for any given set of lot dimensions and street widths. The inventory indicated that the efficiency factors of the recorded plats inventoried over the 50-year inventory period ranged from a low of about 61 to a high of about 92 percent, consistently approximating 75 percent, regardless of the type of subdivision design type.

The maximum possible yield for any given set of lot dimensions was computed by geometrically analyzing a block of lots of given dimensions, assuming appropriate minimum permissible street widths and maximum permissible block lengths. An example of the results of such geometrical analysis is given in Table 14.1. This table indicates that, although lot area is the main determinant of yield, the ratio of lot width to lot depth affects yield as to block length and street width. More recent modeling tools and GIS have enhanced the ability to perform such geometric analyses dynamically, enabling designers to evaluate lot yield potential under a variety of zoning, topographic, and design scenarios.

COMMON ISSUES OF CONCERN

Certain concerns related to land subdivision design are relatively common and are found to recur frequently in land division layouts submitted to approving and reviewing authorities. These concerns, together with some suggested solutions, are addressed in this section.

Private Streets

Generally, private streets should be avoided. Such streets may not be constructed to municipal standards in terms of right-of-way or pavement width, type of cross-section, and type and design of pavement. Because the streets are privately owned, the residents of the subdivision concerned, generally through a homeowners association, are responsible for maintaining the streets, including snow removal. Owners of homes located along private streets often later request that the municipality assume ownership of the streets, including associated maintenance responsibilities. Such requests create difficulties for the municipality because the streets may not have been designed to readily accommodate emergency equipment such as fire engines, certain types of public works maintenance vehicles and equipment, and solid waste collection and snow removal equipment, may have inadequate space for snow storage, and may have been constructed to lower geometric and structural design standards

TABLE 14.1
Maximum Yield in Lots per Acre

Lot Width (Feet)	Lot Depth (Feet)											
	100	105	110	115	120	125	130	135	140	145	150	155
30	10.64	10.24	9.88	9.54	9.22	8.92	8.64	8.38	8.14	7.91	7.68	7.48
32	9.84	9.48	9.14	8.82	8.53	8.25	7.99	7.75	7.53	7.31	7.11	6.92
34	9.31	8.96	8.64	8.34	8.06	7.80	7.56	7.33	7.12	6.92	6.72	6.54
36	8.78	8.45	8.15	7.87	7.60	7.36	7.13	6.91	6.71	6.52	6.34	6.17
38	8.24	7.94	7.65	7.39	7.14	6.91	6.70	6.49	6.31	6.13	5.96	5.79
40	7.98	7.68	7.41	7.15	6.91	6.69	6.48	6.28	6.10	5.93	5.76	5.61
42	7.45	7.17	6.91	6.67	6.45	6.24	6.05	5.86	5.70	5.53	5.38	5.23
44	7.18	6.91	6.67	6.44	6.22	6.02	5.83	5.65	5.49	5.34	5.19	5.05
46	6.91	6.66	6.42	6.20	5.99	5.80	5.62	5.45	5.29	5.14	5.00	4.86
48	6.65	6.40	6.17	5.96	5.76	5.57	5.40	5.24	5.09	4.94	4.80	4.67
50	6.38	6.15	5.93	5.72	5.53	5.35	5.18	5.03	4.88	4.74	4.61	4.49
52	6.12	5.89	5.68	5.48	5.30	5.13	4.97	4.82	4.68	4.55	4.42	4.30
54	5.85	5.63	5.43	5.24	5.07	4.91	4.75	4.61	4.48	4.35	4.23	4.11
56	5.59	5.38	5.19	5.01	4.8	4.68	4.54	4.40	4.27	4.15	4.03	3.93
58	5.32	5.12	4.94	4.77	4.61	4.46	4.32	4.19	4.07	3.95	3.84	3.74
60	5.32	5.12	4.94	4.77	4.61	4.46	4.32	4.19	4.07	3.95	3.84	3.74
62	5.05	4.87	4.69	4.53	4.38	4.24	4.10	3.98	3.87	3.75	3.65	3.55
64	4.79	4.61	4.44	4.29	4.15	4.01	3.89	3.77	3.66	3.56	3.46	3.36
66	4.79	4.61	4.44	4.29	4.15	4.01	3.89	3.77	3.66	3.56	3.46	3.36
68	4.52	4.35	4.20	4.05	3.92	3.79	3.67	3.56	3.46	3.36	3.27	3.18
70	4.52	4.35	4.20	4.05	3.92	3.79	3.67	3.56	3.46	3.36	3.27	3.18
72	4.26	4.10	3.95	3.81	3.69	3.57	3.46	3.35	3.26	3.16	3.07	2.99
74	4.26	4.10	3.95	3.81	3.69	3.57	3.46	3.35	3.26	3.16	3.07	2.99
76	3.99	3.84	3.70	3.58	3.46	3.34	3.24	3.14	3.05	2.96	2.88	2.80
78	3.99	3.84	3.70	3.58	3.46	3.34	3.24	3.14	3.05	2.96	2.88	2.80
80	3.99	3.84	3.70	3.58	3.46	3.34	3.24	3.14	3.05	2.96	2.88	2.80
82	3.72	3.59	3.46	3.34	3.23	3.12	3.02	2.93	2.85	2.77	2.69	2.62
84	3.72	3.59	3.46	3.34	3.23	3.12	3.02	2.93	2.85	2.77	2.69	2.62
86	3.46	3.33	3.21	3.10	3.00	2.90	2.81	2.72	2.64	2.57	2.50	2.43
88	3.46	3.33	3.21	3.10	3.00	2.90	2.81	2.72	2.64	2.57	2.50	2.43
90	3.46	3.33	3.21	3.10	3.00	2.90	2.81	2.72	2.64	2.57	2.50	2.43
92	3.46	3.33	3.21	3.10	3.00	2.90	2.81	2.72	2.64	2.57	2.50	2.43
94	3.19	3.07	2.96	2.86	2.76	2.68	2.59	2.51	2.44	2.37	2.31	2.24
96	3.19	3.07	2.96	2.86	2.76	2.68	2.59	2.51	2.44	2.37	2.31	2.24
98	3.19	3.07	2.96	2.86	2.76	2.68	2.59	2.51	2.44	2.37	2.31	2.24
100	3.19	3.07	2.96	2.86	2.76	2.68	2.59	2.51	2.44	2.37	2.31	2.24

Source: SEWRPC.

than public streets, resulting in safety hazards and in higher maintenance costs. In addition, private streets may lack the infrastructure for EV charging, smart utility metering, and climate-adaptive stormwater management systems, creating long-term sustainability and equity concerns.

The use of private streets also introduces other concerns, including in addition to the adequacy of the design and improvement of the streets the location and adequacy of the design and construction of public utilities, such as sanitary sewers, water mains, and stormwater drains, and the adequacy of building setbacks from the private streets. A strict prohibition of the use of private streets may, however, act as a constraint on the development of "gated community" subdivisions, which seek to exclude the general public from the development by use of walls, gates, and private security measures. Given the problems attendant to the use of private streets, a sound public policy would be to generally prohibit the use of such streets, and to require proposed gated community subdivisions to be designed and developed as planned unit developments under the local zoning ordinance. This requirement would make the location, configuration, and width, and the improvement standards related to private drives within the subdivision that would function as streets, as well as the setback and other locational attributes of the buildings, subject to local plan commission review and approval. Where private streets are permitted, communities should adopt clear standards for their design, EV-readiness, resilience to extreme weather events, and long-term maintenance responsibilities. These provisions can be codified in subdivision ordinances or design manuals to avoid costly disputes and ensure public safety.

Half Streets

It is not uncommon for developers to propose the dedication of only half of the required right-of-way for a street located on and along the perimeter of a tract proposed to be divided. Developers will often argue that, since the proposed lots front only on one side of the proposed perimeter street, the developer should only have to dedicate one-half of the street right-of-way, and assume responsibility for only one-half of the attendant full street improvement costs. Such dedication of half streets and improvement cost allocations related to them present the local municipality and the lot purchasers and homeowners with awkward construction, use, and maintenance problems. The most expedient solution is to simply prohibit the platting of half streets. If a street must be provided on and along the perimeter of a tract, then the full right-of-way should be dedicated and improved. A one-foot development control strip may be dedicated to the municipality to facilitate the equitable recovery of improvement costs from the developer of the abutting property at such time as the abutting property is proposed for development. Moreover, in areas planned for multimodal corridors, partial dedications may hinder future pedestrian or transit enhancements. Planning authorities should ensure full dedication to support long-term transportation infrastructure and connectivity.

Stub-End Streets

Land division designs commonly include temporary stub-end streets that extend to the boundary of the land division and are intended to provide future street connections to abutting lands. Such streets are intended to provide continuity in the street layout of the neighborhood and to provide access to otherwise landlocked tracts of land. The location, configuration, and dedication of stub-end streets normally will present no problems if the local community has prepared a detailed neighborhood development plan that provides for such stub ends where needed. Problems may, however, be presented with respect to the manner in which improvements of the stub-end streets are provided. At least three options are presented for such provision: require the developer to fully improve the stub-end street; leave the stub-end street unimproved but require the developer to provide a surety for the improvement of the stub end at such time as the abutting property is developed; or leave the stub-end street undeveloped and require the developer of the abutting tract to pay for its improvement at such time as the adjacent tract is developed. Experience has shown that the first alternative is preferable, usually being attendant to the least controversy. This alternative has a particular advantage in making it clearly evident to prospective lot and home purchasers that a street is proposed to be extended to the adjoining property to ensure the continuity and integrity of the street system.

In some cases, provision should be made for a temporary turnabout at the end of stub streets. Such a temporary turnabout may be T- or L-shaped, or circular, and should be designed to fit within the right-of-way of the stub-end street. Provision should be made that at such time as the stub-end street is extended the turnabout will be removed and that the cost of such removal will be borne by the developer of the adjacent tract into which the stub street is to be extended. When planning stub-end streets, additional consideration should be given to pedestrian and bicycle connectivity, ensuring that future trail or sidewalk links are also protected through easements or public right-of-way dedications.

Access Control Restrictions

Access control restrictions are permanent reservations that can be used to deny access to abutting lots from public streets and should never be permitted to remain under private control. Such access control restrictions may be used to legitimately control ingress to and egress from double frontage lots platted along arterial, and in some cases, collector streets, and to and from lots, or portions of lots, located at or near street intersections. The exercise of access control restrictions should always rest with the public, and such restrictions should be clearly identified on the face of the land subdivision plat, and noted in the deed for each lot concerned. Access control should also align with contemporary safety goals, such as Vision Zero and Complete Streets principles, to reduce vehicular conflicts and promote safe multimodal movement at intersections.

Cul-de-Sac Streets

As used in land subdivision design, the term cul-de-sac means a street open at only one end with sufficient space at the closed end for turning back vehicles. Cul-de-sac streets constitute a valuable aid in subdivision design, permitting lots to be extended into awkward remnant areas that may be created by a street layout. Far more importantly, however, cul-de-sac streets provide privacy and minimize vehicular traffic, which results in a safe and quiet residential environment. Lots fronting on cul-de-sac streets are highly desirable as places on which to live, and such lots often command the highest values in a subdivision. The use of cul-de-sac streets in subdivision design, however, requires skill and careful attention to detail. Used improperly, cul-de-sac streets may create problems related to traffic, water supply, emergency access, and drainage. All but the last of these potential problems are related to the density of development to be served and the length of the cul-de-sac street. Because of these potential problems, many land division ordinances specify a maximum length for cul-de-sac streets.

Traffic volumes on land access streets should generally not exceed 1,500 vehicle trips per day, but should be substantially lower for cul-de-sac streets, desirably as low as 250 vehicles per day. Traffic volumes in excess of this amount may impair the ability of the cul-de-sac street to provide the desired safe and quiet residential environment. At this lower limit, assuming an average trip generation rate of 10 vehicle trips per dwelling unit per day, a cul-de-sac should serve a maximum of 25 dwelling units. Thus, a cul-de-sac in an area to be developed for single-family residences with minimum lot widths of 50 feet should not exceed about 500 feet in length, while a cul-de-sac so developed with lot widths of 100 feet should not exceed about 1,000 feet in length. Cul-de-sac streets serving areas to be developed for multi-family residences should be substantially shorter than 500 feet. Adequate access for police, fire, and other emergency vehicles is also affected by the length of cul-de-sac streets if the entrance to the street becomes blocked. Cul-de-sac streets exceeding about 1,000 feet should be provided with an emergency vehicle access way from the rest of the street system to the closed end, with the access way closed to normal traffic. Such emergency access routes may take the form of reinforced turf paths or removable bollards integrated with trail corridors, provided they are designed to accommodate the weight and turning radius of fire apparatus and other emergency vehicles.

Cul-de-sac streets may present pressure, taste, and odor problems in water mains if the water distribution system is not properly designed. Fire hydrant location may also present a potential problem. These problems, however, can all be solved by proper water main configuration. Cul-de-sacs should be laid out so as to drain toward the open-end junction with the rest of the street system, and not toward the closed end. This important principle may be violated in some situations, but only if adequate provisions for drainage, including necessary easements, are made. Finally, it should be noted that the circular turnabout area at the end of the cul-de-sac should not extend to the land division boundary, but should be so located as to provide for the layout of lots around the full periphery of the turnabout.

Various designs may be used for the cul-de-sac turnabout area. Three of the more commonly used designs are illustrated in Figure 14.9. Use of a landscaped island in the center of the turnabout greatly enhances the appearance of the cul-de-sac and the value of the lots fronting on the turnabout. The islands, particularly the large teardrop-shaped island, can often be used to preserve one or more old-growth trees that significantly add value and marketability to the development. Where feasible, landscaped islands should include native vegetation or stormwater bioswales to improve water quality and reduce heat island effects. Maintenance responsibilities for these islands should be clearly established via the homeowners' association or municipal agreements.

The use of cul-de-sacs is sometimes criticized as being incompatible with good pedestrian circulation. This clearly need not be the case as exemplified by the plan for Greendale, Wisconsin, one of the federal model "greenbelt" towns. The plan for Greendale, shown in Figure 3.11 of Chapter 3, is based, in part, on the extensive use of cul-de-sacs, and Greendale is an eminently "walkable" community. Figure 14.10

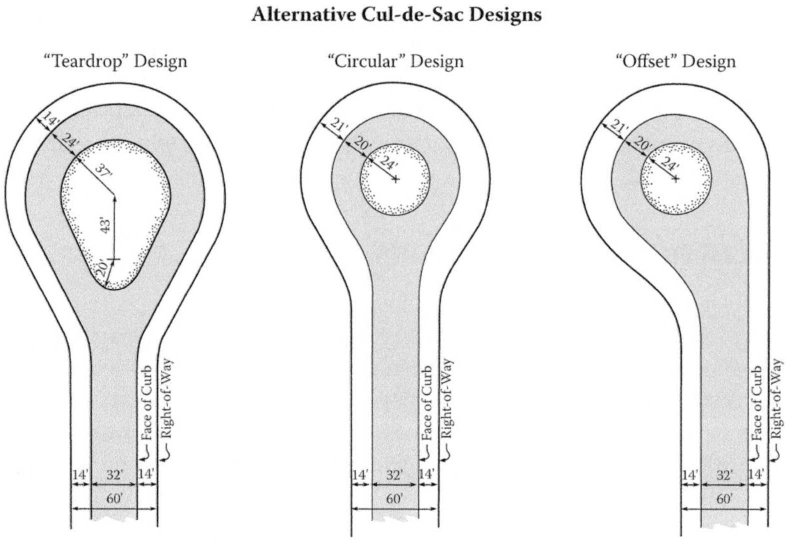

FIGURE 14.9 Illustration of three of the more commonly used designs for cul-de-sac "bulbs." The use of a landscaped island in the center of the turnabout can greatly enhance the appearance of the cul-de-sac street. The area of the teardrop-shaped island approximates 0.12 acre; the area of the circular islands approximates 0.04 acre.

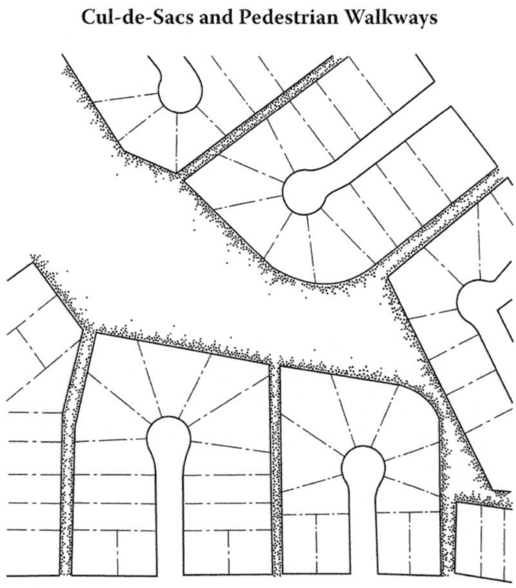

FIGURE 14.10 Illustration of how cul-de-sacs can be used with pedestrian walkways to enhance pedestrian circulation in a neighborhood, providing convenient walkways to parks, schools, and transit stops while separating pedestrians from vehicular traffic.

illustrates the use of cul-de-sacs with pedestrian walkways that can provide direct connections to school and park sites and to transit stops and that separate pedestrian from vehicular traffic. Contemporary cul-de-sac designs should incorporate pedestrian and bicycle cut-through paths to ensure walkability and connectivity. These paths should be publicly accessible, well-lit, and comply with Americans with Disabilities Act (ADA) standards.

OTHER DESIGN CONSIDERATIONS

Many contemporary subdivision plats do not provide for convenient pedestrian circulation within the neighborhood of which the land division is an integral part. Land access street configurations should be supplemented as necessary by the provision of cross-block pedestrian ways to facilitate pedestrian access to local parks and schools, local shopping facilities, and, as needed, to mass transit stops. Blocks having a length in excess of 900 feet should have a 20-foot-wide pedestrian right-of-way provided in mid-block. Careful consideration should be given to integrating bicycle ways into the design of subdivisions. Desirably, such bicycle ways should be located on exclusive rights-of-way, although they may, in some situations, be combined with pedestrian ways.

Streams and watercourses should be located in dedicated parkways and drainageways wide enough to accommodate the stream channel and the adjacent 100-year recurrence interval flood hazard area. Where lots include ravines or other non-buildable terrain, the lots should be required to provide adequate building setbacks from the hazardous features while providing a specified percentage of buildable land. Designers should also consider stormwater best management practices (BMPs), such as vegetated swales, rain gardens, and detention basins, as part of the open space network.

Short blocks should generally be avoided in order to avoid excessive dedication of land for streets and excessive street improvement costs. Acute angle intersections, more than four-way intersections, and jogs in street alignments should also be avoided, as should excessively deep or excessively shallow lots. Other design deficiencies may include the siting of too many homesites along straight streets; lots fronting on arterial streets; failure to properly organize lots into groupings around natural or built features; failure to provide oversize lots at street intersections; lack of streetscape detailing; undersized or oversized street rights-of-way; failure to provide properly sized and located park and school sites; and failure to effectively use site amenities such as entrance landscaping, decorative lighting, and street furniture to complement the design of the land division. Recent best practices also emphasize the inclusion of universal design elements, shade trees along walkways, EV charging readiness in public and private spaces, and integration of passive surveillance principles to enhance community safety.

FURTHER READING

Anderson, Larz T. *Planning the Built Environment*. Planners Press, 2000.
Dewberry, Sidney O. *Land Development Handbook (4th Edition)*. McGraw-Hill, 2019.

15 Street Patterns and Transportation Planning

The end objective of the urban transportation planning process should be informed decisions that lead to the construction, improvement, maintenance, and operation of the physical facilities and services of a total transportation system that is individually designed for the urban area it is to serve and is within the financial means of the area to achieve. The system should make the most effective use of existing facilities and equipment. New facilities as well as expanded services should be as cost-effective as possible.

Transportation Research Record, Issue No. 837
Transportation Research Board, 1982

INTRODUCTION

The street pattern continues to be a critical element in urban planning. It provides the framework for land use and significantly influences mobility, accessibility, and sustainability. The location, width, and alignment of a city's streets are all the more important because once established, they are very difficult to change. The street layout shapes lot configurations, building orientation, neighborhood identity, infrastructure efficiency, and multimodal transportation effectiveness. Contemporary street design incorporates modeling tools, multimodal performance analysis, and design standards supporting safety, equity, and resilience.

PURPOSES OF THE STREET SYSTEM

An urban street system must serve five functions: provide for the free movement of traffic throughout city; provide access to individual building sites; to a great extent, serve as the city's drainage system; provide the rights-of-way for the city's utility systems—the sanitary sewer lines, water supply and gas mains, and electric power and telecommunication cables; and provide light and air to individual building sites. In addition, streets now play a growing role in supporting active transportation (walking, biking), electric vehicle (EV) charging infrastructure, and smart city systems such as traffic sensors and adaptive signals. The efficiency and cost of the utility services, including maintenance and operation as well as construction costs, are determined to a great extent by the layout of the street system. The effectiveness and cost of the stormwater drainage system are also determined to a great extent by that layout.

DOI: 10.1201/9781003500384-15

TYPES OF STREET PATTERNS

Five principal types of street patterns can be identified. Two of these five types—the rectangular grid and the curvilinear—have already been described in Chapter 14, as those patterns are applied to relatively small areas in land subdivision design. These two patterns together with three others—the circumferential and radial, focal point and radial, and organic—can also be applied on an areawide basis to form the street patterns of the city. The cluster pattern described in Chapter 14 is a subdivision design type more than a street pattern type. The collector streets that connect the lot and land access street groups are generally laid out in a curvilinear pattern.

Rectangular Grid Pattern

In its most rigid form, the rectangular grid street pattern consists of a grid-iron of streets of equal width and directness laid over wide areas without regard to topography. A linear street pattern may be regarded as a variation of the rectangular pattern. The rectangular grid street pattern has been widely used as the framework for the development of North American cities. Its application was greatly facilitated by the completion, in advance of settlement, of the U.S. Public Land Survey System, a rectangular survey system. The first network of rural roads in a developing region was generally located on the U.S. Public Land Survey System section and quarter-section lines. These initially rural farm-to-market and land access roads then evolved into a framework of major urban streets located on a one- or a one-half-mile-spaced rectangular grid pattern. This grid pattern of major streets has served remarkably well as an urban arterial network, and has also served to define the boundaries of the neighborhood units and commercial and industrial districts that should constitute a well-planned city.

The grid pattern, however, has a number of disadvantages when used indiscriminately. It is unsuited for application in areas of even gently rolling topography, where it may require excessive amounts of cut and fill for street construction and excessive grading for the development of building sites; it may create costly drainage problems; and it may make the establishment of street and alley grades difficult. As a total pattern, as opposed to a limited arterial street pattern, it may be an unstabilizing influence since it allows heavy traffic to shift readily from street to street. It may be monotonous if applied over large areas without interruption and is considered by some to be aesthetically unsuited to use in residential areas. It is simple to "design" and to lay out, providing efficient lotting. It is well suited for application in commercial core areas because traffic can shift readily from street to street, a desirable feature in such areas. It facilitates the use of one-way streets. It is easy for people to orient to and is readily house numbered. Importantly, as already noted, it provides a sound arterial framework when utilized at a one- or one-half-mile spacing and soundly delineates the boundaries of residential neighborhood units and commercial and industrial districts. The curvilinear or cluster design types can then be applied within the residential neighborhood units. It should be noted that there is some evidence that a closely spaced grid pattern of streets may tend to reduce traffic speeds and increase safety by reducing the severity, if not the number, of accidents.

Additionally, grid patterns can enhance walkability and support transit connectivity in compact urban areas.

Curvilinear Pattern

The use of the curvilinear street pattern is generally confined to the relatively smaller areas contained within the one- to one-half-mile-spaced framework of a larger grid. Applied over a larger area, it becomes known as an irregular pattern. The disadvantages of a curvilinear, or irregular, street pattern include its being difficult to design and to lay out, more difficult for people to orient to, and difficult to house number. The advantages of a curvilinear street pattern include the following. It is readily adjusted to the topography of an area, minimizing grading and facilitating good drainage. It is considered by some to be aesthetically pleasing, providing good architectural settings particularly for residential buildings. And it prevents the shift of through traffic from arterials onto collector and land access streets. Curvilinear streets are often used in conservation subdivisions to preserve natural features and provide a sense of place.

Circumferential and Radial Pattern

A circumferential and radial street pattern is generally applicable only at an areawide scale. It is sometimes used in larger urbanized areas in the form of ring roads and radial arterials. The freeway systems of some major metropolitan areas approximate this pattern, with belt and radial facilities. The principal advantage of this pattern for major streets is the directness of movement afforded. However, overreliance on radial expressways can exacerbate congestion and create barriers to local connectivity.

Focal Point and Radial Pattern

The focal point and radial pattern were introduced as the basis for Christopher Wren's plan to rebuild London and for Pierre L'Enfant's plan for the development of Washington, D.C. Like the circumferential and radial pattern, it is generally applicable only at an areawide scale. The principal advantages of this pattern include directness of movement and, importantly, the provision of good settings for monuments and impressive buildings. It remains a useful concept for ceremonial and civic spaces but must be integrated with modern circulation and accessibility needs.

Organic Pattern

An organic street pattern is marked by segmented, discontinuous, varying-width streets with frequently intervening, irregular-shaped open spaces. It is a pattern that appears to evolve "naturally" as cities develop over time without imposition of a street plan and is a marked and attractive feature of medieval European cities. The pattern is illustrated in Figure 15.1, which shows a part of London and illustrates why taxi drivers in London must serve a lay apprenticeship in order to be municipally licensed. While difficult to replicate, elements of organic design can be intentionally incorporated to create walkable, human-scale environments.

Organic Street Pattern

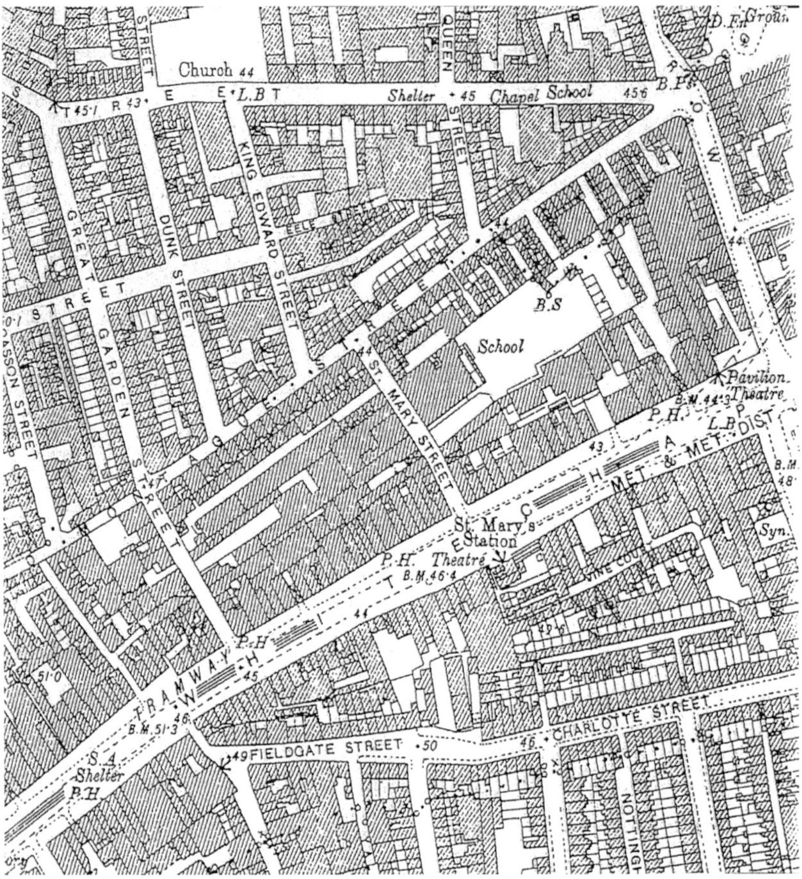

FIGURE 15.1 Illustration of an actual organic street pattern—a part of the Whitechapel area of London.

A good street system might combine a number of these patterns, using a different pattern for the arterial facilities, for the collector facilities, and for the land access facilities, as well as different patterns in residential areas than in commercial and industrial areas. Modern approaches increasingly emphasize multimodal design, flexible street hierarchies, and equitable access, regardless of the chosen geometric pattern.

Functional Classification of Streets

As already noted, an urban street system must generally serve five functions: free movement of traffic; land access; drainage; provision of rights-of-way for utilities; and provision of light and air to building sites. Because two of these five functions—the free movement of traffic and access to individual building sites—are mutually incompatible, the street pattern, for reasons of safety, efficiency, effectiveness, and economy, should be functionally classified, that is, subdivided according to the need and ability to move traffic versus the need and ability to provide good land access. Accordingly, urban streets may be classified by function as arterials, collectors, and land access streets. It should be noted that the recent literature contains some criticisms of the application of the functional classification of streets as applied to residential areas, since it tends to increase traffic on collector and arterial streets while reducing traffic on land access streets. But the concentration of traffic on collectors and arterials and reduction of traffic on land access streets are precisely significant and desirable benefits of the functional classification of streets, and permit street cross-sections and pavement widths to be rationally adjusted to primary function. It also promotes the stability, quiet, and safety—the latter measured in broader terms than traffic safety alone—of residential neighborhoods. Newer frameworks often supplement this model with layered considerations such as pedestrian and bicycle access, environmental sustainability, and "complete streets" principles.

ARTERIALS

Arterials are intended to facilitate the free movement of traffic and may consist of the following types:

1. Freeways—defined as directionally divided arterial highways with full control of marginal access and full grade separation of all intersections. Freeways are thus exempt from one of the five street functions: access to abutting property.
2. Expressways—defined as directionally divided arterial highway with full or partial control of marginal access, and grade separation of some, but not necessarily all, intersections.
3. Major Streets—defined as arterials with intersections at grade and direct access to abutting property. Major streets may or may not be directionally divided and may consist of one-way pairs. Major streets may have partial or—with good land subdivision design—full control of marginal access. Major streets are often spaced at one-half or one-mile intervals. Intersection spacing along major arterials should be 1,000 feet or more to facilitate traffic flow and signalization.
4. Parkways—defined as arterials limited to use by non-commercial traffic; parkways may or may not be directionally divided, and may have full or partial control of marginal access, and are located in a park or a ribbon of park-like development. Parkways may connect larger parks, and often follow streams and lake shorelines. Parkways provide open "green" space in the urban area and can be used to preserve and protect watercourses and shorelines for public use, provide good rights-of-way for trunk sewers, and increase adjacent property values, providing fine settings for residential development. Parkways also provide locations for trail-oriented recreational activities, such as bicycle paths, bridle paths, hiking trails, and pleasure drives. The exclusion of commercial traffic allows construction of

lighter pavements, use of less demanding geometric layouts, and facilitation of maintenance of a park-like character.

5. Boulevards—defined as broad arterials in which through traffic is separated from local traffic by landscaped islands and which are intended to provide an attractive setting for abutting development. Boulevards may carry a busway or light railway facility in the center as well as directionally divided traffic lanes. In recent years, boulevards have reemerged in North American cities as multi-modal corridors supporting transit, cyclists, and pedestrians in addition to vehicles.

Arterials should form an integrated system over an entire substate region or metropolitan area, and should be located and designed to carry the imposed traffic loadings at a desirable level of service. The primary function of an arterial facility should be the expeditious movement of vehicular traffic. Providing access to abutting property may be a secondary function, but should always be subordinate to the primary function of traffic movement. Contemporary arterial design increasingly accounts for transit priority lanes, freight corridors, and smart signalization technologies to support more adaptive, efficient systems.

Collectors

Collector streets are defined as streets intended to collect and distribute traffic to and from land access streets, conveying the traffic to and from arterials. Collector streets often form the principal entrances to residential neighborhood units and may carry transit routes. Collector streets usually fulfill a secondary function of providing access to abutting property. Designs now often include traffic calming features, pedestrian crossings, and bicycle infrastructure, especially in school or park zones.

Land Access Streets

Land access streets are defined as streets intended to provide access to individual building sites: residential, commercial, and industrial. Land access streets include, as subtypes, marginal access streets and alleys. They serve a critical function in ensuring safe, walkable, and community-focused street environments, particularly in low-traffic areas.

Marginal Access Streets

Marginal access streets, or frontage roads, are defined as streets intended to provide access to abutting properties while protecting the capacity and safety of through traffic. While reverse frontage lots remain a tool for minimizing driveway conflicts, marginal access streets can support local circulation near highways or commercial corridors when carefully designed with appropriate buffers and multimodal accommodations.

Alleys

Alleys are defined as secondary means of vehicular access to building sites otherwise fronting on public streets. They are regaining favor in many urban infill developments for supporting accessory dwelling units (ADUs), hiding garages and trash collection, and preserving attractive street frontages.

Design Considerations

Collector and land access streets are usually built to a standard cross-section that is determined by the demands of stormwater drainage, snow storage and removal, utility location, emergency access, and maintenance requirements and municipal service needs, as well as construction and maintenance costs and aesthetic and safety considerations, rather than by the demands of traffic. Anticipated traffic volume is, therefore, not generally a critical factor in the location and design of collector and land access streets, and the location and configuration of collector and land access streets still remain an intuitive design process based on skill and experience. Existing and probable future traffic volumes, however, are critical factors in the location and design of arterial facilities. The location and design of arterials, therefore, must be based on comprehensive traffic studies and analyses, including assignment of existing and forecast traffic volumes by simulation modeling. The arterial network of a city must be designed as an integral part of an areawide system serving the larger region or metropolitan area of which the city is a part. More recently, tools such as dynamic traffic assignment, real-time data feeds, and scenario modeling are being used to optimize network performance while supporting climate goals, equity, and resilience.

Street Cross-Sections

Street cross-sections should support the safe, efficient, and equitable movement of people and goods across multiple modes—vehicles, bicycles, pedestrians, transit, and freight. While vehicles move and park in lanes, and those lanes remain a key element of the street design, modern best practices also give significant weight to pedestrian access, bike safety, transit priority, stormwater management, and landscaping. Roadway widths should be based on a context-sensitive balance of lane needs, including moving and parked vehicles, protected bicycle facilities, wide sidewalks, street trees, and green infrastructure. The traffic loading and the capacity per lane being known, the sum of the widths of the lanes required, including parking lanes, should determine the roadway widths of arterials. As already noted, other considerations should determine the roadway widths of collector and land access facilities. For economy, stability, and beauty, rigid standardization of cross sections should be avoided, even though use of such cross-sections may save engineering time. Flexibility in cross-section design is especially important in mixed-use and residential areas, where human-scale design, accessibility, and aesthetics are paramount.

Figure 15.2 provides a set of typical street cross-sections together with annotations concerning design capacities at level of service C, and attendant approximate unit costs. The cross-sections are designed to facilitate the ready conversion of the rural sections shown to urban sections as may be necessary. For example, rural cross-sections No. 1 and 2 can be readily converted to urban cross-sections No. 9 and 10.

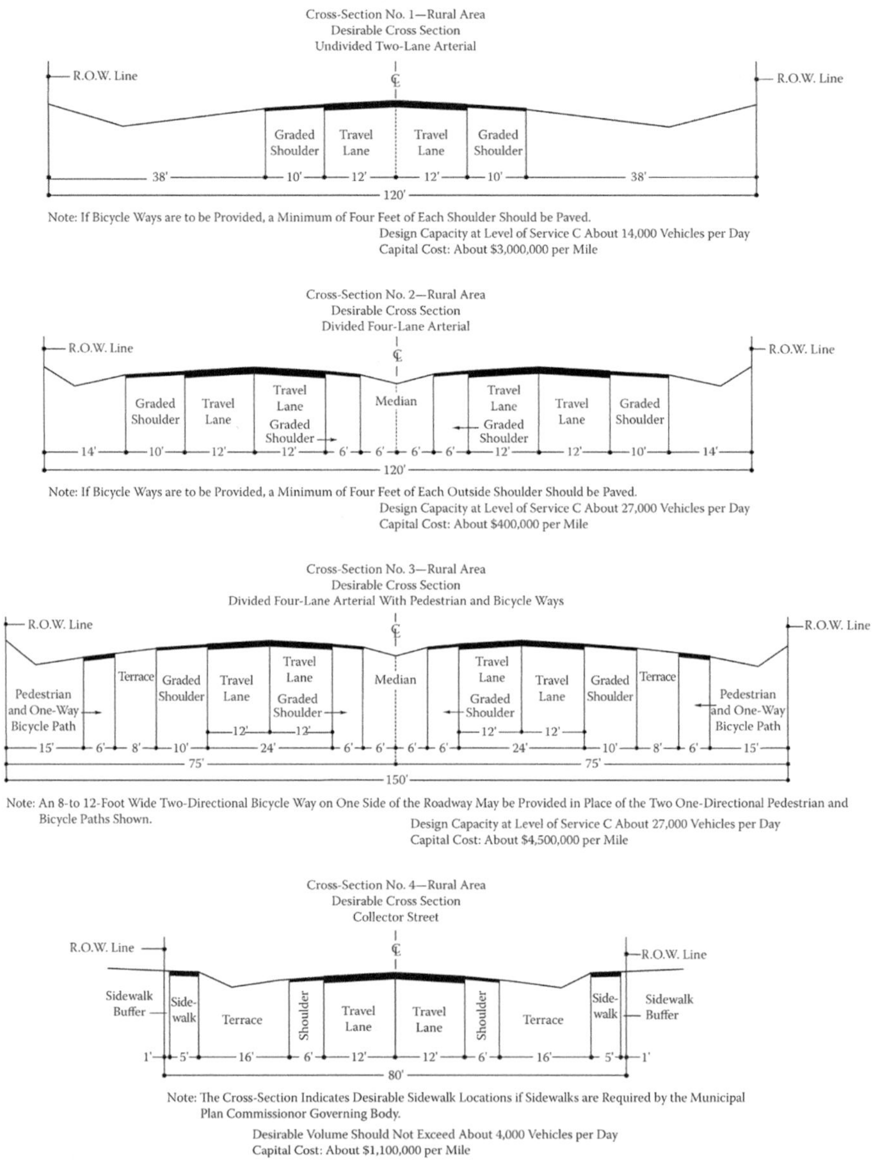

FIGURE 15.2 (a–d) Typical street cross-sections. *(Continued)*

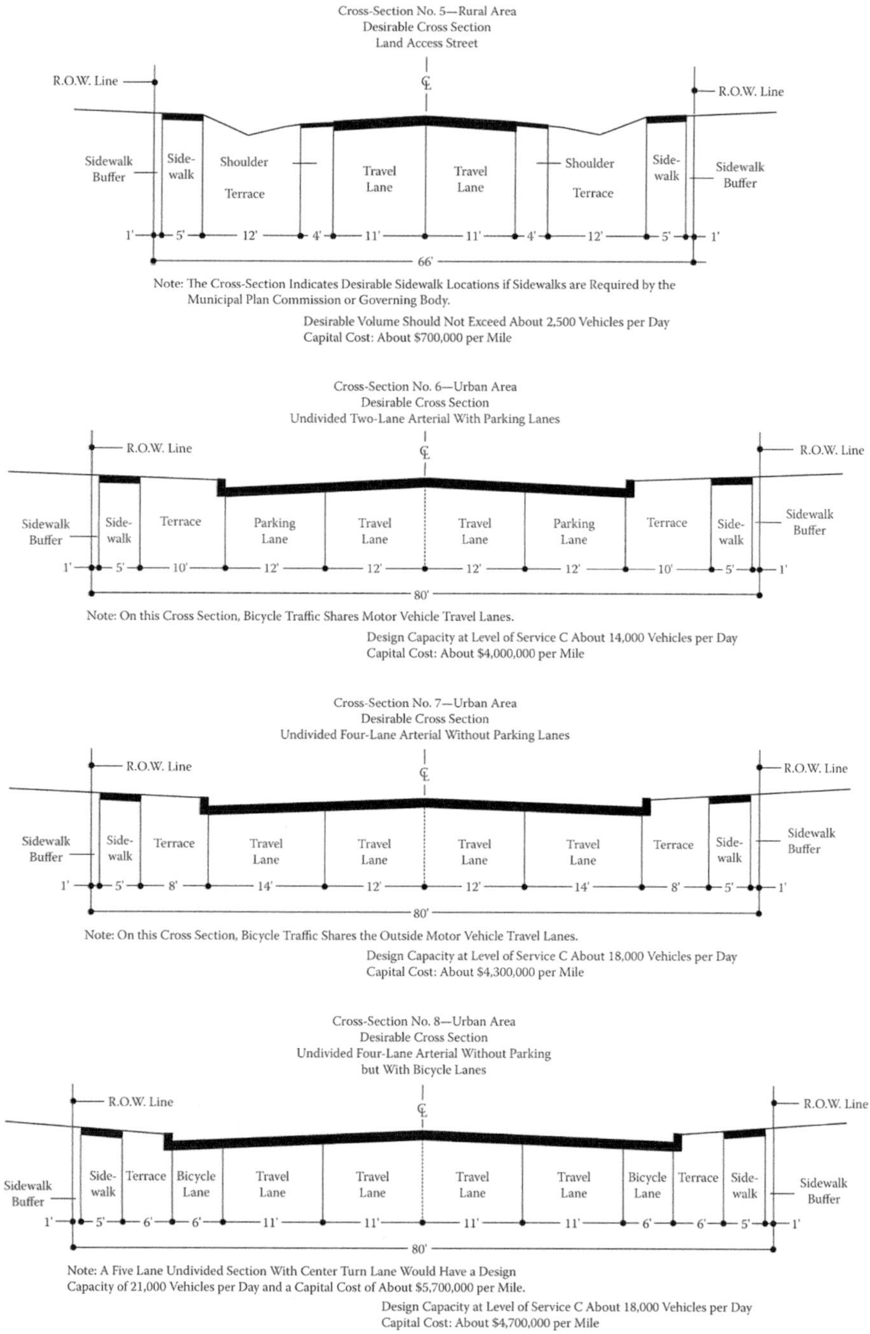

FIGURE 15.2 *(Continued)*

Cross-Section No. 9—Urban Area
Desirable Cross Section
Divided Four-Lane Arterial With Parking Lanes

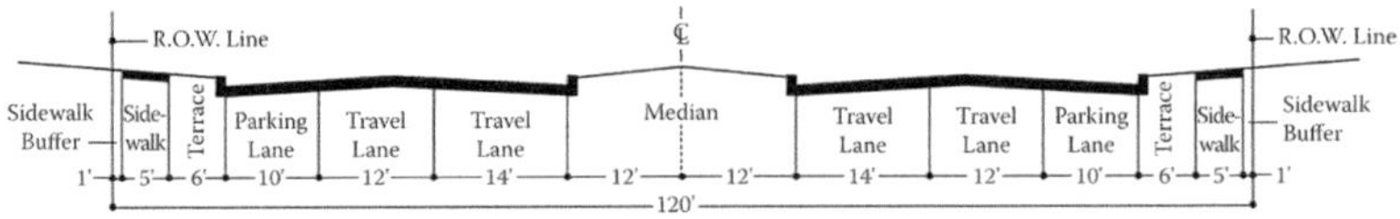

Note: Bicycles and Motor Vehicles Would Share the Outside Travel Lane.
Design Capacity at Level of Service C About 27,000 Vehicles per Day
Capital Cost: About $8,400,000 per Mile

Cross-Section No. 10—Urban Area
Desirable Cross Section
Divided Four-Lane Arterial Without Parking Lanes

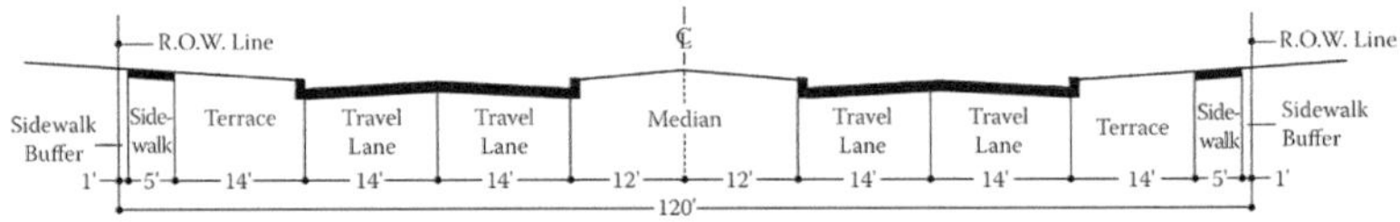

Note: Bicycles and Motor Vehicles Share the Outside Travel Lane.
Design Capacity at Level of Service C About 27,000 Vehicles per Day
Capital Cost: About $6,200,000 per Mile

Cross-Section No. 11—Urban Area
Desirable Cross Section
Divided Six-Lane Arterial Without Parking Lanes

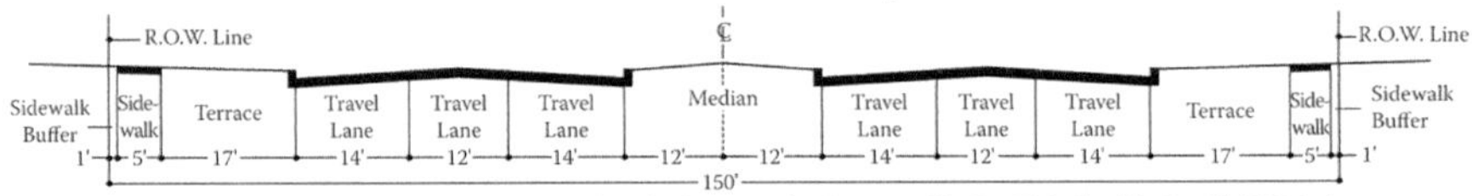

Note: If Bicycle Travel is to be Accommodated, a Bicycle and Pedestrian Path a Minimum of Six Feet Wide Should be Provided Within the Terrace in Lieu of the Sidewalk on Both Sides of the Street.
Design Capacity at Level of Service C About 38,000 Vehicles per Day
Capital Cost: About $9,300,000 per Mile

Cross-Section No. 12—Urban Area
Desirable Cross Section
Collector Street

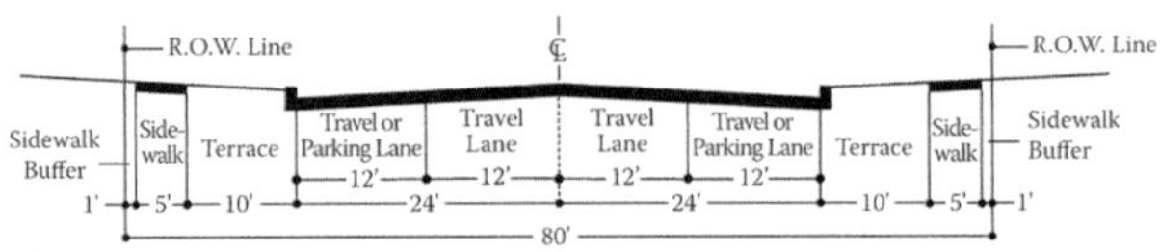

Note: The Pavement Width of an Urban Cross Section Collector Street May Vary from 36 Feet to 48 Feet. The Narrower Width Would Apply to Collector Streets Carrying Average Weekday Traffic Volumes of Under 3,000 Vehicles per Average Weekday and Minimal Truck or Bus Traffic. The Wider Width Would Apply to Collector Streets Carrying Traffic Volumes Exceeding 3,000 Vehicles per Average Weekday and-or Carrying Significant Truck or Bus Traffic.
Desirable Volume Should Not Exceed About 4,000 Vehicles per Day
Capital Cost: About $3,000,000 per Mile

FIGURE 15.2 *(Continued)*

These capacities and costs represent crude approximations and are intended solely to facilitate comparison of the relative potential utility and performance of the cross-sections provided. The annotations should be viewed with caution and reservation. Determination of the actual capacities and costs of given street cross-sections requires detailed engineering studies. Current engineering studies often include additional factors such as pedestrian level of service, microclimate impact, accessibility, and resilience to climate change. Where appropriate, cross-sections should integrate elements such as green infrastructure (bioswales, permeable pavement, and

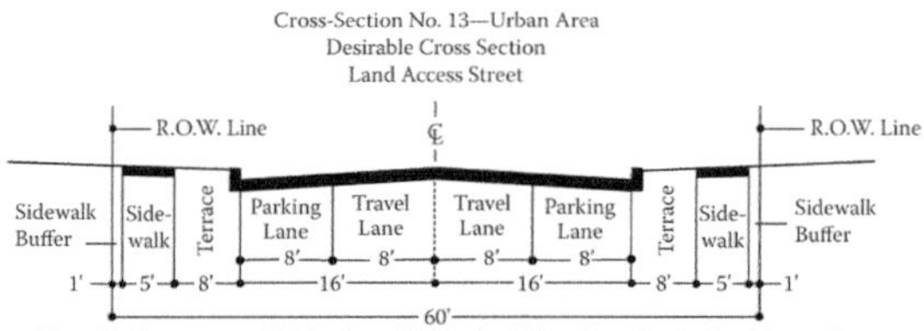

Note: The Pavement and Right-of-way Width of an Urban Cross Section Land Access Street May vary from 32 Feet to 36 Feet of Pavement Width, and from 60 Feet to 66 Feet of Right-of-way Width. Then Arrower Width Would Apply to Land Access Streets Carrying Average Weekday Traffic Volumes of Under 1,500 Vehicles per Average Weekday, With Little Truck or Bus Traffic and Limited Demand for On-street Parking. The Wider Width Would Apply to Land Access Streets With Average Weekday Traffic Volumes of 1,500 or More Vehicles per Average Weekday, Demand for On-Street Parking, and some Truck or Bus Traffic.

Typical Volume Should Not Exceed About 1,500 Vehicles per Day.

Loop Street-each Half Volume Should Not Exceed More than About 750 Vehicles per Day.

Desirable Cul-de-sac Volume Should Not Exceed About 250 Vehicles per Day.

Design Capacity at Level of Service C About 1,500–2,500 Vehicles per Day
Capital Cost: About $2,000,000 per Mile

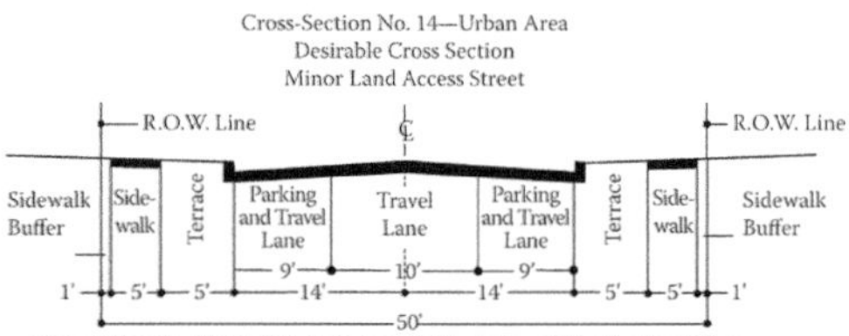

Note: Sidewalks May be Eliminated, at the Discretion of the Municipal Plan Commission or Governing Body, in Areas where such Walks are Not Required Because of the Provision of a Separate Network of Pedestrian Ways, Low Vehicular or Pedestrian Traffic Volumes, or Lot Arrangement.

Desirable Volume Should Not Exceed About 2,500 Vehicles per Day
Capital Cost: About $1,300,000 per Mile

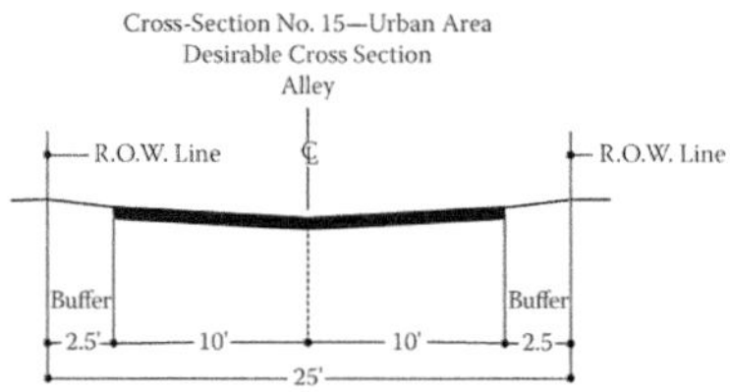

Capital Cost: About $800,000 per Mile

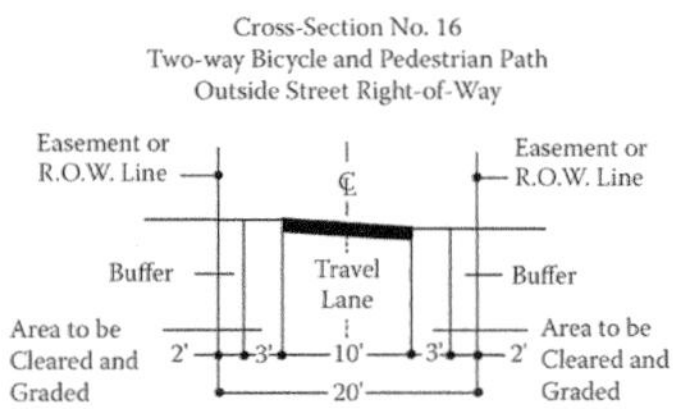

Note: Painted Center Lines are Not Normally Required on Bicycle Path. Where Conditions Such as Limited Sight Distance Make it Desirable to Separate Two Directions of Travel, a Solid Yellow Line Should be Used to Indicate No Traveling to the Left of the Center Line

Capital Cost: About $100,000 per Mile

FIGURE 15.2 *(Continued)*

rain gardens); EV charging infrastructure and smart poles; dedicated bus lanes or light rail corridors; and curb extensions and protected intersections.

ARTERIAL CAPACITY AND LEVEL OF SERVICE

Understanding the capacity values in Figure 15.2 requires familiarity with the terms "capacity" and "level of service" (LOS) as used in modern traffic engineering and planning. Capacity is generally defined as the maximum number of vehicles (or people) that can pass a point on a roadway under prevailing conditions during a given time period. In planning, both theoretical and practical capacities are considered. The practical capacity balances maximum throughput with acceptable operating conditions.

The selection of the proper capacity value for use in planning is one of the most critical steps in the transportation planning process. The concept of level of service describes the operating conditions of a transportation facility from the user's perspective, including speed, freedom to maneuver, delay, and comfort. For planning purposes, maximum lane capacity under ideal conditions is often assumed to be 2,000 to 2,200 vehicles per hour for free-flow conditions on uninterrupted flow facilities. Capacity is reduced by factors such as grade, curves, lane width, heavy vehicle percentage, and adverse weather. On signalized arterials, effective capacity typically drops to 700–900 vehicles per lane per hour, depending on signal timing and spacing.

A commonly assumed practical capacity is 1,500–1,800 vehicles per lane per hour. Thus, a six-lane freeway might accommodate 9,000 to 10,800 vehicles per peak hour. Peak hour volumes typically represent 8–12 percent of daily traffic, so a six-lane freeway might serve 80,000–120,000 vehicles per day. Table 15.1 defines LOS categories based on speed, density, and delay. Although originally developed for rural highways, LOS frameworks now also apply to urban arterials, freeways, and multimodal corridors.

For urban, at-grade arterials, three levels of service are typically used: under design capacity, at design capacity, and over design capacity. Under design capacity conditions, average operating speeds of 25–40 miles per hour, and average delays at signalized intersections are 5–15 seconds. At design capacity, conditions are characterized by average speeds of 20–30 miles per hour, and average delays at signalized intersections of about 25 seconds. Overdesign capacity conditions are characterized by average speeds of 15–20 miles per hour, and average delays at signalized intersections of 35–120 seconds, exceeding one traffic signal cycle. Current best practices increasingly supplement LOS with other metrics such as multimodal LOS (MMLOS) for pedestrians, cyclists, and transit; travel time reliability; and environmental and equity considerations.

ARTERIAL SYSTEM PLANNING

Planning for the development of the collector and land access street system usually rests with the local municipalities concerned and is primarily exercised through the control of the land subdivision process. The design of collector and land access street systems is still an art with the results of the design process dependent upon the skill of the designer. Planning for the development of the arterial street and highway system has, however, evolved into an applied science that involves the conduct of extensive facility and travel inventories, and the formulation and application of travel and traffic simulation models.

The arterial system must be planned on an areawide basis. Travel and traffic patterns develop over entire metropolitan areas, and the transportation system

TABLE 15.1
Arterial Level of Service Definition

Level of Service (LOS)	Description	Upper Limit of Volume Range Vehicles per Lane per Hour
LOS A	Free flow—vehicles unimpeded in ability to maneuver; effects of incidents are easily absorbed.	620–700
LOS B	Almost free flow.	990–1,120
LOS C	Flow at speeds at or near free flow speed but freedom to maneuver noticeably restricted; queues form behind major blockages.	1,450–1,645
LOS D	Speeds begin to decline—freedom to maneuver is restricted; minor incidents will create queuing; little space to absorb disruptions.	1,780–2,015
LOS E	Operation at capacity—volatile; vehicles closely spaced; any disruption can establish a wave that propagates upstream; level of physical and psychological comfort is low; accidents.	1,940–2,200
LOS F	Vehicular flow breaks down—number of vehicles arriving at a point exceeds the number that can move through the point; breakdowns may occur for the following reasons: traffic incidents; points of recurring congestion, such as merging or weaving segments and at lane drops; and chronic congestion on segments where peak period flow rates exceed capacity.	0

Source: SEWRPC.

must form an integrated system over the entire metropolitan area, a system capable of serving the developing travel and traffic patterns. This is recognized by the U.S. Department of Transportation, which requires the preparation of metropolitan transportation system plans as a prerequisite to the receipt of federal funding for street, highway, and transit facility improvement, operation, and maintenance. The capacities of each link in the areawide arterial street and highway and transit systems must be fitted to the existing and probable future traffic loadings, and the effect of each proposed improvement on the remainder of the system quantitatively tested. Not only must arterial street and highway and transit systems be planned together, but transportation planning must be conducted concurrently with, and cannot be separated from, land use planning. The land use pattern determines the amount and spatial distribution of the travel to be accommodated by the transportation system. That system, in turn, is one of the most important determinants of the land use pattern.

Over the last 50 years, an urban transportation planning process has been evolved by which the relationship between land use and trip generation and distribution can be accurately described graphically and numerically, the complex movement of people and vehicles over arterial street and highway and transit systems simulated, and the effects of alternative courses of action with respect to land use and transportation system development evaluated. This process is diagrammed in Figure 15.3. The process is technically complex and costly, and for proper execution requires an experienced planning and engineering staff. It is best carried out at the metropolitan level, and indeed by federal regulation it must be carried out at that level. The arterial street and highway and transit system plans produced should form the basis for the local arterial street and transit elements of the comprehensive city plan. Modern practices increasingly leverage real-time data from connected vehicles, Global Positioning Systems (GPS), and mobile devices to improve the calibration and accuracy of travel demand models. Integration of big data analytics, artificial intelligence, and machine learning has become essential in dynamic transportation system modeling.

Although city planners and engineers are not apt to be directly responsible for the conduct of the necessary areawide transportation planning, a general understanding of the planning process involved is essential since the areawide transportation system plans are intended to form the framework for the local arterial street and transit facility plans. In addition to the socio-economic and land use inventories required for comprehensive planning, areawide transportation system planning requires the conduct of special purpose inventories of the arterial street and highway and transit systems and the conduct of massive travel surveys.

Facility Inventory

There is no ready procedure for identifying the arterial street system of an existing urban area, although traffic simulation modeling may be used to validate, and as necessary adjust, the configuration of an arterial system once identified. The initial identification must be based on the experience and judgment of the officials most intimately familiar with the local street and highway system. This is best done collegially for an entire urbanized area by a committee comprising knowledgeable state and county highway engineers, city engineers, and city planners. Once the arterial system has been delineated, it must be converted to an arterial network for computer simulation by assigning node numbers to all intersections, with each segment between two nodes being defined as a link in the network. Data delineating link lengths, average operating speeds, and traffic capacities must be collected for each link, as well as existing traffic volumes and number of traffic lanes, which determine the capacity for each link. A series of computer programs can then be used to calculate minimum travel time paths through the arterial network and to assign trips to these paths, simulating system utilization.

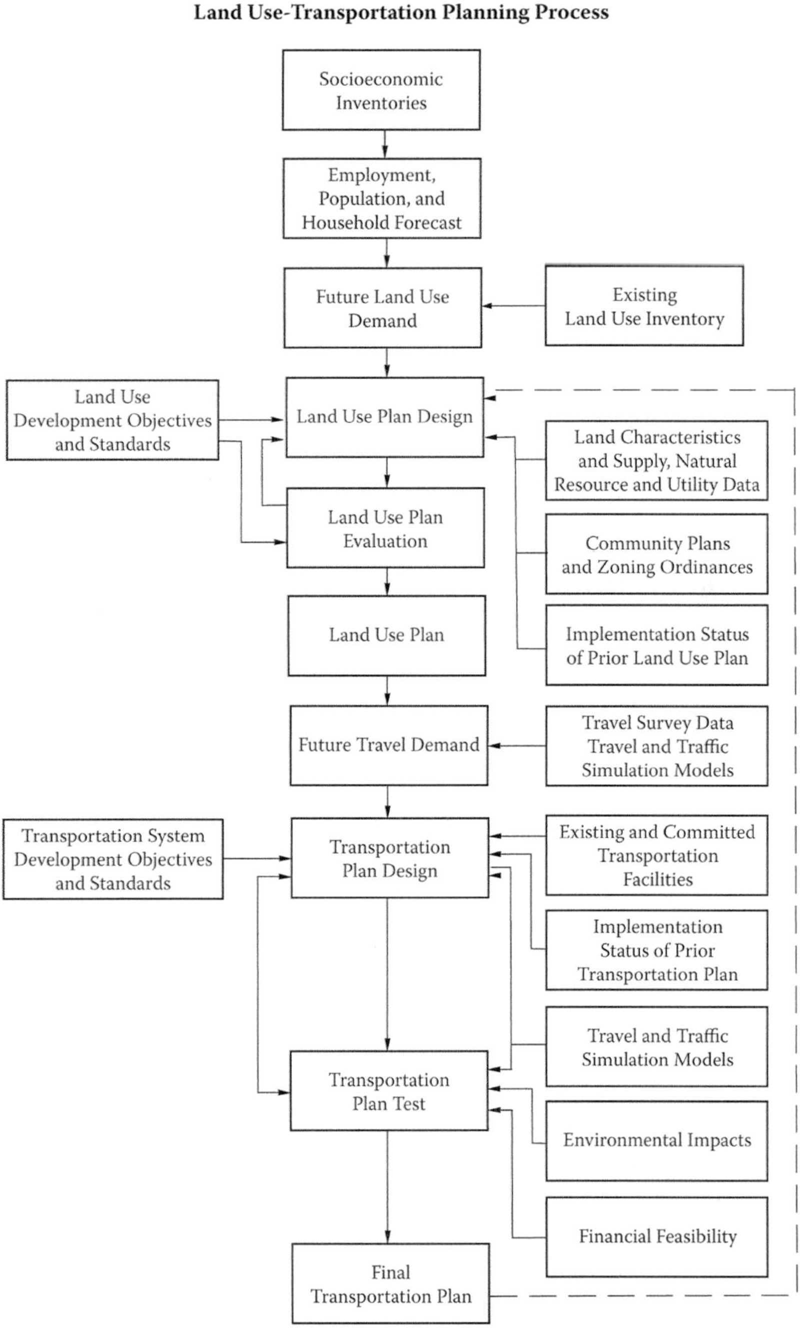

FIGURE 15.3 Illustration of the general procedure to be followed in the preparation of areawide transportation system plans.

Source: SEWRPC.

Travel Surveys

The necessary travel data are obtained by the conduct of travel surveys—often referred to as origin-destination surveys. Travel surveys have evolved significantly, with traditional manual methods increasingly replaced or complemented by passive and anonymized data collection techniques such as location-based services (LBS), GPS logs from mobile apps, Bluetooth and Wi-Fi sensors, automatic traffic recorders (ATRs), and smartphone-based travel diary apps.

Passive data from mobile devices and connected vehicle systems are now widely used to capture trip origins, destinations, times, and travel modes, minimizing the need for manual data collection. Where more detailed surveys are needed, app-based platforms allow participants to self-report with real-time prompts, improving recall accuracy. Privacy-protected automatic license plate readers (ALPR) and camera systems are used where necessary to monitor cross-boundary traffic patterns without physical stopping of vehicles.

The survey data are expanded from the sample to the universe concerned, and a number of checks are made to assure the accuracy and completeness of the data. These checks may include a comparison of the socioeconomic characteristics obtained from the surveys to data obtained from the U.S. Census, American Community Survey (ACS), or commercial mobility datasets, a comparison of vehicle trip volumes as derived from the surveys to actual vehicle volume counts made on selected screenlines, and on-vehicle miles of travel as derived from the survey to actual vehicle miles of travel derived from traffic volume counts. Modern cloud-based data platforms support real-time validation, visualization, and dashboard-based analysis, enabling iterative model development and more transparent public engagement.

Analyses of the survey data are used directly in the planning effort. The principal use of the data, however, is in the formulation, verification, and validation of the mathematical models that are used to simulate existing and probable future volumes of travel and traffic as derived from land use data. The models are based on the concept that travel is an orderly, regular, and measurable occurrence evidenced by travel patterns, as those patterns are definitively established by travel surveys. This orderliness, regularity, and stability are evidenced in Figure 15.4. The remarkable stability of the daily travel patterns over an almost 40-year period is illustrated, along with the peaking characteristics of urban travel patterns—one of the major problems confronted in the design of urban arterial street and highway and transit system plans. The models make it possible to calculate future travel demand as a function of land use development patterns instead of deriving such demand from simple expansions of existing travel patterns. Development and application of the models requires the division of the planning area into traffic analysis areas of relative socioeconomic and land use homogeneity. These areas may range in size from one square mile to one city block.

Simulation modeling involves development and application of four models: trip generation, trip distribution, modal choice, and traffic assignment. The trip simulation model estimates the total number of person trips generated in each traffic analysis area by using relationships established between land use and travel.

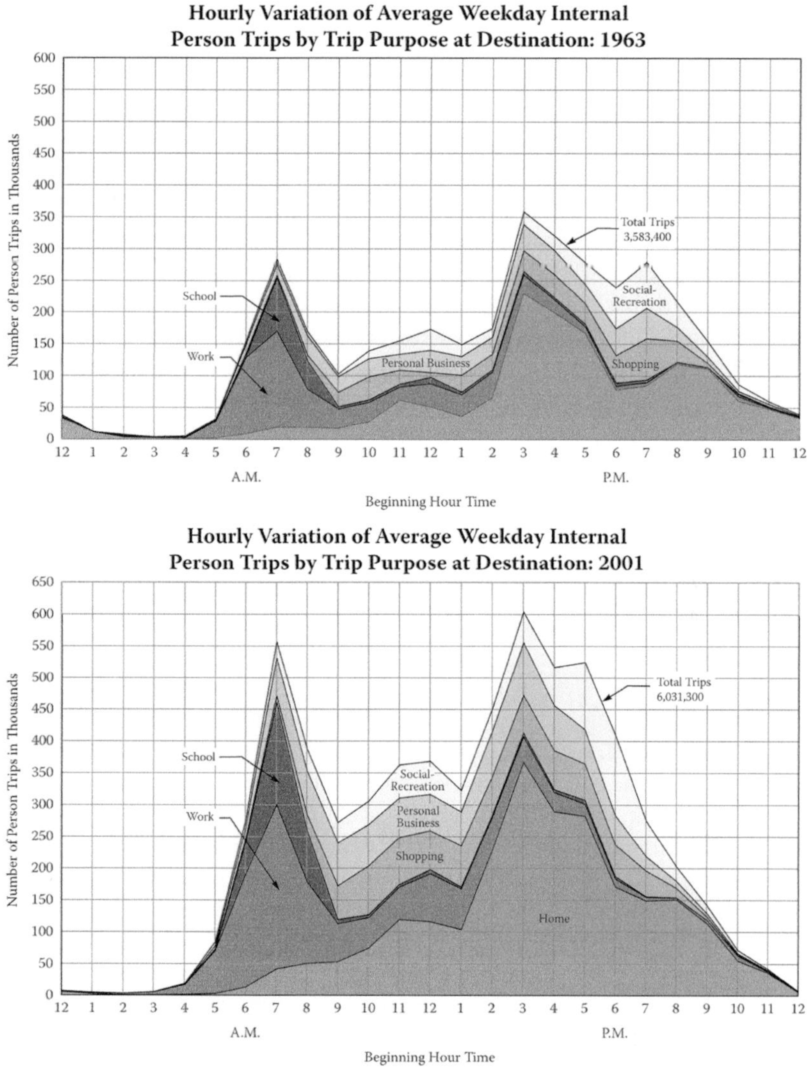

FIGURE 15.4 Illustration of the remarkably orderly and regular as well as stable characteristics of urban travel patterns. Although the peak travel demand in this planning area has increased from about 3.6 million in 1963 to about 6.0 million in 2001, the pattern has remained virtually the same.

Source: SEWRPC.

This model provides the critical link between land use and transportation planning. The data are expressed as trip ends. One trip end is termed the production end and the other end is termed the attraction end. For home-based trips, home is always considered the production end. Two sets of relationships are developed: one for trip production, and one for trip attraction, and usually residential versus non-residential land uses.

Three basic approaches may be used to calculate trip production: cross-classification in which the trip-making characteristics of households are used to establish trip generation rates; regression analysis, in which the dependent variable consists of trips generated in each traffic analysis area and the dependent variables are the socioeconomic and land use characteristics of the traffic analysis area; and the factoring of existing trip productions based on changes in the number of households and/or persons in the traffic analysis area.

Similarly, three basic approaches may be used to calculate trip attractions: trip rates based on socioeconomic characteristics of the traffic analysis area such as households or employment; regression analysis in a manner similar to that used for trip productions; and the factoring of existing trip attractions based on changes in the number of households, persons, or employees in the traffic analysis area.

The trip distribution model links the person trips produced in each traffic analysis area with trip attractions in other such areas. The model consists of the number of person trips made between each pair of traffic analysis areas in the planning area. The most common form of this model is the gravity model, in which the trip interchange between two traffic analysis areas is calculated as a direct function of the number of trip ends in each traffic analysis area and as an inverse function of their spatial separation.

The modal choice model divides the number of trips between each traffic analysis area according to the highway and transit travel modes. The model utilizes factors known to affect choice, such as the characteristics of the trip maker, the characteristics of the trip, and the characteristics of the transportation system. The relationships are developed by trip purpose considering travel times and costs.

The traffic assignment model assigns the transit and highway trips to the transit and highway networks. Assignment is made on the basis of travel times. The model is used to provide average weekday traffic volumes, as well as traffic volumes by period of the day, for each link in the arterial network and for each transit route. Depending on the model inputs, these outputs may represent existing volumes derived from the existing land use or planned design year volumes derived from a planned land use pattern. The models permit both alternative land use plans and alternative supporting arterial street and highway configurations to be proposed and evaluated. Modern assignment models incorporate real-time congestion effects, dynamic route choice, and multi-modal integration, using tools like activity-based modeling and agent-based simulation frameworks.

Experience has shown that the simulation models, if properly calibrated and validated, provide remarkably accurate arterial network traffic volumes and transit route loadings. Ten-year volume forecasts produced by the models may be expected to be within plus or minus 10 percent of actual measured arterial volumes and transit line loadings.

Traffic volumes vary with the hour of the day, the day of the week, and the month of the year. The travel simulation model can provide traffic assignments by time period within the average weekday, for example, peak hourly periods, thus directly yielding design hour volumes. In those cases where only average weekday traffic volumes are provided by the assignment process, two factors are used to convert the average weekday volumes provided by the simulation model applications to design hourly volumes. A K factor, defined as the ratio of design hourly volume to average weekday volume, is used to obtain design hourly volumes. These are not peak hourly volume, but, for economic reasons, are approximately the thirtieth highest hour of the year. For urban facilities this factor may range from 7 to 18 percent, and average about 10 percent. A D factor, defined as the ratio of design hourly directional volume to total facility volume, is then applied to obtain directional design hourly volumes. For urban facilities this factor may provide a directional split ranging from 60–40 percent to 50–50 percent in the peak direction.

Objectives and Standards

A set of transportation system objectives and supporting standards must be prepared as a basis for the system design. An example of such a set for transit as well as arterial street and highway system planning is provided in Table 15.2. The set is provided solely as an example and should not be construed as a recommended set.

TABLE 15.2
Example Transportation System Development Objectives, Principles, and Standards

OBJECTIVE NO. 1

An integrated transportation system that will effectively serve the existing land use pattern and promote the implementation of the land use plan, meeting the anticipated travel demand generated by the existing and proposed land uses.

PRINCIPLE

The urban transportation system serves to freely interconnect the various land use activities within the urban area, thereby providing the attribute of accessibility essential to the support of these activities. Through its effect on accessibility, the transportation system can be used to induce development in desired locations and to separate incompatible land uses.

STANDARDS

1. The relative accessibility provided by the transportation system should be adjusted to the land use plan, and areas in which development is to be induced should have a higher relative accessibility than areas that should be protected from development.
2. Arterial street and highway facilities should be located and designed so as to provide adequate capacity, that is, a volume-to-capacity ratio equal to, or less than, 1.0 based on 24-hour average weekday traffic volumes, to meet the existing and potential travel demand between the various land uses consistent with the trip-generating and trip-interaction characteristics of these uses and the resulting forecast of travel.

(Continued)

TABLE 15.2 *(Continued)*
Example Transportation System Development Objectives, Principles, and Standards

OBJECTIVE NO. 2

A balanced transportation system providing the appropriate types of transportation service needed by the various urban subareas at an adequate level of service.

PRINCIPLE

A balanced transportation system consisting of highway and transit transportation and terminal facilities is necessary to provide an adequate level of transportation service to all segments of the population, to properly support essential economic and social activities, and to achieve economy and efficiency in the provision of transportation service. The transit component provides transportation service to that segment of the population that does not for various reasons own and operate an automobile. Furthermore, transit provides added transportation system capacity to alleviate peak loadings on arterial street and highway facilities and assists in reducing the land use demand for parking facilities in the central business district.

STANDARDS

1. Transit service of an appropriate type should be provided for all routes wherein the minimum potential average weekday revenue passenger loading equals or exceeds the following values: 600 per day per bus for local service; 300 per 4 hours per bus for express transit service; and 21,000 passengers per day per preempted freeway lane for bus rapid transit service.
2. Local transit routes should be provided at intervals of no more than one mile in all residential areas.
3. Maximum operating headways for all transit service throughout the daylight hours should not exceed one hour.
4. The average distance between transit stops should not be less than one quarter mile for local bus transit services, no stops between terminal areas for express bus transit service; two miles for rapid bus transit service.
5. Arterial streets and highways should be provided at intervals of no more than one mile in each direction in residential areas.

OBJECTIVE NO. 3

The alleviation of traffic congestion and the reduction of travel time between component parts of the planning area.

PRINCIPLE

To support the everyday activities of business, shopping, and social intercourse, a transportation system that provides for reasonably fast and convenient travel is essential. Furthermore, congestion increases the cost of transportation, including the cost of the journey to work, which is necessarily reflected in higher production costs and thereby adversely affects the relative market advantages of businesses and industries within the planning area.

(Continued)

TABLE 15.2 *(Continued)*
Example Transportation System Development Objectives, Principles, and Standards

STANDARDS

1. The total vehicle-hours of travel within the planning area should be minimized.
2. Adequate capacity and a sufficiently high level of geometric design should be provided to achieve the following overall speeds based on potential 24-hour average weekday traffic volumes for arterial street and highway facilities:

	Overall Operating Speed in Miles per Hour for Various Type Areas		
Type of Facility	**Central**	**Intermediate**	**Outlying**
A. Arterials			
1. Freeway	35–55	40–55	55–65
2. Expressway	25–40	30–45	40–50
3. Standard Arterials			
a. Divided	15–25	25–35	35–45
b. Undivided	15–25	20–35	25–40
B. Collectors	10–20	15–30	20–35
C. Land Access	5–15	10–20	15–25

3. The proportion of total travel on freeway, expressway, and rapid and modified rapid transit facilities should be maximized.

OBJECTIVE NO. 4

The reduction of accident exposure and the provision of increased travel safety.

PRINCIPLE

Accidents take a heavy toll in life, property damage, and human suffering, contribute substantially to overall transportation costs, and increase public costs for police and welfare services; therefore, every attempt should be made to reduce both the incidence and severity of accidents.

1. Traffic congestion and vehicle conflicts should be reduced by maintaining a volume-to-capacity ratio equal to or less than 0.9, based on 24-hour average weekday traffic volumes.
2. Travel on facilities that exhibit the lowest accident exposure, that is, freeways, expressways, and all forms of transit, should be maximized.

OBJECTIVE NO. 5

A transportation system that is both economical and efficient, meeting all other objectives at the lowest cost possible.

PRINCIPLE

The total resources of the planning area are limited, and any undue investment in transportation facilities and services must occur at the expense of other public and private investment; therefore, total transportation costs should be minimized for the desired level of service.

(Continued)

TABLE 15.2 *(Continued)*
Example Transportation System Development Objectives, Principles, and Standards

STANDARDS

1. The sum of transportation system operating and capital investment costs should be minimized.
2. The total vehicle miles of travel should be minimized by reducing trip length, or the total number of trips made, or both.
3. Full use should be made of all existing and committed major transportation facilities, and such facilities should be supplemented only with such additional major facilities as necessary to serve the anticipated travel demand derived from the land use plan at the desired level of service.

OBJECTIVE NO. 6
The minimization of disruption of desirable existing neighborhood and community development and of the deterioration or destruction of the natural resource base.

PRINCIPLE
The social and economic costs attendant to the disruption and dislocation of homes, businesses, industries, and communication and utility facilities, as well as adverse effects on the natural resource base, can be minimized through proper location of transportation facilities.

STANDARDS

1. The penetration of neighborhood units and of neighborhood facility service areas by arterial streets and highways and rapid transit routes should be avoided.
2. The dislocation of families, businesses, and industries should be minimized.
3. The proper use of land for, and adjacent to, transportation facilities should be maximized and disruption of future development minimized through advance reservation of right-of-way for arterial street and highway facilities.

Arterial Street and Highway System Design

The transportation system design process is one of finding successive approximations to the best design solution by proposing specific facility improvements in each iteration, testing and evaluating these through application of the simulation models, analyzing the results, and repeating the process as necessary. Various sources of potential design solutions will exist. Capacity improvement proposals may be originated by experienced professional engineers in the employ of federal, state, county, and local units and agencies of government who have very intimate knowledge of, and long-standing experience with, arterial street and highway traffic conditions and facility systems in the planning area. Another source of potential design solutions may be developed directly from analyses of the traffic assignments, wherein solutions to correct system deficiencies become apparent through the knowledge provided of existing and probable future traffic patterns in the planning area and the manner in which these patterns are distributed on the existing systems. A third source of potential design solutions may be developed indirectly from the land use planning process, where changes in arterial or transit service based on land use development objectives may be advanced. In addition, the increased use of dynamic simulation platforms, scenario planning tools, and integrated land use–transportation models enables more agile and responsive design development. Maximum use should be made of the existing system, and committed facility improvements recognized.

To assist in the design process, the characteristics of the trips causing overloaded links may be analyzed with the help of screenline and selected-link analyses. It is thus possible to identify specific inter-area trips and analyze the feasibility of rerouting some of these trips over other portions of the system. This permits identification of circuitous travel paths and facilities requiring additional capacity to relieve overloadings on more direct routings. With modern origin-destination data and real-time traffic analytics, these evaluations can now be enhanced by insights into actual travel behavior and temporal variations. The design process is intended to produce an arterial street and highway system plan and a complementary transit system plan that provides recommended functional types and capacities for each link in the arterial and transit networks. The resulting final plan is then presented graphically and described in text and tables for adoption. The areawide plan recommendations should provide the framework for the arterial street and highway element of the local municipal comprehensive plan. In the case of smaller communities, the areawide plan recommendations can be directly incorporated into the local plans. In other cases, mutual adjustments may have to be made in both the local and areawide planning. In all cases, the process should be informed by stakeholder input, environmental justice considerations, climate resilience, and multimodal accessibility objectives.

JURISDICTION CLASSIFICATION OF STREETS

The jurisdictional classification of the various segments of the arterial system of a planning area establishes the level of government that should have the responsibility for the design, construction, operation, and maintenance of each such segment. Jurisdictional classification groups arterial facilities into subsystems that should be under the jurisdiction of a given level of government, and is important to plan preparation and plan implementation through the design, construction, operation, and maintenance of the various segments of the total street and highway system. The arterial subsystems should function as integral parts of the total system and be continuous within themselves or in conjunction with other "higher" level subsystems. The jurisdictional classification of the various arterial subsystems should vary with respect to the types of land use areas served, the types of trips served, and the degree of traffic mobility provided. The design, construction, operation, and maintenance of collector and land access streets are clearly the responsibility of the local municipal level of government, and jurisdictional classification is not intended to be applied to segments, or subsystems, of the collector or land access networks.

In most of the United States, three levels of government—state, county, and local municipal—have direct jurisdictional responsibility for the planning, design, construction, operation, and maintenance of arterial street and highway facilities. These levels of government provide the basis for the jurisdictional classification of the total existing and proposed arterial street and highway system into Type 1, state trunk; Type 2, county trunk; and Type 3, local trunk. The Type 3 facilities are intended to be urban facilities, and are usually located entirely within the corporate limits of cities or villages and within adjacent areas of planned urban expansion. The Type 1 and Type 2 facilities may be rural or urban.

Type 1 arterials should include all those routes within a planning area that are intended to provide the highest level of traffic mobility, that is, the highest speeds and lowest degree of traffic congestion, the minimum degree of land access service, and that must have regional or inter-regional system continuity. Ideally, these Type 1 arterials, because of their function and statewide and nationwide importance, should comprise the state trunk highway system and its underlying federal aid routes and designations.

Type 2 arterials should include all those routes within the planning area that are intended to provide an intermediate level of traffic mobility and an intermediate level of land access service, and that must have intercommunity system continuity. Ideally, these Type 2 arterials, because of their function and subregional importance, should comprise the county trunk highway system of an area and its underlying federal aid routes and designations.

Type 3 arterials should include all those routes within a planning area that are intended to provide the lowest level of arterial traffic mobility and the highest degree of arterial land access service, and that must possess intra-community system continuity. These Type 3 arterials are intended to comprise the local arterial system.

Table 15.3 provides an example of a set of criteria used to prepare a jurisdictional highway system plan for an urbanizing county. It is important to note that the criteria are specific to the county concerned and should not be regarded as models for use elsewhere. The criteria were developed collegially by a committee of experienced highway, traffic, and municipal engineers and land use planners, much in the same way objectives and standards for any planning effort might be developed. The criteria relating to trip service and traffic volume were developed with the aid of analyses such as those summarized in Figure 15.5. Completion of the curves shown in this figure required data provided by facility, travel, and traffic inventories and by simulation modeling. A part of the attendant jurisdictional classification plan for the county concerned is shown in Figure 15.6.

New developments in this area include the integration of smart transportation infrastructure responsibilities across levels of government and the consideration of automated vehicle corridors and charging infrastructure. Jurisdictional classification may increasingly need to reflect not just physical maintenance, but also digital systems such as traffic management platforms, fiber optic networks, and connected vehicle data coordination.

TABLE 15.3
Criteria for Jurisdictional Classification of Arterial Streets and Highways—County A

	Criteria	Arterial Type		
		I (State Trunk)	II (County Trunk)	III (Local Trunk)
Trip Service	Average Trip Length (Miles)	Urban 11.0 or more Rural 41.0 or more	Urban 8.0 to 10.9 Rural Less than 41.0	Urban Less than 8.0
Land Use Service	Transportation Terminals	Urban[a] and Rural[b] Connect and serve interregional rail, bus, and major truck terminals and air-carrier airports	Urban[a] and Rural[b] Connect and serve freeway interchanges, general-aviation airports, pipeline terminals, major intraregional truck terminals, and rapid transit and modified rapid transit system loading and uploading points not served by Type I arterials	Urban[a] Connect and serve truck terminals generating 250 or more truck trips per average weekday and off-street parking facilities having a minimum of 150 parking spaces not served by Type I and II arterials
	Recreational Facilities	Urban and Rural Connect and serve all state parks having a gross area of 500 acres or more	Urban and Rural Connect and serve regional parks and special recreational use areas of countywide significance	Urban Connect and serve community parks not served by Type I and II arterials
	Commercial Centers	Urban and Rural Connect and serve major retail and service centers	Urban and Rural Connect and serve community retail and service centers not served by Type I arterials	Urban Connect and serve neighborhood retail and service commercial centers not served by Type I and II arterials
	Industrial Centers	Urban and Rural Connect and serve major regional industrial centers	Urban and Rural Connect and serve major community industrial centers not served by Type I arterials	Urban Connect and serve minor community industrial centers not served by Type I and II arterials

(Continued)

TABLE 15.3 *(Continued)*

Criteria for Jurisdictional Classification of Arterial Streets and Highways—County A

Criteria		Arterial Type: I (State Trunk)	Arterial Type: II (County Trunk)	Arterial Type: III (Local Trunk)
Land Use Service	Institutional	Urban and Rural Connect and serve universities, county seats, and state institutions	Urban and Rural Connect and serve county institutions; accredited, degree-granting colleges; public vocational schools; and community hospitals not served by Type I arterials	Urban Connect and serve city and village halls and high schools not served by Type I and II arterials
	Urban Areas	Rural Connect and serve urban areas of 2,500 or more population	Rural Connect and serve developed areas of 500 or more population	
Operational Characteristics	System Continuity	Urban and Rural Interregional or regional continuity comprising total systems at the regional and state level	Urban and Rural Intermunicipality and intercounty continuity comprising integrated systems at county level	Urban Intracommunity continuity comprising an integrated system at the city or village level
	Spacing	Urban and Rural Minimum 2 miles	Urban and Rural Minimum 1 mile	Urban Minimum 0.5 mile
	Volume	Urban Minimum 8,000 vehicles per average weekday Rural Minimum 4,000 vehicles per average weekday	Urban 4,000 to 8,000 vehicles per average weekday Rural Fewer than 4,000 vehicles per average weekday	Urban Fewer than 4,000 vehicles per average weekday

Operational Characteristics	Traffic Mobility	Urban Average overall travel speed[c] 30 to 70 miles per hour Rural Average overall travel speed 40 to 70 miles per hour	Urban Average overall travel speed[c] 25 to 50 miles per hour Rural Average overall travel speed 30 to 60 miles per hour	Urban Average overall travel speed[c] 20 to 40 miles per hour
	Land Access Control	Full or partial control of access	Partial control of access	Minimum control of access

Source: SEWRPC.

[a] Urban arterial facilities are considered to "connect and serve" given land uses when direct access from the facility to roads serving the lanc use area is available within the following maximum over-the-road distances from the main vehicular entrance to the land use to be served: Type I arterial facility, 1 mile; Type II arterial facility, 0.5 mile, Type III arterial facility, 0.25 mile.

[b] Rural arterial facilities are considered to "connect and serve" given land uses when direct access from the facility to roads serving the lanc use area is available within the following maximum over-the-road distances from the main vehicular entrance to the land use to be served: Type I arterial facility, 2 miles Type II arterial facility, 1 mile.

[c] Average overall travel speed is defined as the sum of the distances traveled by all vehicles using a given section of highway during an weekday divided by the sum of the actual travel times, including traffic delays.

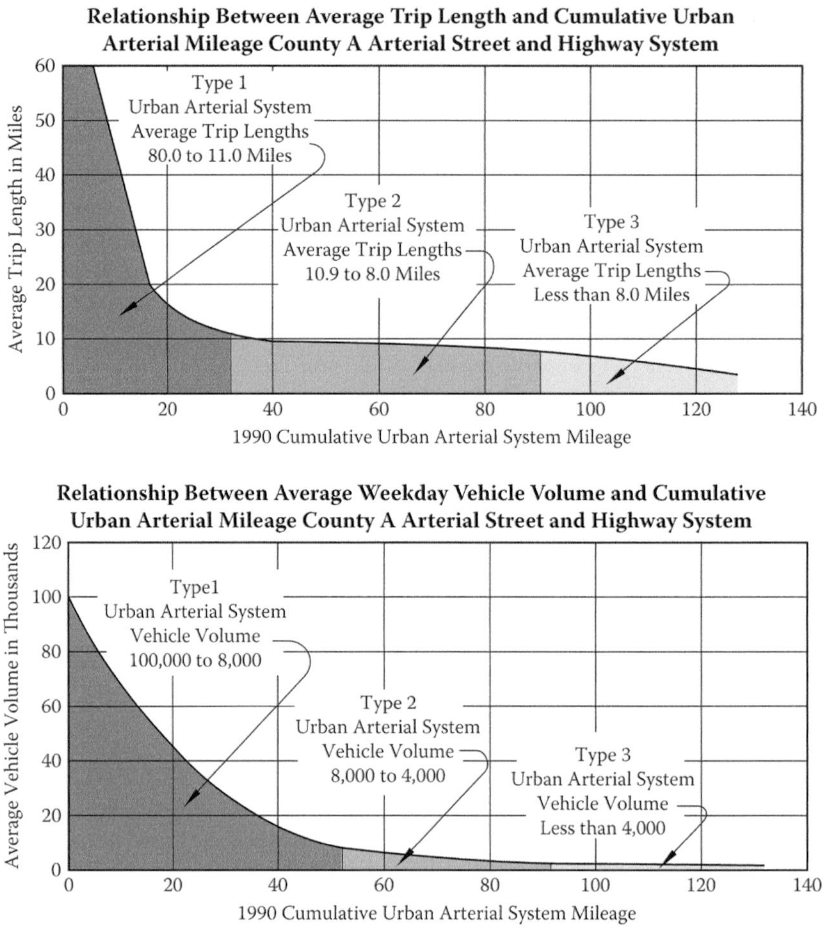

FIGURE 15.5 Illustration of the type of analyses that may be made to identify certain criteria for the jurisdictional classification of arterial facilities. The upper curve provides the basis for the trip length related criteria; the lower for the traffic volume related criteria given in Table 15.3.

Source: SEWRPC.

MASS TRANSIT PLANNING

Mass transit may be defined as the transportation of relatively large groups of people by publicly or quasi-publicly owned and operated vehicles. Mass transit service may be classified as interregional, rural, or urban.

Intercity or interregional mass transit provides service across regional boundaries. It includes commercial air carrier, interregional railway passenger train, and interregional bus service. Interregional mass transit service is usually considered in city transportation planning only to the extent that the terminals—such as airports, railway stations, and bus stations—comprise major trip generators and important land uses.

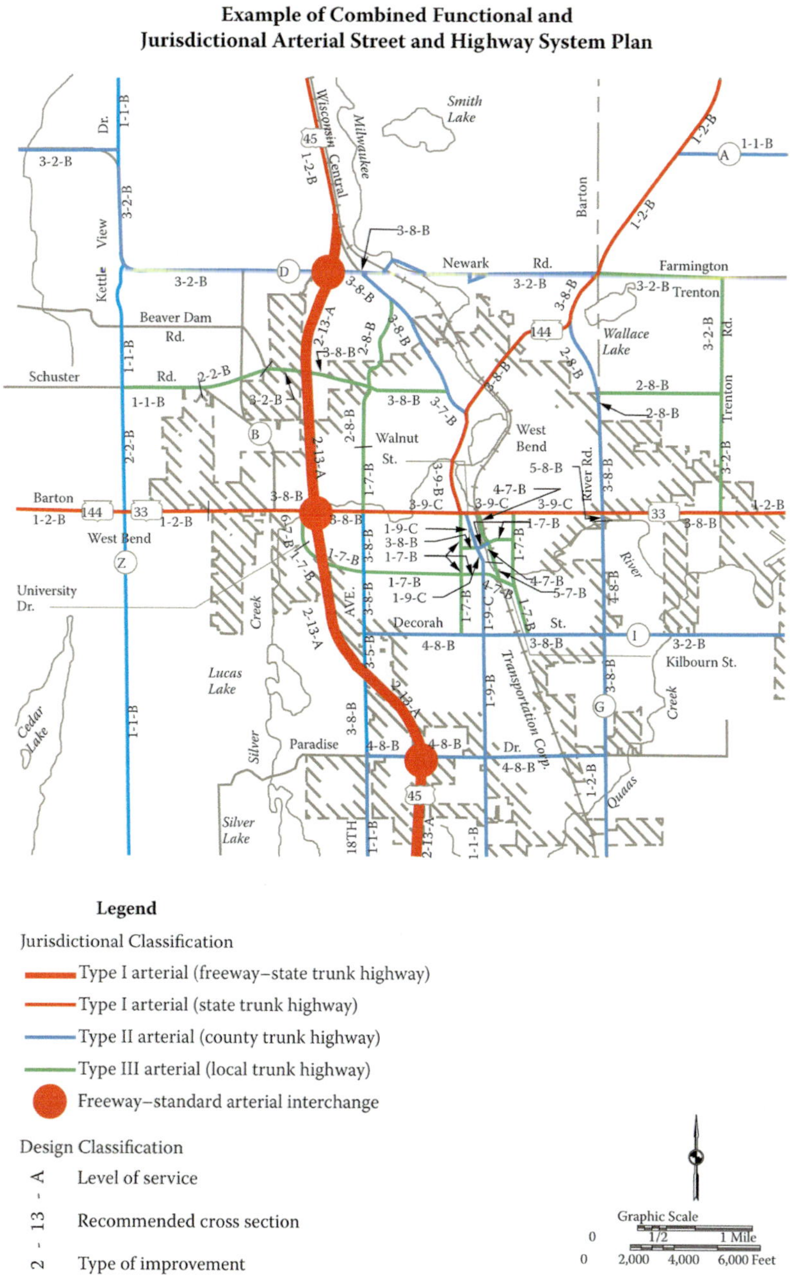

FIGURE 15.6 Illustration of a combined functional and jurisdictional arterial street and highway system plan. The plan recommends required design year capacities for each arterial segment, together with the type of improvement required—resurfacing or reconstruction and the recommended jurisdiction.

Source: SEWRPC.

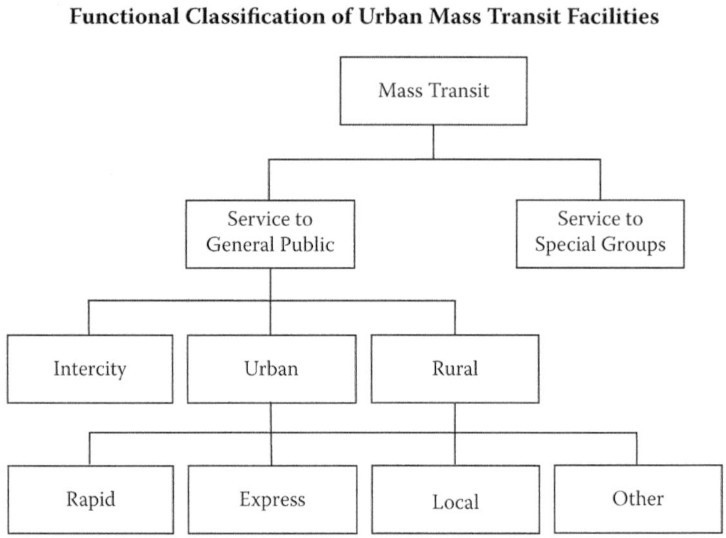

FIGURE 15.7 Illustration of the functional classification of mass transit facilities, a classification conceptually comparable for planning purposes to the functional classification of street and highway systems. The "Service to special groups" category includes yellow school bus and paratransit van service. The "Other" category includes taxi service.

Trips made on interregional mass transit facilities usually constitute less than five percent of all mass transit travel made in a planning area on an average weekday.

Rural mass transit provides service in and between rural communities and areas, and between rural areas and urban areas. Rural mass transit service is typically provided by demand-responsive van service. Federal and state programs increasingly support rural transit through microtransit, mobility-as-a-service (MaaS), and app-based demand-response systems. Rural mass transit service is usually considered only indirectly in urban transportation planning.

Urban mass transit service is essential to meet the travel needs of persons unable to use personal transportation and to provide an alternative mode of travel to the automobile in urban areas, and particularly in heavily traveled corridors. Urban mass transit service can be functionally classified in a manner similar to the functional classification of street and highway facilities. Such a classification system is illustrated in Figure 15.7. It divides urban mass transit service into rapid transit, express transit, and local transit service. An "other" category, consisting primarily of taxi service, is also provided.

Functional Classification

Rapid transit is intended to provide relatively fast transportation along heavily traveled corridors and between major activity centers and high-density residential areas. It is marked by high operating speeds, wide station spacing of one to two miles, and operation on regular schedules over exclusive, fully grade-separated rights-of-way or over exclusive freeway lanes. It provides the highest level of transit service. Express transit is intended to provide an intermediate level of service. It operates on regular schedules over reserved lanes of surface arterials or in mixed traffic over freeways and expressways. Station spacing may range from one-half to one mile with stops at intersecting transit routes. It provides a greater degree of accessibility but lower operating speeds than rapid transit. Local transit is intended to provide a high degree of accessibility at the price of low operating speeds. It operates on regular schedules in mixed traffic over prescribed surface arterials. Stops may be spaced one quarter to one-half mile.

Modes

The three types of service can be provided by four modes or vehicle types: motor bus, light rail, heavy rail, and commuter rail. The commuter rail mode consists of locomotive-hauled trains of two to eight coaches. Modern systems may be electrified via overhead catenary or third rail or use dual-mode locomotives for improved flexibility. Boarding may be from either high-level or low-level platforms at stations, and fare collection is on board. Operation is over mainline railway trackage. Commuter rail is intended to serve trips within metropolitan areas between outlying suburbs and major urban centers. Station spacing will average two to five miles, and access generally is by automobile and feeder bus.

Heavy rail service is provided by multiple-unit, electrically powered subway or elevated railway cars operated in trains of four to ten cars. Power is provided by a third rail. Operation must be over fully grade-separated right-of-way. Boarding and deboarding is from station platforms at floor level of cars. Fare collection is off-board at relatively elaborate stations. Heavy rail is intended to serve trips made within densely developed urbanized areas. Stations are spaced from one-half to two miles apart. Access is by walking, feeder bus, or automobile.

Light rail service is provided by usually articulated, electrically powered, light rail vehicles—modern successors to the streetcar. Vehicles can be operated as multiple-unit trains of from one to three cars. Power is provided by overhead wire. Operation may be in mixed traffic on major streets, over reserved lanes with signal prioritization on such streets, or over an exclusive partial or fully grade-separated rights-of-way. Boarding and deboarding may be at curb level or from high-level platforms at stations. Fare collection is usually on board. Light rail is intended to serve trips within densely developed urbanized areas. Stations and stops are spaced one-quarter to one mile apart. Access is primarily by walking but also by feeder bus and automobile.

Bus service is usually provided by diesel, hybrid, or battery-electric motor coaches operating in mixed traffic on major streets, over reserved lanes with signal prioritization on such streets, or over exclusive partially or fully grade-separated busways. Battery-electric and hydrogen fuel cell buses are increasingly replacing traditional diesel and trolley buses to reduce emissions. Boarding and deboarding is usually at curb level. Fare collection is usually on board. Buses are intended to serve trips throughout an urban area. For local service, stops may be spaced one-quarter to one-half mile apart. Buses are highly flexible as transit vehicles both with respect to routing and suitability for different types of service.

Buses can be used to provide true rapid transit service over exclusive partially or fully grade-separated rights-of-way. When operated in such service, specially designed, often articulated buses are used that have multiple doors for quick boarding and deboarding from station platforms at the floor level of the buses. Stations may be elaborate and spaced one-half to one mile apart. Fare collection is off-vehicle at stations. The buses may be equipped with lateral magnetic guidance systems using electric cables buried in the pavements, which facilitate close docking at station platforms.

Some pertinent characteristics of urban transit modes are provided in Table 15.4. The data are intended for comparative purposes only and should be regarded with caution and reservation. The capacity and cost data particularly represent crude approximations and are intended solely to facilitate comparison of the relative potential utility and performance of the modes represented. Costs will vary widely with site-specific applications, and accurate estimates require detailed engineering study. Comparative data for arterial street and highway facilities are instructive.

Objectives and Standards

As already noted with respect to arterial street and highway system planning, a set of objectives and standards must be prepared as a basis for system design. An example of such a set for both arterial street and highway system planning and transit system planning is provided in Table 15.2.

Transit System Design

The transit system design process is similar to the arterial street and highway design process, particularly with respect to the use of the travel and traffic simulation models. Existing transit facility capacity, however, is not apt to be as meaningful a factor in transit system utilization as it is in arterial street and highway system utilization, and no parallel deficiency analysis may be possible—the potential capacity of the existing systems often far exceeding the demand. Moreover, unlike arterial street and highway service, transit service may not be ubiquitous within the planning area so that certain trips cannot be made solely by transit, making the potential demand for transit service in areas not presently served difficult to assess.

Sources of design solutions may include transit service changes advanced by transit operators serving the planning area. A second source may consist of transit service proposals postulated on the basis of analyses of the socio-economic and existing travel characteristics in the planning area. Such proposals may be focused on major corridors of transportation movement that analyses indicate may have a high potential for transit use. A third source of design solutions may be developed directly from analyses of the traffic assignments that may indicate the need for transit service as an alternative mode in major corridors of transportation movement. In every case, the transit system improvements proposed must be tested by simulation modeling to determine whether the potential passenger traffic demand will justify incorporation into a final transportation system plan.

TABLE 15.4
Characteristics of Urban Transit Modes

Mode	Units per Lane or Train per Track per Hour	Train Length	Headways	Average Speed (Miles per Hour)	Maximum Speed (Miles per Hour)	Passengers per Lane per Hour	Route Length (Miles)	Station Spacing Miles	Capital Costs Millions of Dollars per Mile
Bus on Arterial Street	80–200	—	15–45 seconds	5–12	25–30	4,000–10,000	Indefinite	1/4 to 1/2	$3.1 new lane $0.4 resurfaced lane
Bus on Busway	200–700	—	18–60 seconds	30	70	10,000–32,000	5–20	1/2 to 1	$8.0 new lane $0.4 resurfaced lane
Light Rail	24–92	1–3 cars	40 seconds to 3 minutes	5 to 12 on streets 30 on right-of-way	50	4,000–18,000	5–20	1/2 to 1	$20.0 to $45.0 new – 2 track
Heavy Rail	6–32	4–10 cars	2–10 minutes	18–47	70	10,000–60,000	5–20	1–2	$36.0 to $68.0 new – 2 track
Commuter Rail	3	2–8 coaches	20 minutes	35	79	3,000–5,000	20–50	2–5	$5.0 to $10.0 adapt existing

The design process is intended to produce a transit system plan that complements the arterial street and highway system plan and provides recommendations concerning the configuration of the needed transit facilities by routing and type and level of service to be provided by each route. These recommendations should provide the framework for the transit element of the local municipal comprehensive plan. Of particular interest to the city planner should be type and routing of the proposed transit facilities and, for express and rapid transit facilities, the type and location of stations.

NEW FEATURES OF TRANSPORTATION PLANNING

Recent advances in transportation planning reflect a growing integration of technology, sustainability, and data analytics. Modern planning frameworks now incorporate real-time data from GPS, mobile apps, and connected vehicle systems, enabling dynamic traffic management and more accurate demand forecasting. EV infrastructure planning, micromobility integration (e.g., scooters and bike-share), and mobility-as-a-service (MaaS) platforms are increasingly prioritized to reduce emissions and expand modal options. Artificial intelligence and machine learning tools are being used to enhance simulation models and optimize route planning. Additionally, transportation equity and environmental justice have become core elements, ensuring that system benefits are distributed fairly across all communities. These features collectively signal a transition from static, infrastructure-heavy approaches toward adaptive, data-driven, and human-centered planning strategies.

CONCLUDING COMMENTS

A need has always existed to move people and goods between the various land uses of an urban area, and to deliver certain services for those uses. An inefficient transportation system will decrease personal mobility, will be costly and time-consuming, will make the urban economy less efficient, and will make the overall quality of life in an area less attractive. Problems currently afflicting urban transportation systems include travel and traffic peaking, facility congestion, fragmented and sprawling land use patterns driven by automobile-centric development, lack of modal choice, and constraints placed on lifestyles. The arterial street and highway and the complementary mass transit systems and changes in these systems have a major impact upon the urban environment.

The city planner has a particular interest in certain aspects of these transportation systems. As a major determinant of relative accessibility among areas within the community, the arterial street and highway system determines to a large extent the pattern of land use and of population and employment distribution. Any measure of the desirability of a specific arterial street improvement must, therefore, consider the changes in land use and in population and employment distribution it may bring about as against those that might result from an alternate improvement or none at all. Arterial streets and highways act as natural barriers to divide the urban land use pattern into cellular units. This divisive effect may be used to advantage, as in separating incompatible land uses, or may be a disadvantage as in developing a residential neighborhood into isolated parts. The arterial street and highway system affects the quality and use of mass transit facilities and services and the need for off-street parking facilities. In the larger metropolitan areas, specific system improvements may have far-reaching repercussions in this respect. The location and configuration of the arterial street and highway system affect property values and the efficiency of utility and community services, so that a complex and often subtle relationship exists between the type and spacing of arterials and the overall cost of the urban plant as well as the amenity associated therewith. Since the arterial system forms the basic transportation facility for an urban area, the efficiency of the system in carrying traffic and the cost thereof must always be a basic concern of the city planner. The arterial street and highway system of a city must be adequate to meet not only existing but also probable future traffic needs; must be in harmony with, and an integral part of, the comprehensive plan for city growth and development; and must be compatible with broader metropolitan, regional, state, and national transportation needs.

The highway engineer also has a particular interest in the planning of urban transportation systems. Urban arterials may form important segments of the county, state, and national transportation systems and are essential to the proper functioning of those systems. It is significant to note, in this respect, that urban bypass effectiveness is highly dependent on city size, design, and regional context, but the majority of commuter and freight traffic still targets urban centers directly. The efficiency and effectiveness of the state and national highway system may often be determined by the efficiency and effectiveness of the urban arterial street and highway systems that make up integral parts of the state and national systems.

FURTHER READING

Appleyard, Donald. *Livable Streets*. University of California Press, 1981.

Creighton, Roger L. *Urban Transportation Planning*. University of Illinois Press, 1976.

Hutchinson, B. G. *Principles of Urban Transportation System Planning*. McGraw-Hill, 1974.

Mohan, Dinesh and Tiwari, Geetam. *Sustainable Approaches to Urban Transport*. CRC Press, 2020.

Vuchic, V. R. *Urban Public Transportation Systems and Technology*. Prentice-Hall, 1981.

16 Other Plan Elements

I have yet to see any problem, however complicated, which, when you looked at it in the right way, did not become more complicated.

Paul Anderson
New Scientist
September 25, 1969

INTRODUCTION

Chapters 12 and 15 provided what are, quite simply, introductions to land use and transportation planning and to the preparation of the land use and transportation elements of a comprehensive city plan. These are only two—albeit two of the most important—of the various elements of a comprehensive city plan. The preparation of a number—although not all—of the other elements is as complex and costly a task as the preparation of the land use or transportation elements. For that reason, as well as for the sheer number of elements involved, even introductory descriptions of these other elements lie beyond the scope of this text. Some remarks that set these other plan elements within the context of this text may, however, be in order.

COMMON BASE

By their very nature, the elements of a comprehensive plan, as listed in Chapter 10, vary in their complexity and in terms of the technical skills required for their proper preparation. All of the elements, however, share a common foundation in the basic inventories conducted as a part of any good city planning effort: the maps, the demographic and economic studies, projections and forecasts, the land use and supporting infrastructure inventories, and the natural resource base inventories. These data are now typically organized in geospatial databases or parcel-based land information systems, and their availability greatly expedites the preparation of any and all of the various elements of a comprehensive plan.

Each element of a comprehensive plan is intended to address the development of a specific part of the total infrastructure—or physical plant—that provides the basis for urban life: the arterial street and highway, mass transit, sanitary sewerage, water supply, stormwater drainage, park, and other community facilities. Each of the individual plan elements is intended to deal with an identified developmental or environmental problem or need. It is important to understand that the individual plan elements are coordinated by being related to a common land use plan. Thus, the land use plan is the most basic element of any comprehensive plan, an element on which all other elements are based. No facility plan can be properly prepared in

DOI: 10.1201/9781003500384-16

the absence of a land use plan. It is the land use pattern that determines the loadings in the arterial street and highway, mass transit, sanitary sewerage, water supply, and stormwater drainage facilities and determines the capacities and configurations of these systems. The land use pattern also determines the need for and the size and location of parks and open spaces, and for the type and location of facilities such as fire stations, schools, and libraries.

If all the elements of a comprehensive plan are prepared at one time and presented in a single plan report, the relationship of each element to the land use element becomes apparent. If, however, the various elements are prepared individually over time, as is most often the case, the important relationship of the element concerned to the land use element must be made clear. Therefore, planning reports presenting each individual plan element may be expected to contain a description of the land use plan on which the individual element is, in the final analysis, based.

CHARACTERISTICS OF INDIVIDUAL ELEMENTS

While linked to a common land use plan, and while utilizing common map, demographic, economic, and natural resource base data in its preparation, each plan element will usually require the conduct of some special inventories and the application of specialized analytical and design techniques. All will require the formulation of an appropriate set of objectives, principles, and standards.

Relatively simple graphic analyses and design technologies may suffice for the preparation of some plan elements, such as for the location of fire stations or elementary schools. The proper preparation of other plan elements—for example, sanitary sewerage, water supply, and stormwater drainage—may require the formulation and application of simulation models for plan design, test, and evaluation. Model formulation, calibration, and validation may in turn require the conduct of detailed inventories. For example, the formulation, calibration, and validation of an aquifer performance simulation model for use in planning the development of a water supply system that depends upon groundwater as the source of supply may require the conduct of extensive and costly hydrogeologic studies. The potential complexity involved in the preparation of some plan elements is illustrated by Figure 16.1.

The preparation of some plan elements may also require the development of a special report on the state of the art of the technology involved. The preparation of a sanitary sewerage system element of a comprehensive plan for a large community or urbanized area might require such a report. The scope of the report might include such topics as gravity and pressure flow sanitary sewer collection and transmission, including methods of hydraulic analyses; collection, transmission, and treatment capacity determination; flow and load reduction; treatment processes such as nutrient removal technologies for nitrogen and phosphorus in treatment plant effluents; land application of treated effluent; effluent polishing and receiving water treatment; treatment facility schematics; and, importantly, unit capital and operating cost data. The information presented in such a report, and particularly the unit-cost data, should constitute direct inputs to the formulation of objectives and standards, alternative plan preparation and evaluation, and final plan selection. In this respect, it should be noted that the preparation of such state-of-the-art reports need not be limited

Example Work Flow Diagram for a Sanitary Sewerage System Planning Program

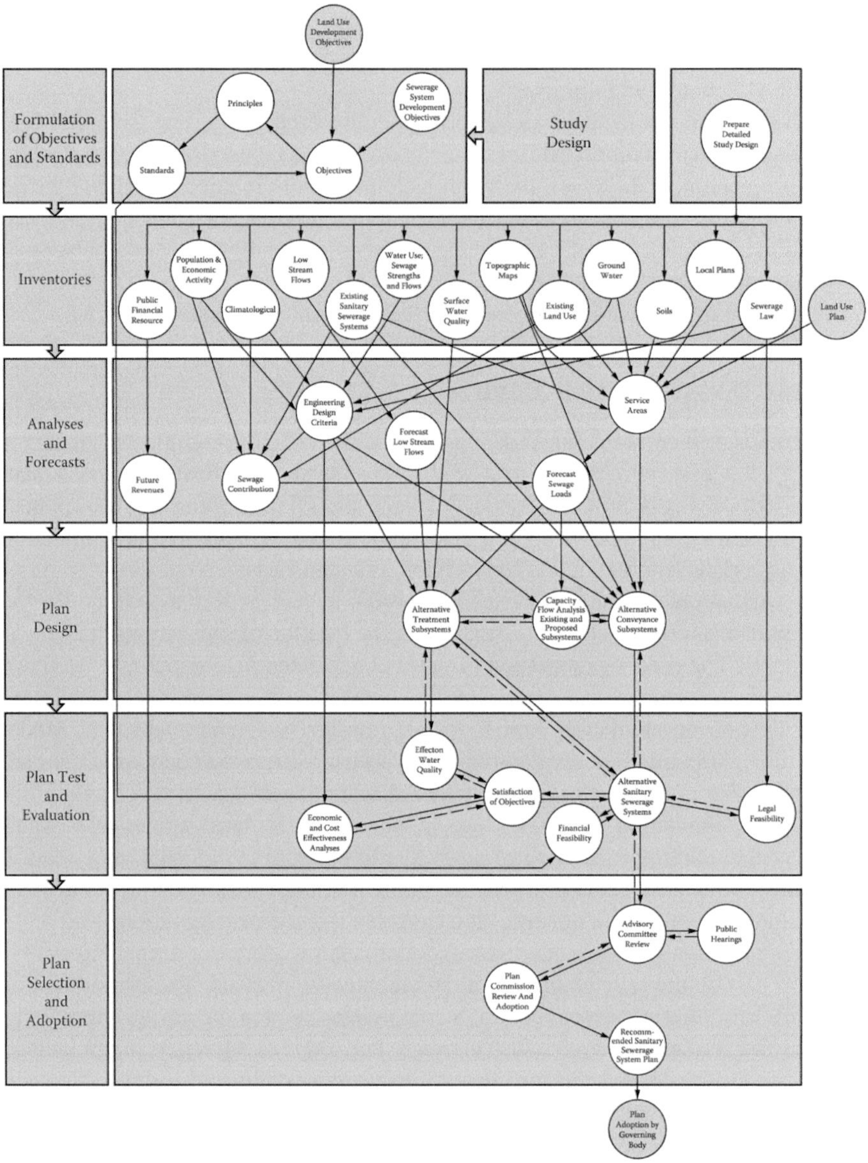

FIGURE 16.1 Illustration of the complexity and inter-relationships of the tasks involved in the preparation of a sanitary sewerage system plan. The preparation of water supply and stormwater management plans represents similarly complex planning efforts.

Source: SEWRPC.

to application in the more "science"-oriented planning efforts, but may extend to application in the more "art"-oriented planning efforts. For example, a state-of-the-art report might be developed for use in the preparation of the land use element of a comprehensive plan, addressing such issues as the application of form-based or context-sensitive concepts in the planning effort.

NEED FOR STUDY DESIGN

The initiation of any program to prepare a comprehensive plan, or an element of such a plan, should be preceded by the preparation of a study design. The study design should establish the need for the planning program; specify the scope and content of the work required to prepare the proposed plan or plan element in a technically sound manner; recommend the most effective means for establishing, organizing, and accomplishing the required work; recommend a practical time sequence and schedule for the work; and provide sufficient cost data to permit preparation and approval of a budget for the program.

The ultimate responsibility for accomplishing the planning work should rest with the plan commission and planning director. The necessary staff work may be accomplished entirely with planning department staff; by a staff team including assignment of staff to the team by other departments such as public works, engineering, or parks, as may be appropriate; or by a planning or engineering consulting firm engaged for this purpose. Increasingly, partnerships with academic institutions, community-based organizations, or technical assistance programs may also support this work. An interdepartmental advisory committee may be established to guide the work and recommend a final plan to the plan commission for consideration and adoption.

17 Plan Implementation—Land Subdivision Control

We have learned that planning without implementation is futile! We know too well that implementation without planning is chaos.

Dennis O'Harrow
Plan Talk and Plain Talk, 1981

INTRODUCTION

Three important plan implementation powers are generally available to municipalities in the United States: taxation and appropriation, eminent domain, and the police power. The first of these—taxation and appropriation—consists of the power to levy a tax usually and primarily on real property, including special assessments, and to expend the monies raised for public purposes. The second—eminent domain—consists of the power to take private property for a public purpose upon payment of fair market price. The power is exercised through the legal process of condemnation. The third plan implementation power available to municipalities is the police power. This is the power to make and enforce regulations to promote the public health, safety, and general welfare without payment of compensation to private property owners who may be affected by the regulations. The scope of this power has evolved over time, particularly through state and federal court interpretations, to include emerging concerns such as environmental protection and climate resilience. This chapter deals with land subdivision control. Such control is exercised through ordinances adopted under the police power.

HISTORICAL BACKGROUND

The use of land division plats for the purpose of describing real property ownership parcels, for facilitating the transfer of title to such parcels, and for the development of both rural and urban areas in North America extends back into the early colonial era.

The use of land division plats was continued after the end of the colonial era. In the post-colonial period, the land division process was often grossly abused by land speculators and promoters of urban developments. Developers often subdivided land into very small lots, 25 feet in width by 100 feet in depth, and promoted the sale of unimproved lots to unwary, absentee, and speculating buyers. Many individuals lost lifetime savings in land speculation, and local governments were often left with large areas of platted, but unimproved, lots that imposed an obsolete and costly development pattern on the community. Often such plats were located in areas that should not have been developed at all.

DOI: 10.1201/9781003500384-17

The most serious problems attendant to this uncontrolled land division were often the result of the land being divided without the installation of necessary improvements, such as sanitary sewers, water mains, stormwater drains, street pavements, curbs and gutters, and sidewalks. In many cases, no provision was made for adequate school and park sites or for other necessary public facilities. Other manifestations of the misuse of land in the past have included the lack of adequacy and uniformity in street width, alignment, and continuity, and a generally inferior arrangement of blocks and lots with inadequate consideration being given in the platting layout to the topography, drainage pattern, and natural assets of the site, such as woodlands, wetlands, floodplains, streams, and watercourses. In many cases, streets were laid out with little thought as to function, resulting in unnecessary traffic conflicts and also in an unnecessarily excessive area being devoted to streets. Poor surveying and survey monumentation practices gave rise to disputes over land ownership and public versus private rights with respect to land use regulations.

The net results of this widespread misuse of land resources, which occurred during the Industrial Revolution in the late nineteenth and early twentieth centuries, are evidenced today in many older, deteriorating urban areas. In terms of costs to the individual taxpayer, the damage has been incalculable. The community has often been left with the responsibility for installing and maintaining improvements that may never be fully used. Poor land division and the resulting development have often made areas generally undesirable as places in which to live and work, with the result that the areas contribute less and less over time to the municipal tax base by virtue of premature obsolescence, lower property values, high cost of municipal services, and high rates of vacancies and tax delinquencies. The social cost of poor land division and development may also be significant where that physical deterioration has led to social deterioration. The results have been reflected in increased fire and police protection and public welfare costs.

The federal government in 1928 promulgated a Standard City Planning Enabling Act. This act, among other provisions, shifted the emphasis of land division regulation from its narrow focus as a means for facilitating the description of real property boundaries and the transfer of land ownership to a broader one of shaping urban growth in the public interest.

In a 1932 report entitled "Report of a Presidential Conference on Home Building and Home Ownership," the prominent civil engineer and city planner Harland Bartholomew presented key subdivision design principles, suggested design standards for street widths and alignment based on functional need, and demonstrated the cost savings inherent in good subdivision design. This report is sometimes viewed as the beginning of effective modern municipal land subdivision regulation in the United States. Subsequent to its publication, and in accordance with the report, municipalities across the country began to adopt subdivision control ordinances. However, it was not until the massive land development and housing construction period that followed the end of World War II that modern land division regulations were widely adopted by municipalities, particularly in metropolitan areas.

The Federal Housing Administration (FHA), through its mortgage insurance program, and the Federal Housing and Home Finance Agency, through its planning grant program, were important proponents of local subdivision regulations. These agencies published recommended subdivision regulations during the 1950s that increased public understanding of the benefits of good subdivision design.

In the 1960s, infrastructure improvement requirements were commonly incorporated into land subdivision control ordinances. Municipalities began to require developers to install on-site improvements, such as street pavements, sanitary sewers, water mains, and storm drains. Land dedication requirements for school and park sites were also incorporated into local ordinances. Some municipalities required developers to provide off-site improvements as well. The timing and location, as well as the design and improvement, of subdivisions were also addressed in some municipal regulatory programs. Subdivisions were approved only if a finding was made by a cognizant public agency that the existing public facilities and services were adequate to serve the needs created by the proposed subdivisions. In the 1980s, subdivision regulations were further expanded to address certain environmental protection needs, such as the conservation of woodland, wetland, and floodplain areas. The use of impact fees and financial exactions, such as the payments in lieu of dedication for municipal off-site infrastructure and social needs, particularly schools and parks, was also introduced to assist municipalities in meeting the costs attendant to urban development. In more recent years, subdivision control has continued to evolve to address climate adaptation, equity in infrastructure investment, green stormwater infrastructure, and smart growth principles through form-based codes and performance zoning.

REASONS FOR PUBLIC REGULATION

Land subdivision has far-reaching impacts upon the community. A variety of interests are involved in, concerned with, and affected by the subdivision of land for development. These include homeowners, mortgage lending institutions, realtors, land developers, builders, public and private utilities, special purpose governmental districts; general-purpose, municipal, county, and state governments; and sometimes the federal government.

The purchase of a home is a major investment for the average citizen, and the debt normally incurred with such a purchase is usually amortized over many years. Good land subdivision design and improvement will help protect the security of that investment for all concerned. In protecting the investment of the homeowner against premature obsolescence, good land subdivision practice also provides protection against inaccurate and ambiguous titles and costly boundary disputes, indicates to buyers of individual building sites in undeveloped areas what the area will be like when it is fully developed, and discourages later changes in streets and public places incorporated in a land subdivision plat.

Good land subdivision practices benefit realtors, developers, and builders by requiring compliance by all with established layout and improvement standards, thereby avoiding arbitrary regulation. Moreover, good land subdivision practices prevent poorly designed and inadequately improved subdivisions from deteriorating and causing nearby well-designed developments to depreciate.

Most importantly, good land subdivision practices benefit the community as a whole. The act of land subdivision establishes the pattern for future community development. The community is required to furnish public facilities and services to newly platted areas. Therefore, the community should be able to require that streets are wide enough to accommodate firefighting, emergency, solid-waste collection, and snow-removal equipment; that street configurations, alignment, and grades are adequate for the safe movement of traffic, access to building sites, and snow storage; and that building sites are adequately drained and not subject to flooding. If public sanitary sewerage facilities are not to be provided, the community should be able to require that lots are of sufficient size and have soils that can accommodate on-site sewage disposal facilities without creating a public health problem. Proposed land subdivisions may have impacts extending beyond the site boundaries of the land subdivision itself, and the community should be able to require that the developer bear a fair and proportionate share of required offsite improvement costs. This involves consideration of the impacts of the proposed development on arterial streets and highways; on sewage conveyance, storage, and treatment facilities; on water supply, storage, transmission, and treatment facilities; on stormwater management facilities; and on school and park facilities. These infrastructure impacts also translate into fiscal effects on municipal budgets, service delivery, and long-term maintenance. These impacts may be positive or negative depending upon whether the revenues obtained from a new development exceed the costs of the facilities and services to be provided or are less than those costs. In the latter case, some developments may result in higher incremental costs of services than the taxable value of the development may provide in revenues. Municipalities increasingly require subdivision proposals to submit a fiscal impact analysis and sustainability or resilience assessments.

Clearly, land subdivision has far-reaching impacts, particularly upon municipal and county governments. Many of the perplexing problems that face communities, such as traffic congestion, poor drainage, flooding, high street and utility maintenance costs, inadequate park and school sites and facilities, high fire, emergency, and police protection costs, and rural and urban deterioration may be directly attributable to the manner in which the areas of the community concerned were originally subdivided. Just as important in this respect as the design of land subdivisions is the issue of whether or not the areas concerned should have been subdivided at all. Scattered, leapfrog-style subdivisions lacking access to essential community services increase car dependency, energy use, and the financial burden of extending infrastructure. The viability of agricultural areas may also be destroyed through scattered land subdivision and development. Also important in this respect is the issue of whether an excess of building sites is being created at any one time, thereby leaving the community with widely scattered, partially developed neighborhoods. Public regulation is therefore critical to ensure subdivision patterns align with long-term growth management goals, infrastructure capacity, and environmental stewardship.

PURPOSE OF PUBLIC REGULATION

Because land subdivision affects a broad spectrum of interests and, particularly, the welfare of the community in so many respects, its regulation has become widely accepted as a function of municipal, county, and state government. Land subdivision regulation is the exercise of control by the community over the conversion of undeveloped land into buildable lots. It is through such regulation that the public interest in land subdivision is expressed and protected.

Land subdivision regulation is intended to accomplish a number of public purposes. The first of these is the original reason for such regulation, that is, to provide a basis for clear and accurate real property boundary line records. In this text it should be noted that there are three basic means by which real property boundaries may be described: by reference to adjoining lands or waters, by courses and distances, and by reference to a plat. When numerous smaller parcels of land are involved, particularly in urban areas, legal description by reference to a recorded plat becomes the most practical means of describing ownership parcels. Although not commonly recognized as such, legal descriptions of properties expressed in terms of U.S. Public Land Survey system sections and aliquot parts of such sections are actually descriptions by reference to a plat. The U.S. Government Surveys represent the first division of the federal lands made in the public land survey states for the purpose of transferring ownership from the public to the private sector. The later preparation of plats for the further division of land into smaller parcels, primarily for the purpose of urban development, has, therefore, become known as land subdivision.

In areas covered by the U.S. Public Land Survey system, good practice would dictate that subdivision plats be referenced to at least two monumented corners set in the government survey. Subdivision plats must be recorded and filed in the office of the register of deeds in the county concerned. Legal descriptions of individual parcels can then be readily expressed in terms of numbered or lettered blocks and numbered lots and outlots in a titled subdivision plat, and the resulting building sites are easily identified, readily marketed, and efficiently taxed.

Land subdivision regulation is intended to ensure that proposed land subdivisions will fit harmoniously into the existing land use pattern and will serve to implement the comprehensive plan and its various components for the physical development of the community. Such regulation is intended to ensure that adequate provision is made for necessary and planned community and neighborhood facilities, including parks, accessways to navigable waters, schools, and shopping areas, so that an attractive and efficient environment results. Such regulations are also intended to ensure that sound standards for the development of land are met, with particular attention to such factors as street layouts, widths, and grades; bicycle and pedestrian circulation; park and open space requirements; block configurations; lot sizes; and street, utility, stormwater management, and transit improvements. Land subdivision regulation is also intended to ensure the fiscal stability of the community, minimizing the cost of public facilities and services and protecting against the development over time of blighted areas. Importantly, such regulation is intended to provide for the public health, safety, and general welfare. Modern approaches further emphasize climate resilience, environmental protection, and equitable access to infrastructure.

Land subdivision regulation should be regarded as an important means of implementing community comprehensive plans. Accordingly, land subdivision regulations should be prepared and administered within the context of, and be consistent with, such a plan. This integration of land subdivision regulation with comprehensive planning assists the community in avoiding ad hoc and irresponsible development decisions. Such integration may, indeed, be required by the state planning enabling legislation.

Historically, in communities that do not have a comprehensive plan, the land subdivision control ordinances, together with the zoning ordinances, have often been substituted for such a plan and thus have had to bear the full weight of guiding and shaping the physical development of the community. Zoning relates to the type of building development that can be placed on the land, whereas land subdivision control relates to the way in which land is divided and made ready for building development. The substitution of these plan implementation devices for the plans themselves has been particularly common in smaller communities, and in such situations the preparation and administration of good land subdivision control regulations is all the more important. Although land subdivision control is far more effective if based on a comprehensive plan, substantial benefits may be derived from enacting a land subdivision control ordinance even in the absence of such a plan. In the absence of a plan, such regulations may either promote or retard change, an influence that may be either good or bad, depending upon the validity of the assumptions on which the regulations are based. It should be noted in this respect that, under the new state planning legislation in Wisconsin, communities must have an adopted plan in order to enforce land subdivision and zoning regulations.

In considering public regulation of land subdivision, it must also be recognized that land subdivision design is a dynamic art. New ideas and continually emerging community concerns must be integrated into the land subdivision design and infrastructure improvement process. Emerging topics include sustainability, climate adaptation, smart infrastructure, and green infrastructure integration.

If land subdivision design is to be amenable to new direction, then the land subdivision control regulations imposed must be reasonably flexible. Land subdivision control regulations should not be regarded as set formulae from which developers can cheaply and mechanically extract the best solution to a given development problem. Good land subdivision design is never a process of simply following a standard set of regulations. The concern should be with the effect of the land subdivision produced, rather than with the regulations per se. Accordingly, the intent of a land subdivision control ordinance should be to ensure compliance with at least minimum standards for new development, and to prevent further occurrences of the abuses in land development that occurred in the past, while at the same time facilitating the best subdivision design possible. The quality of each land subdivision design will be a reflection of the developer's ingenuity in creating a functional development that meets or exceeds the requirements of public regulation while respecting the natural characteristics of the site.

STATUTORY AUTHORITY FOR LAND SUBDIVISION CONTROL

Land subdivision control is in most, if not all, states of the United States authorized and regulated by state statute. For example, in Wisconsin, land subdivision is regulated by Chapter 236 of the Wisconsin Statutes, entitled "Platting Lands and Recording and Vacating Plats." Under this Statute, cities, villages, towns, and counties are given the authority to regulate the subdivision of land. The Statute is of the mandatory type, rather than the enabling type, and requires that all subdivisions, as defined in the Statutes, be approved by specified governmental agencies in order to be entitled to be recorded. The Statute requires that any division of land that results in a subdivision shall be, and provides that any other division may be, surveyed and a plat thereof approved and recorded. The Statute makes exceptions for cemetery plats, assessors' plats, and the sale or exchange of parcels of public utility or railway right-of-way to adjoining property owners, provided certain conditions are met.

The Wisconsin Statutes defines a subdivision as "a division of a lot, parcel or tract of land by the owner thereof or the owner's agent for the purpose of sale or of building development, where: a) the act of division creates five or more parcels or building sites of 1.5 acres each or less in area; or, b) five or more parcels or building sites of 1.5 acres each or less in area are created by successive divisions within a period of five years." A plat is defined as "a map of a subdivision." City, village, town, and county subdivision control ordinances can use more restrictive definitions of the term subdivision than that used in the Statute concerned, and can, indeed, require the preparation and recording of a subdivision plat for any land division.

REQUIRED PLAT APPROVALS

State law and local ordinances often provide for two categories of review authorities with quite different functions in the platting process: approving authorities and objecting agencies. Approving authorities must demonstrate their acceptance of a proposed subdivision by signing the plat before it can be recorded. Objecting agencies are provided the opportunity to review the proposed plat and are allowed to object to the subdivision if it fails to comply with extant regulations. For subdivisions located within a city or village, the governing body of the municipality will usually be the approving authority. If within an unincorporated area, the approving authorities may include the town board and the county planning agency.

Various state and county agencies may be identified as objecting agencies. These may include state departments of administration, transportation, and natural resources. If the state statutes include detailed regulations governing the land subdivision process, a department of administration may be charged with reviewing all proposed plats for compliance with, for example, statutory surveying and layout requirements, the data required to be shown on final plats, and the certificates required to accompany a final plat. A department of natural resources may have responsibility for reviewing proposed plats that will not be served by sanitary sewer with respect to lot size, elevation, and soil suitability. A department of transportation may have responsibility for reviewing proposed land subdivision plats abutting state trunk highways for compliance with the access regulations intended to provide for the safety of entry onto and exit from abutting state trunk highways, for the protection of arterial capacity, and for right-of-way reservation.

Counties may be objecting agencies to subdivision plats located within cities and villages, as well as approving authorities for plats located in unincorporated areas. The county agencies may review proposed plats for conflicts with existing or proposed parks, parkways, highways, airports, drainage-ways, or other planned county public improvements. In some states, regional planning commissions may also play a review or coordination role.

FINAL PLAT DATA

The state statutes may provide detailed specifications for the preparation of a final subdivision plat. In the absence of such state requirements, the county or local municipal subdivision control ordinance should provide such detailed specifications. Such specifications may include the size and type of media on which the plat is to be proposed; the margins to be used; the minimum allowable scale of the plat; and provision of a graphic scale, the name, and, in U.S. Public Land Survey system states, the location of the plat by U.S. Public Land Survey system quarter-section, section, township, range, and county, and by bearing and distance referenced to a boundary line of a quarter section in which the subdivision is located, the monumentation at the ends of the boundary line being used as well as the state plane coordinate system grid bearing and distance between those ends. Survey data requirements typically include accurate definitions of all exterior boundaries and interior lines for lots, easements, public spaces, and infrastructure.

Because a land subdivision plat provides the vehicle for transferring ownerships, certain certificates should be provided on the face of a plat to entitle the plat to be recorded. These should include a surveyor's certificate stating by whose direction the subdivision plat was made, providing a description of the land surveyed, divided, and mapped; a statement that the plat is a correct representation of the exterior boundaries of the land surveyed and the subdivision made; and a statement that the surveyor has fully complied with the provisions of the state statutes and applicable local ordinances in surveying, dividing, and mapping the land. The certificates should also include an owner's certificate stating that the owner caused the land described on the plat to be surveyed, divided, mapped, and dedicated as represented on the plat and listing the approving authorities that must act in order to entitle the plat to be recorded. The certificate must be executed by the owner, and all persons holding an interest in the fee of record, or by possession and, if the land is mortgaged, by the mortgagee of record.

The certificates should also include a certificate of the clerk or treasurer of the municipality in which the subdivision is located and a certificate of the treasurer of the county in which the subdivision is located, stating that there are no unpaid taxes or unpaid special assessments on any of the lands included within the plat. The plat should also show on its face certificates relative to the actions of all of the approving and objecting authorities concerned. Importantly, these certificates should include a certified copy of the action of the common council, village board, or town board required to approve the plat and accept the dedications provided by the plat, and, as may be necessary, a certificate of the county clerk or other designated county official certifying approval actions by the county agencies concerned. The certificates must also include statements by the city, village, or town clerks concerned that copies of the plat were forwarded to the concerned state and county objecting agencies, and that either no objections to the plat were filed or all objections filed were met. In some cases, a certificate restricting direct vehicular ingress or egress from specified lots and blocks to and from state trunk highways may be required. Special certificates relating to the use of any outlets on the plat may also be required. Some jurisdictions may also require right-of-way reservations, easement declarations, and digital plat submissions.

RECORDING OF PLATS

Usually state law will require that the final plat be recorded and filed in the office of the register of deeds of the county in which the subdivision is located. Upon recording, the lots in the plat are described by the name of the plat and the lot and block descriptions for all purposes, including assessment, taxation, devise, descent, and conveyance. The act of recording the plat also, in effect, conveys all of the dedications to the public noted on the plat to the city, village, or town concerned, to be held by those bodies in trust for the uses and purposes intended by the plat. In jurisdictions with digital land records, electronic recording systems may also apply, enabling faster and more secure document handling.

LOCAL LAND SUBDIVISION REGULATION

The state planning enabling acts, or accessory acts, usually specifically empower cities, villages, towns, and counties to adopt ordinances governing the division of land that are more restrictive than the state statutes. This empowerment may in some cases be limited to cities, villages, towns, and counties that have established planning programs.

For example, the Wisconsin Statutes state that the purpose of local subdivision regulations shall be to: "promote the public health, safety, and general welfare of the community …; to lessen congestion in the streets and highways; to further the orderly layout and use of land; to secure safety from fire, panic, and other dangers; to provide adequate light and air …; to prevent the overcrowding of land; to avoid undue concentration of population; to facilitate adequate provision for transportation, water, sewerage, schools, parks, playgrounds and other public requirements; to facilitate the further resubdivision of larger tracts into smaller parcels of land." The Statutes further provide that local ordinances are to be made "with reasonable consideration, among other things, of the character of the municipality, town, or county with a view of conserving the value of buildings placed upon land, providing the best possible environment for human habitation, and for encouraging the most appropriate use of land throughout the municipality, town, or county."

Given the provisions of the state statutes that regulate the procedural and technical aspects of land division, county and local land division control ordinances should be viewed primarily as a means of implementing local comprehensive or master plans. As such, county and local ordinances should be particularly concerned about good location, good design, adequate improvements, and design criteria for such improvements. Design requirements may include those for streets, lots and blocks, bicycle and pedestrian facilities, public sewerage and water supply systems, and stormwater management facilities. Modern ordinances may also include considerations for renewable energy readiness, climate resilience, and electric vehicle infrastructure.

Before adopting a land division control ordinance or any amendments thereto, the governing body concerned should receive the recommendation of the plan commission and should hold a public hearing on the proposed ordinance or amendment. Land division control ordinances should always be reevaluated and, if necessary, updated at the time a community revises its comprehensive plan, or elements thereof. Periodic review ensures alignment with emerging best practices, statutory updates, and evolving community needs.

EXAMPLE LOCAL SUBDIVISION CONTROL ORDINANCE

An example of a land subdivision control ordinance is provided in Appendix A to this text. This ordinance is not to be interpreted as a so-called "model" ordinance, but simply as an example of a relatively good ordinance. The example ordinance is specifically based on Wisconsin law and practice. Although the laws and practices governing land subdivision will differ from state to state, the ordinance provided presents good practices generally widely applicable. General provisions, submittal requirements, and review and approval procedures for subdivision plats are presented in Sections 1.00 through 6.00 of the model ordinance. Section 7.00 presents design standards for lots, blocks, streets, pedestrian and bicycle ways, protection of natural resources, and requirements for the dedication or reservation of public sites. Section 8.00 presents requirements for improvements, including roadways, sidewalks, street trees; and lamps; sewer, water, and stormwater management facilities; and landscaping and erosion control facilities. Section 9.00 presents construction standards, which primarily refer to the standard specifications adopted by a municipality and the judgment of the municipal engineer to ensure high-quality improvements. Section 10.00 sets forth fees for the various components of the land division and development process, which reflect local practices. Section 11.00 includes definitions of terms used in the ordinance, and Section 12.00 provides for the adoption of the ordinance. Some comments concerning specific requirements of the ordinance follow.

Plat Review and Approval Procedures

The procedure for dividing land should entail three basic steps. These will provide for a smooth and expeditious development process that will be satisfactory to both the municipality and the subdivider, and will avoid poorly planned and executed land development. Following these three steps will help ensure the evolution over time of an efficient, healthful, and attractive community.

The first step, which is discretionary, should consist of a pre-application meeting of the subdivider with the plan commission or its staff to discuss in a preliminary manner the proposed land division. The second step should consist of the submission of a preliminary plat to the plan commission for review and recommendation to the governing body for approval, approval with conditions, or rejection. The third step should consist of the submission of a final plat to the plan commission and governing body concerned for review and approval or rejection. Local review and approval procedures for preliminary and final plats must comply with the requirements set forth in the state statutes concerned.

Under the ordinance, the local plan commission has primary authority for the review of proposed land divisions and the application of the design standards set forth in the ordinance. Local plan commissions are, however, advisory agencies and must forward commission recommendations regarding the land division to the governing body for action. Final plats involve the dedication of streets or other public lands and should, therefore, be approved by the governing body. The ordinance, in Section 11.02, defines a subdivision in terms of the number of parcels or building sites created and the time over which they were created. Other ordinances are also often included in the definition of the maximum size of parcels or building sites created.

It should be noted that the appended ordinance recommends that the city, village, or town engineer concerned be delegated the responsibility of making certain professional judgments and decisions concerning the design and funding of required improvements. Some communities employ, on either a full-time or a part-time consulting basis, a municipal engineer duly appointed as an officer of the municipality and authorized to exercise certain powers, duties, and functions as prescribed in the state statutes and in local ordinances. Every municipality should employ a registered professional engineer on staff or on a consulting basis as the municipal engineer. The municipal engineer should be duly appointed by the governing body and authorized to exercise the prescribed powers, duties, and functions of that office.

Pre-Application Conference

As already noted, the land division development process should be initiated by the subdivider contacting the local plan commission to discuss the proposed land division plat informally with the commission and its staff. The subdivider may prepare a conceptual or "sketch" plan to illustrate proposed street and lot layouts, open space areas, and other development concepts. This meeting allows the plan commission an opportunity to advise the subdivider as to the platting procedures to be followed and the regulations governing the platting of land in the municipality concerned. It also affords the plan commission to provide the subdivider with pertinent information concerning the adopted comprehensive plan for the community. The plan commission or its staff may also furnish the subdivider with a checklist of requirements relating to the format and content of the preliminary and final plats. This step should be, in all respects, an informal one and is intended to save both the subdivider and local government much time in moving through the subsequent steps in the land development process.

Preliminary Plat Submission

The submission of a preliminary plat is the first formal step toward land division. It enables review for compliance with the community's comprehensive plan, especially neighborhood unit design policies. For the subdivider, it offers the opportunity to obtain general acceptance of the proposed land division by the affected governmental agencies before large expenditures of time and money are made.

The preliminary plat should include all of the contiguous lands owned or controlled by the subdivider even though only a small part of such lands may be intended to be immediately subdivided and developed. This is necessary in order to avoid the creation of an undesirable overall development pattern, to avoid the creation of landlocked parcels, and to ensure that any single development, no matter how small, will properly form an integral part of the whole. The plat must show on its face all of the information pertinent to the proper review of the proposed development as specified in the local land division control ordinance. An example preliminary plat is shown in Figure 17.1. Under the model ordinance, a site analysis must accompany a preliminary plat if the area within the plat is not included within a neighborhood unit development plan.

FIGURE 17.1 Illustration of a well-prepared preliminary plat for a proposed land subdivision. It should be noted that the only precise survey data given on the plat are for the lengths and bearings of the exterior boundaries, which lengths and bearings are referenced to the State Plane Coordinate System. All other survey data—such as the width and depth of lots—are approximations. *(Continued)*

Source: SEWRPC.

LAKE HIGHLANDS

BEING A SUBDIVISION OF A PART OF THE SOUTHWEST ONE QUARTER OF THE SOUTHWEST ONE-QUARTER OF SECTION 35, TOWNSHIP 8 NORTH, RANGE 9 EAST

CITY OF LAKE, LAKE COUNTY, WISCONSIN

SURVEYOR'S CERTIFICATE

STATE OF WISCONSIN)
COUNTY OF LAKE) SS

I, John W Doe, registered land surveyor, being first duly sworn, on oath hereby depose and say:

THAT I have surveyed, divided, and mapped "LAKE HIGHLANDS," being a Subdivision of a part of the SW ¼ of the SW ¼ of Section 35, T 8 N, R 9 E, in the City of Lake, Lake County, Wisconsin, which is bounded and described as follows:

Beginning at the Southwest corner of said ¼ Section; running thence North 03° 43'00" East along the West line of said ¼ Section 953.50 ft. to a point; thence South 86°17'00" East at right angles to the West line of said ¼ Section 273.00 ft. to a point; thence South 77° 09' 38" East 153.00 ft. to a point; thence South 68° 21' 22" East 507.00 ft. to a point; thence South 82° 49' 10" East 192.00 ft. to a point; thence South 37° 28' 50" East 212.00 ft. to a point; thence South 23° 23' 06" West 293.00 ft. to a point; thence North 66° 36' 54" West 389.78 ft. to a point; thence South 23° 23' 06" West 368.58 ft. to a point; thence South 02° 38' 00" West, at right angles to the South line of said ¼ Section 98.00 ft. to a point in the South line of said ¼ Section; thence North 87° 22' 00" West along the South line of said ¼ Section 650.00 ft. to the point of beginning. EXCEPTING therefrom those parts heretofore dedicated for road and highway purposes.

The coordinates of the southwest corner of the said ¼ section are on the Wisconsin Plane Coordinate System, South Zone NAD 27: 414557.85 feet North, 2185 076.50 feet East, and all description bearings are referred to said coordinate grid.

THAT I have made such survey, land division and plat by the direction of HIGHLANDS INVESTMENT CORP., owner of said land.

THAT such plat is a correct representation of all the exterior boundaries of the land surveyed and the land division thereof.

THAT I have fully complied with all the provisions of Chapter 236 of the Wisconsin Statutes and all the provisions of the Land Division Ordinances of the City of Lake and of Lake County in surveying, dividing and mapping the same.

_______________________ (SEAL)
John W Doe, Registered Wisconsin Land Surveyor
No. S-157
Signed this ____ day of __________, 2000

GENERAL CURVE DATA					
CURVE	RADIUS	ARC	CHORD	CHORD BEARING	CENTRAL ANGLE
A-N. LINE	810.00	278.05	276.69	S 76°26'57" E	19°40'06"
A-C_L	785.00	269.47	268.15	S 76°26'57" E	19°40'06"
A-S. LINE	760.00	260.89	259.61	S 76°26'57" E	19°40'06"
B-N. LINE	135.00	165.72	155.51	N 58°33'03" E	70°19'54"
B-C_L	160.00	196.40	184.30	N 58°33'03" E	70°19'54"
B-S. LINE	185.00	227.09	213.10	N 58°33'03" E	70°19'54"
C-N. LINE	625.00	214.56	213.50	N 76°26'57" W	19°40'06"
C-C_L	600.00	205.97	204.96	N 76°26'57" W	19°40'06"
C-S. LINE	575.00	197.38	196.42	N 76°26'57" W	19°40'06"
D-SAC	60.00	217.97	116.40	S 23°23'06" W	208°09'04"
E-FILLET	25.00	39.27	35.36	N 21°36'54" W	90°00'00"
CURVE DATA BLOCK - 1					
LOT - 3	810.00	63.05	63.03	S 84°09'13" E	4°27'38"
4	810.00	65.00	64.98	S 78°31'29" E	4°35'52"
5	810.00	65.00	64.98	S 74°55'36" E	4°35'52"
6	810.00	65.00	64.98	S 70°19'44" E	4°35'52"
7	810.00	20.00	20.00	S 67°19'21" W	1°24'54"
11	60.00	52.98	51.28	S 59°22'38" E	50°35'38"
16	60.00	51.00	49.48	S 08°44'45" E	48°47'10"
PED. WAY	60.00	10.01	10.00	S 23°22'56" W	9°33'32"
17	60.00	47.00	45.81	S 50°36'21" W	44°52'58"
18	60.00	52.00	50.39	N 82°07'26" W	49°39'28"
19	60.00	4.98	4.98	N 54°58'32" W	4°45'20"
26	185.00	39.09	39.02	S 29°26'18" W	12°06'24"
27	185.00	64.00	63.68	S 45°24'06" W	19°49'18"
28	185.00	62.00	61.71	S 64°54'51" W	19°12'08"
29	185.00	62.00	61.71	S 64°06'37" W	19°12'06"
CURVE DATA BLOCK - 2					
LOT - 2	760.00	9.00	9.00	S 85°56'39" E	00°40'42"
3	760.00	77.00	76.97	S 82°42'09" E	5°48'18"
4	760.00	77.00	76.97	S 76°53'51" E	5°48'18"
5	760.00	73.00	72.98	S 71°14'38" E	5°30'12"
6	760.00	24.89	24.89	S 67°33'12" E	1°52'36"
9	25.00	39.27	35.36	S 21°36'54" E	90°00'00"
10	625.00	46.96	48.95	N 68°51'33" W	4°29'18"
11	625.00	70.00	69.96	N 74°18'43" W	6°25'02"
12	625.00	71.00	70.96	N 80°46'30" W	6°30'32"
13	625.00	24.59	24.59	N 85°09'33" W	2°15'14"
CURVE DATA BLOCK - 3					
LOT - 3	575.00	7.42	7.42	S 85°54'49" E	00°44'22"
4	575.00	74.00	73.95	S 81°51'25" E	7°22'26"
5	575.00	74.00	73.95	S 74°28'59" E	7°22'26"
6	575.00	40.96	41.95	S 68°42'30" E	4°10'52"
7	135.00	28.35	28.30	S 29°24'02" W	12°01'52"
8	135.00	121.00	110.39	S 61°05'36" W	51°21'16"
9	135.00	16.37	16.36	N 89°45'23" W	6°56'46"

CORPORATE OWNER'S CERTIFICATE OF DEDICATION

HIGHLANDS INVESTMENT CORP., a corporation duly organized and existing under and by virtue of the laws of the State of Wisconsin, as owner, does hereby certify that said Corporation caused the land described on this plat to be surveyed, divided, mapped and dedicated as represented on this plat.

HIGHLANDS INVESTMENT CORP., as owner, does further certify that this plat is required by S. 236.10 and S. 236.12 to be submitted to the following for app roval or objection:

(1) COMMON COUNCIL OF THE CITY OF LAKE
(2) COUNTY PLANNING AGENCY
(3) WISCONSIN DEPARTMENT OF ADMINISTRATION

IN WITNESS WHEREOF, the said HIGHLANDS INVESTMENT CORP. has caused these presents to be signed by JOSEPH C. SMITH, its President, and countersigned by JOHN BROWN, its Secretary, at Lake, Wisconsin, and its Corporation seal to be hereunto affixed this ____ day of ________________, 2000.

In the presence of: HIGHLANDS INVESTMENT CORP.

/s/ JOHN A. GREEN ____________

/s/ JANE B. GREEN ____________

/s/ JOSEPH C. SMITH ____________ (SEAL)
Joseph C. Smith, President

COUNTERSIGNED

/s/ JOHN D. BROWN ____________ (SEAL)
John D. Brown, Secretary

STATE OF WISCONSIN)
COUNTY OF LAKE) SS

On this the ____ day of ____________, 2000, before me, __________________, the undersigned officer, personally appeared JOSEPH C. SMITH, who acknowledged himself to be the President of the abovenamed corporation and also personally appeared JOHN D. BROWN, who acknowledged himself to be the Secretary of the abovenamed corporation, and that they, as such, being authorized to do so, executed the foregoing Corporate Owner's Certificate for the purposes therein contained, by signing the name of the corporation by themselves as President and Secretary.

IN WITNESS WHEREOF I hereunto set my hand and official seal.

____
Notary Public, JANE E. GREY, Lake County, Wisconsin
My Commission Expires December 31, 2002

LOT AREA

BLOCK NUMBER	LOT NUMBER	AREA IN SQUARE FEET	BLOCK NUMBER	LOT NUMBER	AREA IN SQUARE FEET
1	1	8,960	2	1	10,603
	2	8,639		2	9,385
	3	8,407		3	8,866
	4	8,538		4	7,742
	5	8,475		5	7,492
	6	8,374		6	7,406
	7	9,015		7	7,919
	8	8,334		8	9,174
	9	8,475		9	9,714
	10	9,017		10	8,677
	11	9,435		11	8,839
	12	9,463		12	9,606
	13	9,148		13	9,191
	14	11,086		14	9,225
	15	15,900		15	10,802
	16	17,533			
	17	21,738	3	1	10,260
	18	13,587		2	8,601
	19	16,108		3	8,625
	20	10,718		4	8,078
	21	10,374		5	7,841
	22	9,097		6	7,260
	23	8,790		7	8,148
	24	8,498		8	7,734
	25	9,104		9	8,085
	26	10,411		10	8,799
	27	12,629		11	8,829
	28	16,709		12	9,191
	29	12,033		13	10,411
	30	9,032			
	31	9,001			
	32	8,699			
	33	10,019			
	34	11,805			

ACCESS RESTRICTION

HIGHLANDS INVESTMENT CORP., a corporation duly organized and existing under and by virtue of the laws of the State of Wisconsin, as owner, does hereby restrict Lots 28 through 34 of Block 1, in that no owner, possessor, user, licensee nor any other person shall have any right of direct vehicular ingress or egress with County Trunk Highway A (Jefferson Avenue), as shown on the plat. This access restriction is expressly intended to constitute a restriction for the benefit of the public according to Section 236.293 of the Wisconsin Statutes and shall be enforceable by the City of Lake and by Lake County.

In the presence of: HIGHLANDS INVESTMENT CORP.

/s/ JOHN A. GREEN ____________

/s/ JANE B. GREEN ____________

/s/ JOSEPH C. SMITH ____________ (SEAL)
Joseph C. Smith, President

COUNTERSIGNED

/s/ JOHN D. BROWN ____________ (SEAL)
John D. Brown, Secretary

CERTIFICATE OF CITY TREASURER

STATE OF WISCONSIN)
COUNTY OF LAKE) SS

I, JOHN F. BLACK, being duly appointed, qualified and acting City Treasurer of the City of Lake, Wisconsin, do hereby certify that the records in my offi ce show no unpaid taxes or unpaid special assessments as of __________________ on any of the lands included in the plat of LAKE HIGHLANDS.

____
Date

/s/ JOHN F. BLACK ____________
John F. Black, Treasurer, City of Lake, Wisconsin

CERTIFICATE OF COUNTY TREASURER

STATE OF WISCONSIN)
COUNTY OF LAKE) SS

I, JOSEPH G. BLACK, being the duly elected, qualified and acting Treasurer of the County of Lake, Wisconsin, do hereby certify that the records in my off ice show no unredeemed tax sales and no unpaid taxes or special assessments as of __________________ affecting the lands included in the plat of LAKE HIGHLANDS.

____
Date

/s/ JOSEPH G. BLACK ____________
Joseph G. Black, Treasurer, Lake County, Wisconsin

COMMON COUNCIL RESOLUTION

RESOLVED, that the plat of LAKE HIGHLANDS, being a Subdivision of a part of the SW ¼ of the SW ¼ of Section 35, T8N, R9E, in the City of Lake, Lake County, Wisconsin, having been approved by the Plan Commission is hereby approved by the Common Council of the City of Lake, Wisconsin, on this ____ day of _____ ________, 2000.

____
Date

/s/ JOHN H. GRAY ____________
John H. Gray, Mayor, City of Lake, Wisconsin

I, JOAN J. CLAY, do hereby certify that I am duly appointed, qualified and acting City Clerk of the City of Lake and the foregoing is a true and correct copy of a resolution passed and adopted by the Common Council of the City of Lake, Lake County, Wisconsin, this ____ day of ______________, 2000.

AND I DO further certify that copies of this plat were forwarded as required by S. 236.12 (2) on the ____ day of ______________, 2000, and that no objections to the plat have been filed or, if filed, have been met.

AND I DO further certify that the City Plan Commission has approved the use of 50-foot-wide streets in the Lake Highlands subdivision, and that such street widths comply with the City subdivision ordinance.

____
Date

/s/ JOAN J. CLAY ____________
Joan J. Clay, Clerk, City of Lake, Wisconsin

There are no objections to this plat with respect to Secs. 236.15, 236.16, 236.20 and 236.21 (1) and (2), Wis. Stats., or by the County Planning Agency.

Certified ______________________, 20____

Department of Administration

CERTIFICATE OF COUNTY REGISTER OF DEEDS

RECEIVED for record this ____ day of ______________, 2000, at ____ o'clock __M., and recorded in Volume ____ of Plats on Page ____.

/s/ JOSEPH K. WHITE ____________ (SEAL)
Joseph K. White, Register of Deeds, Lake County, Wisconsin

SHEET 2 OF 2

FIGURE 17.1(b) *(Continued)*

An application for approval should accompany the preliminary plat. Platting fees based on the number of lots in the proposed plat, or on the acreage of the proposed plat, may be required by the municipality concerned at this time in order to help defray the cost of review.

Upon submission of the preliminary plat, copies should be distributed by the local government concerned to other approving agencies, to the objecting agencies, and to other concerned local officials, agencies, and organizations for review and comment.

A public hearing may be held before the plan commission. Based on the findings of the public hearing, on the recommendations of the reviewing agencies, on the relationship of the proposed plat to the community's comprehensive plan, including the neighborhood unit development element of such a plan, and on other local considerations, the plan commission determines whether it should recommend to the governing body that the preliminary plat be approved, be approved conditionally, or be rejected. The plan commission then forwards its recommendation to the governing body, which must act on the plat within 90 days of plat submission. If the plat is rejected, the subdivider may resubmit a revised preliminary plat for approval at a later date. If the preliminary plat is approved, or approved conditionally, the subdivider may proceed to prepare and submit for approval a final plat.

Final Plat Submission

An example final plat is provided in Figure 17.2. The area included within the final plat is for a part of the area included within the preliminary plat shown in Figure 17.1. In the submission of a final plat, the subdivider must, under the ordinance, follow one of two alternative procedures regarding the installation of improvements. Each of these alternatives is intended to achieve a high standard of development and a low initial cost to the local government concerned.

Under the first alternative, the subdivider elects to install the required improvements in accordance with the approved preliminary plat and to municipal specifications. The subdivider then submits the final plat along with accompanying documents, fees, and evidence of approval of the installed improvements by the affected governmental agencies to the clerk of the local municipality. This alternative presents some risks to both the developer and the municipality concerned. These risks relate to potential changes in factors affecting the timing of the development and the layout of the plat, and in market conditions, during the time between preliminary and final plat approval. Moreover, supplemental survey control that is otherwise, in effect, provided by the final plat may be required if the street and utility improvements are to be properly constructed in the precise intended locations. Such precision, important to the creation of automated public works management and parcel-based land information systems, cannot be achieved if improvements are permitted to be installed solely on the basis of the information shown on a preliminary plat.

The plan commission then determines whether the plat is to be recommended for approval or rejection, and forwards the plat, together with its recommendations, to the governing body. The governing body must approve or reject the final plat within 60 days after the plat has been submitted to the local government. If approved, the municipal clerk must so certify, and the subdivider can then present the plat to the county Register of Deeds for recording. After recording, lots may be sold.

Example Final Plat

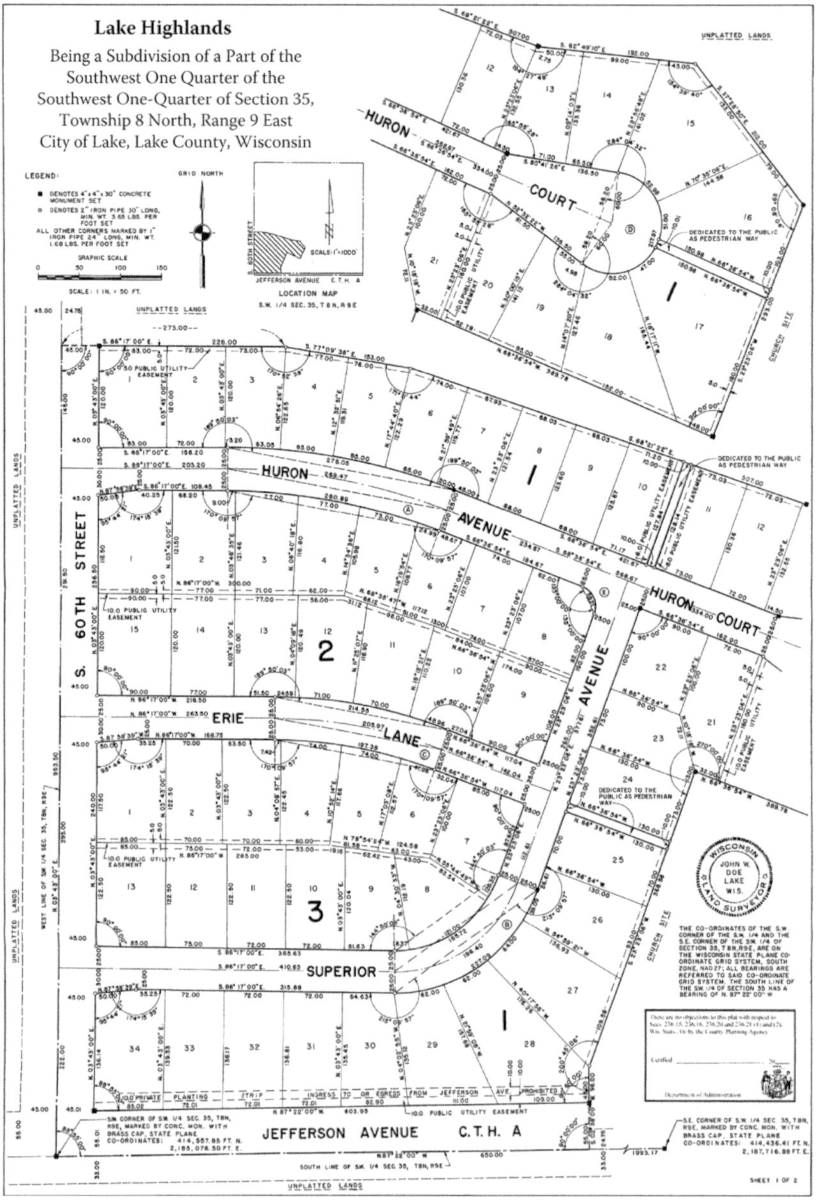

FIGURE 17.2 **(a, b)** Illustration of a properly prepared final land subdivision plat. The final plat is for a part of the area covered by the preliminary plat shown in Figure 17.1. Note that the exterior boundaries of the plat are referred to the State Plane Coordinate System, and since the blocks and lots form closed figures, State Plane Coordinates can be computed for all block and lot corners. *(Continued)*

Source: SEWRPC.

LAKE HIGHLANDS

BEING A SUBDIVISION OF A PART OF THE SOUTHWEST ONE QUARTER OF THE SOUTHWEST ONE-QUARTER OF SECTION 35, TOWNSHIP 8 NORTH, RANGE 9 EAST

CITY OF LAKE, LAKE COUNTY, WISCONSIN

SURVEYOR'S CERTIFICATE

STATE OF WISCONSIN)
COUNTY OF LAKE) SS

I, John W. Doe, registered land surveyor, being first duly sworn, on oath hereby depose and say.

THAT I have surveyed, divided, and mapped "LAKE HIGHLANDS," being a Subdivision of a part of the SW ¼ of the SW ¼ of Section 35, T 8 N, R 9 E, in the City of Lake, Lake County, Wisconsin, which is bounded and described as follows:

Beginning at the Southwest corner of said ¼ Section; running thence North 03°43'00" East along the West line of said ¼ Section 953.50 ft. to a point; thence South 86°17'00" East at right angles to the West line of said ¼ Section 273.00 ft. to a point; thence South 77° 09' 38" East 153.00 ft. to a point; thence South 68° 21' 22" East 507.00 ft. to a point; thence South 82° 49' 10" East 192.00 ft. to a point; thence South 37° 28' 50" East 212.00 ft. to a point; thence South 23° 23' 06" West 293.00 ft. to a point; thence North 66° 36' 54" West 389.78 ft. to a point; thence South 23° 23' 06" West 368.58 ft. to a point; thence South 02° 38' 00" West, at right angles to the South line of said ¼ Section 98.00 ft. to a point in the South line of said ¼ Section; thence North 87° 22' 00" West along the South line of said ¼ Section 650.00 ft. to the point of beginning. EXCEPTING therefrom those parts heretofore dedicated for road and highway purposes.

The coordinates of the southwest corner of the said ¼ section are on the Wisconsin Plane Coordinate System, South Zone NAD 27: 414557.85 feet North, 2185 076.50 feet East, and all description bearings are referred to said coordinate grid.

THAT I have made such survey, land division and plat by the direction of HIGHLANDS INVESTMENT CORP., owner of said land.

THAT such plat is a correct representation of all the exterior boundaries of the land surveyed and the land division thereof.

THAT I have fully complied with all the provisions of Chapter 236 of the Wisconsin Statutes and all the provisions of the Land Division Ordinances of the City of Lake and of Lake County, in surveying, dividing and mapping the same.

______________________ (SEAL)
John W. Doe, Registered Wisconsin Land Surveyor
No. S-157
Signed this ____ day of __________, 2000

GENERAL CURVE DATA					
CURVE	RADIUS	ARC	CHORD	CHORD BEARING	CENTRAL ANGLE
A-N. LINE	810.00	278.05	276.69	S 78°28'57" E	19°40'06"
A-C$_L$	785.00	269.47	268.15	S 78°28'57" E	19°40'06"
A-S. LINE	760.00	260.89	259.61	S 78°28'57" E	19°40'06"
B-N. LINE	135.00	165.72	155.51	N 58°33'03" E	70°19'54"
B-C$_L$	160.00	196.40	184.30	N 58°33'03" E	70°19'54"
B-S. LINE	185.00	227.09	213.10	N 58°33'03" E	70°19'54"
C-N. LINE	625.00	214.55	213.50	N 78°28'57" W	19°40'06"
C-C$_L$	600.00	205.97	204.96	N 78°28'57" W	19°40'06"
C-S. LINE	575.00	197.39	196.42	N 78°28'57" W	19°40'06"
D SAC	60.00	217.97	116.40	S 23°23'06" W	208°08'04"
E-FILLET	25.00	39.27	35.36	N 21°36'54" W	90°00'00"
CURVE DATA BLOCK – 1					
LOT – 3	810.00	63.05	63.03	S 84°03'12" E	4°27'36"
4	810.00	65.00	64.98	S 79°31'28" E	4°35'52"
5	810.00	65.00	64.98	S 74°55'38" E	4°35'52"
6	810.00	65.00	64.98	S 70°19'44" E	4°35'52"
7	810.00	20.00	20.00	S 67°19'21" W	1°24'04"
15	60.00	52.88	51.48	S 55°23'38" E	50°35'36"
16	60.00	51.00	49.48	S 05°44'45" E	48°42'10"
PED. WAY	60.00	10.01	10.00	S 33°33'08" W	9°33'32"
17	60.00	47.00	45.81	S 60°36'21" W	44°52'58"
18	60.00	53.00	50.39	N 82°07'28" W	40°38'28"
19	60.00	4.98	4.98	N 54°55'02" W	4°45'20"
26	185.00	39.09	39.02	S 28°28'18" W	12°06'24"
27	185.00	64.00	63.68	S 45°24'09" W	19°49'15"
28	185.00	62.00	61.71	S 64°54'51" W	19°12'06"
29	185.00	62.00	61.71	S 84°06'57" W	19°12'05"
CURVE DATA BLOCK – 2					
LOT – 2	760.00	9.00	9.00	S 85°58'33" E	00°40'42"
3	760.00	77.00	76.97	S 82°43'09" E	5°48'18"
4	760.00	77.00	76.97	S 76°53'51" E	5°48'18"
5	760.00	73.00	72.98	S 71°14'36" E	5°30'12"
6	760.00	24.89	24.89	S 67°33'12" E	1°52'36"
9	25.00	39.27	35.36	S 21°36'54" E	90°00'00"
10	625.00	48.96	48.95	N 68°51'33" W	4°29'18"
11	625.00	70.00	69.96	N 74°18'43" W	6°25'02"
12	625.00	71.00	70.96	N 80°46'30" W	6°30'32"
13	625.00	24.59	24.59	N 85°09'23" W	2°15'14"
CURVE DATA BLOCK – 3					
LOT – 3	575.00	7.42	7.42	S 85°54'48" E	00°44'22"
4	575.00	74.00	73.95	S 81°51'25" E	7°22'28"
5	575.00	74.00	73.95	S 74°28'58" E	7°22'28"
6	575.00	49.98	41.95	S 68°42'20" E	4°10'52"
7	135.00	28.35	28.30	S 29°24'02" W	12°01'52"
8	135.00	121.00	116.99	S 61°05'36" W	51°21'16"
9	135.00	16.37	16.36	N 89°45'23" W	6°56'48"

CORPORATE OWNER'S CERTIFICATE OF DEDICATION

HIGHLANDS INVESTMENT CORP., a corporation duly organized and existing under and by virtue of the laws of the State of Wisconsin, as owner, does hereby certify that said Corporation caused the land described on this plat to be surveyed, divided, mapped and dedicated as represented on this plat.

HIGHLANDS INVESTMENT CORP., as owner, does further certify that this plat is required by S. 236.10 and S. 236.12 to be submitted to the following for app roval or objection:

(1) COMMON COUNCIL OF THE CITY OF LAKE
(2) COUNTY PLANNING AGENCY
(3) WISCONSIN DEPARTMENT OF ADMINISTRATION

IN WITNESS WHEREOF, the said HIGHLANDS INVESTMENT CORP., has caused these presents to be signed by JOSEPH C. SMITH, its President, and countersigned by JOHN BROWN, its Secretary, at Lake, Wisconsin, and its Corporation seal to be hereunto affixed this ____ day of ______________, 2000.

In the presence of: HIGHLANDS INVESTMENT CORP.

/s/ JOHN A. GREEN
/s/ JANE B. GREEN

/s/ JOSEPH C. SMITH (SEAL)
Joseph C. Smith, President

COUNTERSIGNED

/s/ JOHN D. BROWN (SEAL)
John D. Brown, Secretary

STATE OF WISCONSIN)
COUNTY OF LAKE) SS

On this the ____ day of ____________, 2000, before me, ________________, the undersigned officer, personally appeared JOSEPH C. SMITH, who acknowledged himself to be the President of the abovenamed corporation and also personally appeared JOHN D. BROWN, who acknowledged himself to be the Secretary of the abovenamed corporation, and that they, as such, being authorized to do so, executed the foregoing Corporate Owner's Certificate for the purposes therein contained, by signing the name of the corporation by themselves as President and Secretary.

IN WITNESS WHEREOF I hereunto set my hand and official seal.

____
Notary Public, JANE E. GREY, Lake County, Wisconsin
My Commission Expires December 31, 2002

LOT AREA

BLOCK NUMBER	LOT NUMBER	AREA IN SQUARE FEET	BLOCK NUMBER	LOT NUMBER	AREA IN SQUARE FEET
1	1	8,900	2	1	10,609
	2	8,639		2	9,365
	3	8,407		3	8,886
	4	8,539		4	7,742
	5	8,476		5	7,492
	6	8,374		6	7,406
	7	8,035		7	7,918
	8	8,334		8	9,174
	9	8,475		9	9,714
	10	8,617		10	8,677
	11	9,435		11	8,838
	12	8,403		12	8,895
	13	8,148		13	9,191
	14	11,086		14	8,235
	15	15,900		15	10,653
	16	17,533			
	17	21,736	3	1	10,280
	18	13,587		2	8,981
	19	10,108		3	8,625
	20	10,716		4	8,678
	21	10,374		5	7,841
	22	8,417		6	7,260
	23	8,290		7	8,146
	24	8,496		8	7,734
	25	8,104		9	8,086
	26	10,411		10	8,799
	27	12,828		11	8,820
	28	16,709		12	9,191
	29	12,023		13	10,411
	30	9,532			
	31	8,801			
	32	8,608			
	33	10,018			
	34	11,605			

ACCESS RESTRICTION

HIGHLANDS INVESTMENT CORP., a corporation duly organized and existing under and by virtue of the laws of the State of Wisconsin, as owner, does hereby restrict Lots 28 through 34 of Block 1, in that no owner, possessor, user, licensee nor any other person shall have any right of direct vehicular ingress or egress with County Trunk Highway A (Jefferson Avenue), as shown on the plat. This access restriction is expressly intended to constitute a restriction for the benefit of the public according to Section 236.293 of the Wisconsin Statutes and shall be enforceable by the City of Lake and by Lake County.

In the presence of: HIGHLANDS INVESTMENT CORP.

/s/ JOHN A. GREEN
/s/ JANE B. GREEN

/s/ JOSEPH C. SMITH (SEAL)
Joseph C. Smith, President

COUNTERSIGNED

/s/ JOHN D. BROWN (SEAL)
John D. Brown, Secretary

CERTIFICATE OF CITY TREASURER

STATE OF WISCONSIN)
COUNTY OF LAKE) SS

I, JOHN F. BLACK, being duly appointed, qualified and acting City Treasurer of the City of Lake, Wisconsin, do hereby certify that the records in my offi ce show no unpaid taxes or unpaid special assessments as of __________________ on any of the lands included in the plat of LAKE HIGHLANDS.

____
Date

/s/ JOHN F. BLACK
John F. Black, Treasurer, City of Lake, Wisconsin

CERTIFICATE OF COUNTY TREASURER

STATE OF WISCONSIN)
COUNTY OF LAKE) SS

I, JOSEPH G. BLACK, being the duly elected, qualified and acting Treasurer of the County of Lake, Wisconsin, do hereby certify that the records in my office show no unredeemed tax sales and no unpaid taxes or special assessments as of __________________ affecting the lands included in the plat of LAKE HIGHLANDS.

____
Date

/s/ JOSEPH G. BLACK
Joseph G. Black, Treasurer, Lake County, Wisconsin

WISCONSIN
JOHN W. DOE
LAKE
WIS.
LAND SURVEYOR

COMMON COUNCIL RESOLUTION

RESOLVED, that the plat of LAKE HIGHLANDS, being a Subdivision of a part of the SW ¼ of the SW ¼ of Section 35, T8N, R9E, in the City of Lake, Lake County, Wisconsin, having been approved by the Plan Commission is hereby approved by the Common Council of the City of Lake, Wisconsin, on this ____ day of __________, 2000.

____
Date

/s/ JOHN H. GRAY
John H. Gray, Mayor, City of Lake, Wisconsin

There are no objections to this plat with respect to Secs. 236.15, 236.16, 236.20 and 236.21 (1) and (2), Wis. Stats., or by the County Planning Agency.

Certified ______________________, 20____

Department of Administration

I, JOAN J. CLAY, do hereby certify that I am duly appointed, qualified and acting City Clerk of the City of Lake and the foregoing is a true and correct copy of a resolution passed and adopted by the Common Council of the City of Lake, Lake County, Wisconsin, this ____ day of ____________, 2000.

AND I DO further certify that copies of this plat were forwarded as required by S. 236.12 (2) on the ____ day of ____________, 2000, and that no objections to the plat have been filed or, if filed, have been met.

AND I DO further certify that the City Plan Commission has approved the use of 50-foot-wide streets in the Lake Highlands subdivision, and that such street widths comply with the City subdivision ordinance.

____
Date

/s/ JOAN J. CLAY
Joan J. Clay, Clerk, City of Lake, Wisconsin

CERTIFICATE OF COUNTY REGISTER OF DEEDS

RECEIVED for record this ____ day of ____________, 2000, at ____ o'clock __.M., and recorded in Volume ____ of Plats on Page ____.

/s/ JOSEPH K. WHITE (SEAL)
Joseph K. White, Register of Deeds, Lake County, Wisconsin

SHEET 2 OF 2

FIGURE 17.2(b) *(Continued)*

The subdivider may elect to proceed with final platting by using the preferred alternative set forth in Section 9.08 of the ordinance. Under that second alternative, the subdivider may enter into a development agreement with the local government and submit a letter of credit, a performance bond, or a certified check covering the total cost of required improvements. Then the same procedure is followed, as in the first alternative, for the recording of the final plat. After recording, the subdivider must install the required improvements to municipal specifications within specified time limits. If the improvements have not been installed at the end of the specified time limits, the local government concerned may obtain the funds from the letter of credit or performance bond, or use the funds in escrow, to install the improvements. In no case, under either approach, should installation of improvements commence before construction plans have been approved by the municipality, a development agreement entered into, and financial guarantees provided.

Required Information for Plats

The data required to be shown on preliminary and final plats are set forth in Sections 4.03 and 5.02, respectively, of the ordinance. In addition to the information required to be shown on a preliminary plat, Section 4.02 of the ordinance requires that a site analysis be submitted for the area included within the preliminary plat, unless the area within the plat is included within a neighborhood unit development plan.

Topography and Steep Slopes

Section 4.03B of the ordinance requires that existing topographic contours be shown on a preliminary plat. Topographic information helps ensure that the proposed land division is appropriate for the site's terrain. Moderately sloped sites are best suited to urban development. If the site is too level, drainage difficulties may be encountered that will require careful design to overcome; and if the site is too steep, it will be difficult to construct streets and buildings without extensive grading. Generally, urban development should not occur on slopes of 12 percent or more.

Floodplains

Sections 4.03C and 5.02G of the model ordinance require that preliminary and final plats, respectively, include the boundaries of the 100-year recurrence interval floodplain and related regulatory stages, as determined by the Federal Emergency Management Agency. Where such data are not available, the floodplain boundaries and related stages are to be determined by a registered professional engineer retained by the subdivider, and the engineer's report providing the required data is to be submitted with the plat for review and approval by the municipal engineer. Floodplain areas are generally not well suited to intensive development, not only because of the flood hazard but also because of the presence of high water tables and, often, of poor soils. Ideally, floodplain areas should be retained in natural, open space uses.

Wetlands

Wetlands perform an important set of natural functions, including stabilizing lake levels and stream flows, reducing stormwater runoff by providing areas for floodwater impoundment and storage, providing groundwater recharge and discharge areas, trapping soil particles and soil nutrients in runoff, thus reducing stream sedimentation and the rate of enrichment of surface waters, and providing habitat for many plant and animal species. Wetlands also have severe limitations for residential, commercial, and industrial development. Generally, these limitations are due to the erosive character, high compressibility and instability, low bearing capacity, and high shrink-swell potential of wetland soils, along with the inherent high water table. Wetlands should, therefore, be protected from development. A number of local, state, and federal regulations have been adopted to protect wetlands, particularly wetlands within shoreland areas and wetlands associated with navigable streams.

Section 4.03F of the ordinance requires that the location of wetlands be shown on preliminary plats. It also requires that the name of the person, agency, or firm identifying the wetlands, and the date of the field survey be included on the plat or certified survey map.

Design Considerations

Section 7.00 of the model ordinance sets forth design standards for proposed land divisions. The design standards are consistent with, and would serve to implement, the principles of good design presented in Chapter 14, including standards related to street widths, lot size and configuration, access control restrictions, pedestrian and bicycle ways, provision for park and school sites, and protection of drainageways and environmentally significant lands. The design standards included in the ordinance will accommodate any and all of the three alternative subdivision designs described in Chapter 14, namely the curvilinear, cluster, and new urbanism design types. Ordinance provisions intended to accommodate such designs include Subsection 7.06B, which would allow the plan commission to vary the design of lots to approve a nonconventional subdivision layout. The ordinance allows alleys in residential subdivisions to support the flexibility needed for new urbanism features.

Required Improvements

Section 8.00 of the ordinance includes requirements for improvements related to survey monuments; street improvements, including surfacing, curb and gutter, sidewalks, lamps, signs, and trees; sewage disposal; water supply; stormwater management; other utilities; erosion and sedimentation control; and landscaping.

Applicable state statutes should allow the governing body of a city, village, or town within which a land subdivision is located to require that the subdivider make and install those public improvements reasonably necessary to serve the proposed land division, and may further require that such improvements be installed at no cost to the local government. It is normally expected that the subdivider will bear the costs of installing all improvements needed to serve the land subdivision. The details related to required improvements should be specified in a development agreement between the subdivider and local government.

Oversized Streets and Utilities

In some cases, a subdivider may be required to install oversized streets, sanitary sewers, water mains, or storm sewers in order to assure implementation of municipal transportation, sanitary sewerage, water supply, and stormwater drainage system plans. The ordinance addresses the allocation of the costs associated with oversized facilities by requiring the municipality, or other unit or agency of government having jurisdiction, to pay for facilities larger than the size that would normally be expected to be adequate to serve a land division. The municipality may then recover the cost of the larger facilities from the owners of the properties benefiting at the time of development.

Off-Site Improvements

The courts have generally decided the legality of the imposition of off-site improvement costs on a subdivider on the basis of one of three tests. The first of these tests, the reasonable relationship test, allows the most liberal interpretation. It permits an exaction if there is a relationship, even though general and long-term, between the identified improvement and a proposed new development. The second test, the rational nexus test, requires a closer demonstrated association between the proposed development and the identified improvement, to the extent of showing a reasonable linkage between the two. The third test, the specific and unique test, is the most stringent. Under this test, a subdivider may be made to pay a share of off-site improvement costs only if an actual need can be shown and specifically attributed to the proposed development.

If an off-site improvement is needed solely because of a proposed new development, and that development will be the sole beneficiary of the improvement, then it should be clear that the subdivider should pay the full cost of the improvement. In most cases, however, the needed improvement will benefit more than just the proposed development, and a fair means of allocating the cost of the improvement among the beneficiaries is required, with the proposed new development paying only its proportionate share. County and local governments commonly use impact fees to equitably recoup the cost of off-site improvements.

Dedication or Fees in Lieu of Dedication for Park and School Sites

Counties, cities, villages, and towns usually have the authority to require a subdivider to dedicate land for park or school purposes, or to pay a fee in lieu of land dedication as a condition of plat approval. The courts have determined that the amount of land to be dedicated or the fee paid in lieu of dedication, must bear a reasonable relationship to the increased demand for such sites attributable to the land division.

Municipal governments have historically relied upon dedication or fees in lieu of dedication to acquire park and school sites needed to serve new land subdivisions. It should be noted that dedication or fee-in-lieu payments for school sites may be required in land subdivision ordinances, but the imposition of impact fees for school facilities may be specifically prohibited under some state statutes.

Section 7.10 of the ordinance requires that a proposed public school site designated on the local government's official map or comprehensive plan and encompassed within a proposed land subdivision be made a part of the plat and reserved at the time of final plat approval for a period not to exceed three years, unless extended by mutual agreement, for acquisition by the school board at a price agreed upon and set forth in the development agreement. The ordinance does not require a dedication or fee-in-lieu payment for school sites.

Section 7.10 further requires that a proposed public playground, park, parkway, trail corridor, public open space site, or other public lands designated on a locally adopted official map or comprehensive plan and encompassed within a proposed land subdivision be made a part of the subdivision plat and dedicated to the public by the subdivider. Should the value of the land to be dedicated be less than the value of the public site fee that would be required for the subdivision under Section 10.06, the ordinance would require the subdivider to pay the local government the difference between the value of the land dedicated and the public site fee. Should the value of the land to be dedicated exceed the public site fee, any lands in excess of the value of the public site fee would be reserved for a period not to exceed three years, unless extended by mutual agreement, for purchase by the local government at the price agreed upon and set forth in the development agreement. If no proposed park lands are located within a proposed land division, the subdivider would be required to pay the public site fee set forth in Section 10.06.

It does not appear fair to require the community at large to purchase sites for neighborhood parks for which only a segment of the entire community population may be expected to derive direct benefit. On the other hand, it does not appear fair to require a subdivider of, say, an 80-acre tract of land to dedicate a park site that will benefit an entire neighborhood comprising in addition to the 80-acre tract, say, 560 acres of surrounding lands. Sometimes, however, the most desirable locations for such needed public facilities may lie within one or two specific tracts, while adjacent tracts may have no such proposed facilities within their boundaries. Therefore, the most equitable approach is to levy a uniform fee per acre or per dwelling unit, in lieu of dedication. Such fees can then be pooled toward the purchase of the park lands necessary to serve a particular neighborhood unit. In this way, each subdivider pays an equitable share of the total cost of the public sites and can pass this cost on to the purchasers of lots that will directly benefit from the facilities.

The legality of a required dedication, or fee in lieu of such dedication, will be enhanced if the dedication or fee bears a reasonable relationship to the stated purpose of the regulation concerned, that is, to the provision by the subdivider of those park sites that are reasonably required for use by future residents of the proposed land division. Accordingly, any fees collected and pooled should be used to serve the neighborhood within which the proposed land division is located.

Surveying Requirements

When a control survey network, such as described in Chapter 5, is in place within a planning area, survey requirements similar to those contained in Section 5.06 of the model ordinance should be enforced.

Protective Covenants

Protective covenants are a way of ensuring long-term quality and consistency in residential neighborhoods by private rather than public action. Such covenants are legal agreements between the subdivider and the lot purchasers in which all parties seek to gain certain advantages: the subdivider to aid his or her development program, and the purchasers to protect their investments. It is an agreement to which all homeowners in the area are bound, and it gives assurance that no owner may use his or her land for any purpose that tends to reduce the environmental quality of the area, lower property values, or create a nuisance. To assure the legality and validity of the covenants, they should be drawn by a qualified attorney, and should then be recorded as a public declaration of restrictions with the county register of deeds in order that they may be transferred by deed from one owner to another. If restrictive covenants are required by a local government to protect and manage common open space within a proposed land division or condominium, the covenant should be reviewed as to form by the municipal attorney.

Protective covenants may be used as supplements to city, village, town, or county zoning regulations. Whereas zoning districts and regulations may be modified at any time by the local government, protective covenants are agreements binding on all parties owning lands in the land division for a given period of time, usually from 20 to 30 years. After such time, the covenants are usually automatically extended for successive periods of a set number of years unless changes are agreed upon by a majority of the property owners within the restricted land division. In order to achieve design and development objectives, the protective covenants can be more restrictive than the local zoning ordinance and land division regulations.

Protective covenants should be required by local governments to protect common open space or other common areas retained in private ownership within a subdivision or condominium. A homeowner or condominium association should be created by the subdivider to manage and maintain the common open space and any facilities located within common areas. Section 2.06 of the ordinance sets forth regulations governing the creation of homeowner and condominium associations. Importantly, as set forth in Subsection 2.06J, the local government should reserve the right to maintain common open space and facilities and to assess that cost to the owners of property within the land division, should the homeowner or condominium association fail to properly maintain such areas and facilities.

IMPROVEMENTS

The creation and maintenance of attractive and stable urban areas require that good land division design be accompanied by the proper installation of adequate public utilities and facilities. Requiring such improvements at the time of land division benefits the future residents of the land division. The installation of high-quality, durable improvements also has a direct bearing on future operation and maintenance costs, which will be incurred by the community virtually in perpetuity. Requiring developers to install adequate improvements also tends to discourage excessive land division, which in the past has created serious problems for many local governments.

Improvement standards should vary with the type of development, differing for rural and urban areas, and for single-family residential, multi-family residential, commercial, and industrial areas. All improvements should be designed and installed in accordance with good municipal engineering standards, and the design and installation of what will become an integral part of the public infrastructure system should be subject to the approval of the city, village, town, or county engineer concerned.

Minimum improvements in urban residential areas should include survey monuments, street grading to the full street width in accordance with community-approved cross-sections and to established street grades, permanent roadway pavements, adequate stormwater management facilities, and public sanitary sewers and public water supply distribution mains. In higher-density urban areas, concrete curb and gutter and piped storm sewers may be required. Curbs may be vertical-faced, roll-faced, or mountable. Improvement standards may also require the installation of sidewalks, street lights, street signs, the planting of street trees and seeding of terraces, and other landscaping. A land division ordinance may also require the installation of sanitary sewer, stormwater, and water supply laterals to the lot lines in order to protect newly installed roadway pavements.

In low-density residential areas, on-site sewage disposal systems and private wells may be installed in lieu of public sanitary sewer and water supply facilities under suitable topographic and soil conditions. Careful attention must, however, be given to lot size in order to properly accommodate the necessary on-site sewage treatment and disposal facilities. In low-density residential areas, in lieu of the use of curb and gutter and storm sewers, the stormwater management system may take the form of roadside ditches and parallel and cross culverts discharging to open drainage channels.

Street Improvements

The creation of a desired appearance or character in the finished land division requires careful attention to the details of proposed street improvements and landscaping. The appearance of the land division will be affected by, for example, the use of asphalt as opposed to Portland cement concrete for roadway and sidewalk pavements; the use of vertical face as opposed to rolled face or mountable curb and gutter; the use of decorative as opposed to utilitarian street lighting fixtures; the design of street name signs; and the type and spacing of street trees and other landscaping.

Street improvements should be carefully related to the function that each street in a land division is intended to perform. Roadway widths should be determined on the basis of the functional classification of the street, the anticipated traffic volume, the number of parking and traffic lanes to be accommodated, the type of bicycle improvements to be provided, and the anticipated volume of bus or truck traffic. Right-of-way widths should be established on the basis of the roadway width required and attendant areas needed for terraces and sidewalks in urban areas and roadside drainage swales in rural areas, as well as requirements for storm and sanitary sewers, water distribution lines, and other utilities to be installed in the street right-of-way. Standard locations within the street rights-of-way for underground locations should be required based on locally adopted utility placement standards or applicable state Department of Transportation (DOT) guidelines. In addition to the street cross-section, careful attention should be given to street alignments and grades and to the treatment of intersections with respect to the configuration of the curb returns and the treatment of pavement crowns.

The arterial street and highway system is intended to provide a high degree of travel mobility, serving the through movement of traffic between and through urban areas. The right-of-way width and pavement widths and configurations of arterial streets and highways should be based on adopted county and municipal transportation system plans. Collector and land access streets should be designed to carry only locally generated traffic, and cross-sections for such streets should be specified in the local land division ordinance.

Within the constraints of good engineering practice, it is generally desirable to hold pavement widths to a minimum. Use of minimum pavement widths reduces capital and maintenance costs, reduces the amount and rate of stormwater runoff, and reduces nonpoint source water pollution. Collector and land access street right-of-way and pavement widths should be determined on the basis of careful consideration of the street pattern and abutting development. A detailed neighborhood unit development plan provides the best source of information required to identify specific traffic and parking conditions related to each proposed land access and collector street, and permits attendant desirable right-of-way and pavement widths to be properly selected.

Particular attention should be given to the improvement of cul-de-sac streets. A landscaped island in the center of the turnabout greatly enhances the appearance of the cul-de-sacs. Maintenance responsibilities for the center island should be specified in the local land division ordinance. Landscaping within the center island may include new plantings or may involve preservation of existing vegetation, including mature trees, and should be installed or preserved in accordance with a landscaping plan approved by the local government concerned. Where feasible, cul-de-sac designs should allow pedestrian or bicycle connections to adjacent streets or paths to improve connectivity and reduce vehicle miles traveled.

In urban areas, terraces—sometimes known as curb lawns—should be provided between the curb and the inside edge of the sidewalk. Such terraces provide separation between vehicular and pedestrian traffic and thereby a more pleasant environment for pedestrian traffic. Such terraces also provide an area for the location of street signposts, streetlights, utility poles, fire hydrants, buried utility lines and mains, and mailboxes, for street trees and other landscaping, for driveway aprons, and for snow storage. Terraces that are to contain trees should be at least six feet wide, and desirably eight feet or wider, to allow sufficient space for the tree root system and to minimize damage by tree growth and adjacent pavements, especially sidewalks. Consideration should be given to soil volume and species selection to ensure long-term tree health in narrow terraces.

STREET LAMPS

Generally, street lamps should be required at each street intersection and at such interior block spacing as may be required by the municipal engineer. Developers should be given the option of installing street lamps designed to complement the neighborhood or type of development proposed or to help establish the character of the land division, as an alternative to standard utilitarian street lights, subject to the approval of the municipality concerned. The municipality may also allow the developer to require by covenant the installation of private post lighting in the front yard of each residence, in lieu of or to complement public street lamps. Private post lighting may contribute to neighborhood character, but municipalities should ensure consistency in lighting quality and coverage. Care should be taken to avoid overly intensive lighting, and fixtures should be selected that direct light downward onto pavement and walk surfaces, reducing glare and avoiding what has been termed "light pollution." Lighting plans should consider energy-efficient fixtures and dark-sky-compliant designs.

Street Name Signs

Street name signs should be installed at all street intersections. Such signs should meet municipal design specifications and should be uniform along arterial streets and highways and, generally, throughout a municipality. The municipality may permit use of variations in street signage on the interior streets of land divisions to lend character to a development, provided such signs are clearly legible as well as attractive. A municipality may also require the use of a special street sign design approved for use throughout the municipality. Care should be given to the selection of street names and to the assignment of street addresses within land divisions. Desirably, local municipalities should have adopted street naming and address numbering plans to guide such selection and assignment. Emergency services coordination should be considered in all naming conventions to avoid confusion or duplication.

Street Trees

Generally, the planting of trees along new land division streets should be required at a minimum average spacing of one tree per 50 feet of street frontage. Municipalities may require such trees to be planted at regular intervals or allow the trees to be grouped in irregular or informal patterns, in accordance with a landscaping plan submitted by the developer. Street trees should be planted between the curb and sidewalk along streets having an urban cross-section and between the drainage swale and the outside edge of the right-of-way for streets having a rural cross-section. Requirements for street tree spacing may be waived on the basis of an approved landscaping plan that provides substantial alternative landscaping, including trees. Care should be used to ensure that trees and other landscape materials do not interfere with the location and maintenance of public and private utility lines. Tree species should be selected to be climate-resilient, non-invasive, and compatible with local soil conditions and urban infrastructure.

Landscaping

Landscaping within a land division enhances the overall attractiveness of the land division and the community as a whole. Landscaping should be provided within or adjacent to easements used to control access to arterial and collector streets, along the rear of double frontage lots along arterial streets to provide screening between residences and the arterial street, and within street medians and islands of cul-de-sac turnabouts. Landscaping should also be required at entrances to land divisions and within common open space areas. Landscaping should be considered an important element of required improvements. The provision of trees and other plant materials by the developer, in accordance with a landscaping plan approved by the local government, should be required by the local land division ordinance. The landscaping plan should provide for the maintenance of vision clearance areas at street intersections. Requirements regarding the quantity, spacing, maintenance, stormwater compatibility, and type of plant materials should be set forth in the land division ordinance. The use of land division entrance signs and attendant landscaping has become common. Care should be taken that such signs are made attractive and not located in public rights-of-way, and that arrangements for proper maintenance are provided. Landscaping plans should prioritize low-maintenance native species and support pollinator-friendly environments.

Grading

A land division ordinance should require that plans for the rough and finish grading of a land division be submitted by the developer for approval by the municipal engineer. Such grading plans should provide for a proper relationship of the elevation of finished building pads to street grades and environmental features, thereby creating a good architectural setting for finished buildings. Grading plans must also carefully consider such details as desirable driveway grades and good drainage of each building site. Great care must be taken in the preparation of grading plans to assure adequate overland flow paths where such paths are required as an integral part of the major stormwater management system. Such paths must provide effective routes to receiving watercourses and have adequate capacity to carry peak rates of discharge without flooding adjacent buildings. Every effort should be made to minimize grading in site preparation and to preserve existing desirable trees, which can lend value to the finished lots and greatly enhance the appearance of the land division. Grading of areas to be used for on-site sewage treatment and disposal should be avoided, as well as the operation of heavy construction equipment across them. Certification by a registered land surveyor of conformance of the finished grading to an approved grading plan may be required by the local government prior to the issuance of any building permits. Where stormwater management relies on low-impact development (LID) approaches, grading should support the natural hydrology and maintain infiltration zones.

Stormwater Management

Stormwater management facilities should be adequate to serve a proposed land division. Such facilities may include curbs and gutters, catch basins and inlets, storm sewers, parallel and cross culverts, stormwater storage facilities for both quantity and quality control, roadside swales or ditches, overland flow paths, other open channels, and infiltration facilities. Drainageways should incorporate naturalized, vegetated channels wherever feasible to support water quality goals and habitat creation. Drainage facilities should be of adequate size and grade to accommodate runoff from the specified design rainfall event and should be configured so as to prevent and control soil erosion and sedimentation. Stormwater infrastructure should be designed as part of a multi-scale, integrated system that manages both minor and major storm events. Stormwater management facilities within a subdivision must comply with the requirements of any adopted county or municipal stormwater management system plans and stormwater management ordinances. Where applicable, facilities should be aligned with regional green infrastructure or resilience plans.

Drainage Considerations

Some communities have adopted a policy of employing rural cross-sections for street improvements in what are really urban areas. Such rural cross-sections employ roadside ditches together with parallel and cross culverts for drainage as opposed to curb and gutter and storm sewers. Although such rural cross-sections generally have a lower initial cost than typical urban sections, they can lead to higher long-term maintenance costs, reduced pedestrian safety, and limited stormwater quality control. The use of rural cross-sections may be justified, however, in residential areas that have a relatively low density of population and that desire to maintain a rural appearance for aesthetic reasons. When so employed, the use of rural cross-sections presents design problems that are not encountered in the use of the urban cross-section and that require careful engineering for proper resolution. These problems relate to the establishment of street grades and the design of attendant drainageways and structures. Since the use of a rural cross-section usually dictates that all stormwater must be carried and disposed of by means of surface drainage channels, the proper establishment of street grades becomes more critical than when storm sewers are available, and it is absolutely essential that the street grades be established in accordance with an overall plan encompassing the entire drainage area involved. The necessary areawide street grade study is best accomplished on the basis of a detailed neighborhood unit development plan or planning layout, which shows all planned streets in the drainage area, together with topographic contour lines having a vertical interval of no more than two feet.

Failure to carefully consider drainage in the land division design and improvement may result in costly property damages when major rainfall events occur. Runoff may accumulate in low spots in the street system and in the center of blocks, may flow across low-lying lots and blocks, and sometimes into and through buildings. Developers should be required by the land division ordinance to provide the municipal engineer with a thorough analysis of how both the minor and major stormwater drainage facilities are expected to function during major rainfall events. This analysis should include climate-adjusted rainfall projections, peak flow modeling, and evaluation of critical drainage paths and flow routing under future development scenarios.

Stormwater Storage and Infiltration Facility Considerations

Stormwater storage facilities may be designed as "dry ponds," "wet ponds," or infiltration basins. Dry ponds temporarily store and later release stormwater runoff, draining dry between rainfall events in order to reduce peak rates of runoff. Wet ponds, while functioning like dry ponds to control the rate of stormwater runoff, maintain a permanent pool to also provide water quality control. Infiltration basins, which store runoff for infiltration and evaporation, may be designed to provide for no release of runoff except under excessive rainfall events, and, like wet ponds, provide both water quality and quantity control. The application of any of these types of facilities requires careful systems and project development engineering and sound long-term arrangements for proper maintenance.

Where stormwater storage or infiltration facilities are required for control of the rate and quality of stormwater runoff, the construction plans and specifications should be subject to review and approval by the municipal engineer. The general shape of required storage of infiltration ponds should blend into the surrounding topography. Care should be taken to provide an adequate safety shelf around the perimeter of storage and infiltration ponds.

Although developers may be required to bear the capital and maintenance costs attendant to stormwater storage and infiltration facilities, it should be recognized that such facilities may present future problems with respect to water quality and water level conditions and may create public safety and health hazards. Serious problems may be created if the landowners do not properly maintain the facilities. Accordingly, legal and physical provisions should be made for maintenance by the municipality, permitting the assessment of the costs entailed to the facility owners concerned. Such owners may consist of homeowners associations created as a part of the land subdivision process. Maintenance agreements should include contingency plans, enforceable funding mechanisms, and inspection requirements. Consideration should also be given to integrating green infrastructure such as rain gardens, bioswales, and constructed wetlands where feasible.

Public and Private Utilities

The location and installation of private utilities in a new land division is normally the responsibility of the private utility companies concerned. Utility corridors should be strategically planned with multi-utility coordination to minimize conflicts and future excavation. Gas mains are typically installed within the public right-of-way, while electric power and communication cables are commonly placed within designated utility easements, often located along rear or side lot lines. To promote aesthetics, safety, and climate resilience, the underground installation of electric and communication infrastructure should be required in all but the most exceptional cases.

Sanitary sewers, water distribution mains, and storm sewers can also normally be accommodated within public street rights-of-way. These utilities should, however, be carefully sized in accordance with municipal sanitary sewerage, water supply, and stormwater drainage system plans. Municipalities may require developers to install oversized utility lines to support future growth or system redundancy. Where this occurs, municipalities should ensure transparent cost-sharing mechanisms or reimbursement policies are in place to equitably cover excess infrastructure costs not solely benefiting the development.

Land division improvement costs will vary widely, not only with required standards, but also with location, site characteristics, and the quality of the land division design itself. While high-quality improvement standards are generally desirable, requiring over-improvement may do as much to impede good development as under-improvement. As already noted in this respect, it is manifestly unfair to require developers to install oversized utility lines for the benefit of areas that may develop beyond the limits of a proposed land division without assessing the excess costs of such oversized utilities to the benefited area or to the community at large.

Construction plans and profiles for required sanitary sewers, water mains, and storm sewers may be prepared by municipal engineering staffs, by consulting engineering firms employed by the municipality, or by consulting engineering firms employed by the developer. In any case, the improvement plans and profiles should be prepared to good municipal engineering standards. Careful attention should be given to such details as sewer systems, manhole spacings and locations, manhole rim and inlet grate elevations, and depths to sewer crowns and inverts. For water supply systems, careful attention should be given to such details as the number and location of valves and hydrants, the avoidance of dead-end mains that may result in future water quality and maintenance problems, and the location of any needed air release valves.

Upon completion of the infrastructure improvements, and particularly for the public underground utilities, accurate "as-built" records should be provided to the municipality. The horizontal location of all key facility components such as sanitary and storm sewer manholes, storm sewer inlets and catch basins, water main valves and fittings, and hydrants should be recorded using standardized coordinate systems (e.g., the State Plane Coordinate System or a local geographic information systems [GIS] reference system) aligned with the municipality's parcel-based geospatial information systems. These positions can be readily and accurately determined even in curvilinear layouts utilizing modern survey-grade Global Navigation Satellite Systems (GNSS) technology while the installation trenches are open. The vertical location of those facility components should be similarly determined including rim, grate, and invert elevations of all sewer manholes, inlets, and catch basins, and the crown elevations of all water mains and fittings. The elevations should be referred to the vertical datum used as a basis for the municipality's public works management system, and can also be determined utilizing GNSS technology.

Similar "as-built" data should be required for private underground utilities such as gas mains and power and telecommunication cables. Obtaining accurate "as-built" records utilizing GNSS-based methods will make the "as-built" records, in effect, scale independent and will facilitate the integration of new infrastructure data into the municipality's land and public works information systems. This data should be submitted in digital formats compatible with the municipality's GIS and asset management systems and updated as infrastructure is modified or replaced over time.

STANDARDIZATION OF UTILITY LOCATIONS

Standard underground utility corridors, such as those illustrated in Figure 17.3, should be adopted and maintained by municipalities as part of a broader utility coordination framework. These standardized locations should be integrated into the municipality's comprehensive plan and supported by enforceable local ordinances. Coordination should involve key stakeholders, including municipal engineering departments (e.g., water, sewer, drainage, transportation), public and private utility providers (e.g., gas, electric, broadband), and, where applicable, transit agencies and emergency services. The utility coordination plan should mandate the creation and maintenance of accurate digital "as-built" records for all underground and aboveground utility infrastructure, covering both existing facilities and all future installations. Key components of the utility location standardization plan should include the adoption of consistent geospatial standards (scale, projection, datum) for utility base mapping to ensure GIS interoperability; designation of standardized corridors within public rights-of-way and easements for each utility type to avoid conflict and facilitate maintenance; standard construction specifications for

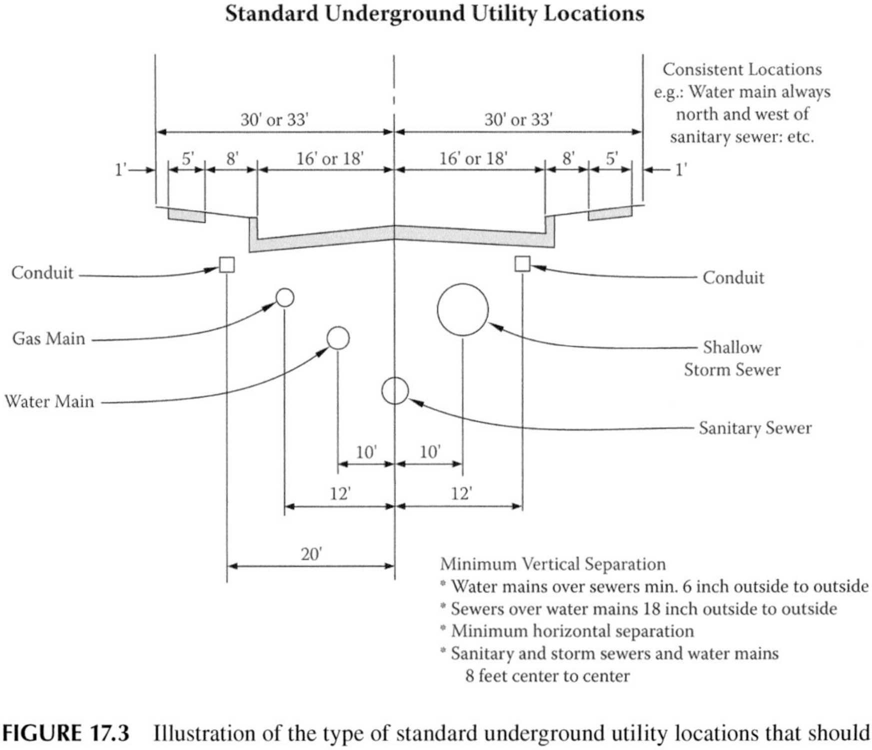

FIGURE 17.3 Illustration of the type of standard underground utility locations that should be promulgated by municipalities as an element of the comprehensive plan relating to utility locations and service areas.

manholes, handholes, vaults, duct banks, and conduits; permitting protocols to regulate utility taps, pavement cuts, and repair responsibilities; a rolling capital improvement plan (minimum five-year horizon) that coordinates public utility and roadway projects to minimize duplication and excavation; and digital infrastructure integration, including utility data layers within a municipal asset management and land information system.

IMPROVEMENT GUARANTEES

Improvement guarantees are critical tools for ensuring the timely and quality installation of required land division improvements, including survey monuments, grading, streets, and public utilities. Municipalities should have the authority to require the installation of improvements before final plat approval.

To address developer financing constraints, the municipality may, as already noted, require the posting of an irrevocable letter of credit. The amount of the letter of credit must be adequate to allow the municipality to design and install the required improvements if the developer defaults. An alternative to the use of a letter of credit is the use of a performance bond. The letter of credit is issued by an independent financial institution, and the municipality concerned is made the beneficiary of the credit. The municipality is entitled to collect on the credit prior to pursuing any claims against a developer who may default on the installation of required improvements. The issuer is obligated to make payment whenever the beneficiary presents the letter of credit and any related documents required by the letter. In many cases, the accompanying documents may consist solely of an affidavit by the municipality stating that the developer is in default. Modern best practices recommend setting guarantee amounts at 110–125 percent of the estimated cost to account for inflation, administrative overhead, and contingency. The guarantees should be released promptly once the improvements are completed, inspected, and accepted by the municipality. Digital tracking of guarantee instruments and completion milestones is now increasingly common in municipal permitting software, enhancing transparency and accountability.

FUTURE LAND SUBDIVISION CONTROL TRENDS

Emerging trends in land subdivision control reflect the growing need to integrate sustainability, resilience, and technological innovation into community planning practices. Increasing emphasis is being placed on incorporating green infrastructure, such as bioswales, rain gardens, and permeable pavements, to support stormwater management and reduce environmental impacts. Greater attention is also being given to long-term maintenance responsibilities and lifecycle costs of infrastructure. In addition, many municipalities are adopting digital tools for subdivision review and permitting, including GIS, online submission platforms, and automated compliance checks. Equity-oriented planning considerations are also shaping subdivision controls to ensure inclusive access to infrastructure and amenities across neighborhoods. As housing needs evolve, local governments may encourage compact and mixed-use development forms through flexible subdivision standards that support walkability, transit accessibility, and affordability. These emerging directions are expected to shape the next generation of subdivision ordinances and associated administrative procedures.

FURTHER READING

American Society of Civil Engineers Manual of Engineering Practice No. 14, "Location of Underground Utilities," 1937.

Zevenbergen, Jaap, et al., *Advances in Responsible Land Administration*. CRC Press, 2016.

18 Plan Implementation—Zoning

In analyzing any case in which it is claimed that land is taken: without just compensation—whether the regulated land is a wetland within a shoreland area, or land within a primary environmental corridor, or an isolated swamp—the test to be applied is the same; public benefit versus public harm. Where land is taken for the public benefit, the taking is compensable. However, if the land is regulated to avoid a public harm, then the regulation is not compensable unless the regulation results in a value diminution to the landowner which is so great as to amount to a confiscation.

Justice William G. Callow
Wisconsin Supreme Court,
M&I Bank v. Town of Somers, 1987

INTRODUCTION

Zoning is one of the major plan implementation devices available to local units of government. A zoning ordinance is typically designed to implement the land use element of a comprehensive plan by regulating the private use of land. The formulation and administration of a zoning ordinance are thus important planning activities. As indicated in Chapter 8, the urban land market tends to group together like kinds of land uses. Zoning may be seen as systematizing and legalizing this market-driven process. Because it so directly affects property rights, zoning is fraught with legal complications, and the preparation of a zoning ordinance requires the assistance of an attorney experienced in land use control law.

DEFINITION

By definition, a zoning ordinance is a public law that regulates and restricts the use of private property in the public interest. Zoning ordinances are based on the police power, that is, on the power of the community to make regulations for the purpose of promoting the public health, safety, and general welfare. The police power permits public restrictions to be placed on the use of private property without payment of compensation. A zoning ordinance divides a community into a number of districts for the purpose of regulating the use of land and buildings, the height and bulk of buildings, and the density of population. In regulating the use of land and buildings, the zoning ordinance attempts to confine certain land uses to those areas of the community that are particularly suited to, and should be set aside for, those particular uses, thereby implementing the community land use plan and encouraging the most

DOI: 10.1201/9781003500384-18

appropriate use of the land throughout the community. In regulating the height and bulk of buildings, the zoning ordinance attempts to ensure adequate light and air to adjacent private properties and to public streets, reduce fire hazard, and provide for off-street parking and loading. In regulating the density of population, the zoning ordinance attempts to assure good sanitation, abate traffic congestion, and prevent the overcrowding of land. These three types of regulations are often referred to as use, height, and area regulations, respectively.

Obviously, a single set of regulations applying to the entire community would not be appropriate since the different areas of a city differ in character and function. In this respect, zoning differs from building or sanitation codes that, in general, apply uniformly to all land and buildings of like use wherever they may be located in a community. Zoning regulations may be different for different types of districts but must be uniform within any one district. Zoning ordinances are evolving to be more inclusive and adaptable, accommodating a mix of uses where appropriate, encouraging flexible design, and supporting social equity goals.

A zoning ordinance then consists of two parts: a written text setting forth the regulations that apply to each of the various districts together with general information about the regulations, and a map delineating the districts to which the various regulations apply. Modern zoning ordinances may also include overlay districts, form-based codes, or performance standards that supplement or replace conventional zoning to achieve desired urban form, sustainability, and livability outcomes. Digital zoning maps and integrated geographic information system (GIS) are increasingly used to enhance accessibility and transparency.

BRIEF HISTORY

The history of zoning is long, complex, and convoluted, involving case as well as statutory law. Land subdivision control regulates the use of private property, but does so only as a property owner proposes to subdivide and develop a specific ownership parcel. Zoning, however, regulates the use of all of the real property within a community and, therefore, is much more apt to be perceived and regarded as an interference with the exercise of private rights. Real property consists of intangible interests—or rights—in land. These interests encompass the right of disposition as the owner may see fit, the right of exclusive use, and the privileges of use. Zoning focuses on the privileges of use portion of the bundle of property rights, and only on those privileges. Monarchs and legislatures have long placed limitations on these privileges of use, demonstrating that real property ownership does not carry with it any absolute right to use as seen fit by the individual owner.

Early examples of crude zoning through regulation of the location of industrial uses include an act of the English parliament in 1581 regulating the location of iron mills; an act of the colonial Massachusetts legislature in 1692 regulating the location of slaughterhouses, still houses, chandleries, and tanneries; and an act of the English parliament in 1835 regulating the location of stationary steam engines. In 1889, the Wisconsin legislature authorized cities to designate zones with varying regulations as to buildings and structures depending upon the fire risk involved. In 1892, Boston introduced height zoning, and in 1910, Los Angeles introduced use zoning.

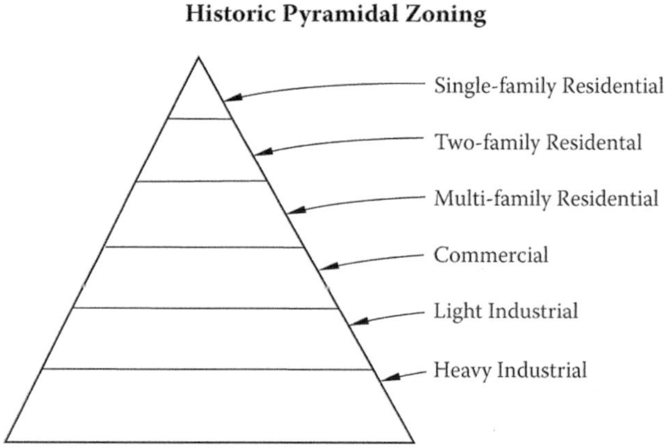

FIGURE 18.1 Illustration of the concepts involved in the historic pyramidal zoning ordinances. Such ordinances envisioned a hierarchy of uses but permitted all "higher" uses in "lower" zoning districts. This permitted, for example, the undesirable intrusion of residential uses into industrial districts and the creation of serious nuisance conditions.

The first comprehensive zoning ordinance in the United States was enacted by New York City in 1916. Interestingly, some early comprehensive zoning ordinances—including the City of Milwaukee ordinance—separately regulated height, use, and density so that the ordinances required the use of three zoning district maps, the district boundaries often having different configurations, resulting in complex overlays and difficulties in administration. Historical zoning ordinances were also often organized in a hierarchy of uses, with single-family residential use being regarded as the "highest" use and heavy industrial use as the "lowest," as illustrated in Figure 18.1. In these "pyramidal" ordinances, the higher uses were permitted in the lower use zoning districts. Experience proved this to be a major mistake in that it destroyed the integrity of the lower use districts, negating one of the purposes of good zoning, namely, to set aside land needed for commercial and industrial development. The practice often resulted in the development of scattered pockets of residential uses in commercial and industrial areas, with resulting complaints by the residents concerned of noise, vibration, glare, and air pollution, and the filing by the residents of nuisance action lawsuits against the commercial and industrial use owners. The basic U.S. Supreme Court case upholding the legality of zoning was decided in 1926 in the case of the *Village of Euclid v. the Amber Realty Company.* Zoning, however, remains the subject of almost continuous litigation, and court decisions actively and continuously serve to modify the principles and practices involved.

The courts have not always agreed with the planning profession on the most basic planning concept involved, namely, that zoning should be a plan implementation device. The courts have considered this issue somewhat indirectly through the definition of the term *master plan* or *comprehensive plan* as related to the preparation of a zoning ordinance. The position of the courts has varied from a finding that a zoning ordinance is in accord with a comprehensive plan if it regulates use, height, and area; to a finding that a zoning ordinance is in accord with a comprehensive plan if it is based on careful and comprehensive study by the enacting agency prior to adoption; to the finding that a zoning ordinance must be based on a "master plan" of land use. The second position is probably the one taken by most courts; the third is the position taken by the planning profession.

Court cases have also frequently involved the issue of whether the zoning regulations, in effect, constitute an unreasonable taking of private property rights without payment of compensation. This issue is currently raised by private property rights advocates who propose "compensatory zoning." Under this concept, municipalities would have to reimburse property owners for any reduction in the value of real property that a zoning ordinance might cause. Aside from the practical problem of determining such a reduction, the concept would, in effect, destroy zoning together with all of its benefits to the community. Most municipalities would be financially unable, or politically unwilling, to assume the unknown—but large—fiscal burden entailed. The concept represents a selfish and greedy disregard for the wisdom on which the police power is based. It would substitute the promotion of private gain for the protection and promotion of the public health, safety, and general welfare for which the police power exists, and would ignore the need to further the well-being of the community as a whole. Two court decisions made more than 100 years apart may be cited as addressing this central issue:

> We think it is a settled principle, growing out of the nature of well ordered civil society, that every holder of property, however absolute and unqualified may be his title, holds it under the implied liability that his use of it may be so regulated that it shall not be injurious to the equal enjoyment of others having an equal right to the enjoyment of their property, nor injurious to the rights of the community …
>
> There are two reasons of great weight for applying this strict construction of the constitutional provision to property in land: 1st, such property is not the result of productive labor, but is derived solely from the State itself, the original owner; 2nd, the amount of land being incapable of increase, if the owners of large tracts can waste them at will without State restriction, the State and its people may be helplessly impoverished and one great purpose of government defeated.
>
> **Chief Justice Shaw**
> *Supreme Court of Massachusetts,*
> *Commonwealth v. Alger, 7 Cush. 53 Mass. (1851)*

> Nothing this court has said or held in prior cases indicates that destroying the natural character of a swamp or a wetland so as to make that location available for human habitation is a reasonable use of that land when the new use, although of a more economical value to the owner, causes a harm to the general public …
>
> While loss of value is to be considered in determining whether a restriction is a constructive taking, value based upon changing the character of the land at the expense of harm to public rights is not an essential factor or controlling …
>
> The land belongs to the people … a little of it to those dead … some to those living … but most of it belongs to those yet to be born …
>
> **Chief Justice E. Harold Hallows**
> *Supreme Court of Wisconsin,*
> *Just v. Marinette Co. (1972), 56 Wis 2d 7*

Benefits of Good Zoning

Good zoning provides a number of benefits to the community. Good zoning assures the maintenance of the integrity of those areas identified on the land use plan as best suited to certain land uses, thereby promoting the most appropriate and desirable use of land throughout the community. Good zoning promotes the permanency of desirable home surroundings, helping assure that residential areas are free from the noise, pollution, and safety hazards of commercial and industrial areas and have abundant light and air. Good zoning exercises a favorable influence on traffic and parking problems. By regulating the intensity and location of development, zoning helps mitigate congestion, enhance pedestrian safety, and promote multimodal transportation. Modern zoning ordinances often incorporate parking maximums, shared parking arrangements, and support for transit-oriented development. Good zoning reduces fire hazards by requiring appropriate setbacks, building separations, and emergency access. Good zoning facilitates the design and provision of public utilities such as sanitary sewerage, water supply, stormwater drainage, and well-designed roadway pavements, enhancing their economy and efficiency. Good zoning also facilitates the orderly selection and preservation of future sites for such community facilities as schools, parks, playgrounds, and fire stations. Form-based codes and overlay zoning districts have also been adopted in many communities to guide the design quality of the built environment and promote neighborhood character. Finally, good zoning contributes to the stabilization of the tax base by helping prevent the deterioration of residential and commercial areas and the attendant destruction of real property values. The benefits of good zoning are readily apparent to lay citizens, and the restrictive zoning of residential areas is often one of the more popular and strongly supported actions of local municipal government. Without a good zoning ordinance, implementation of the comprehensive plan—including housing, transportation, environmental, and economic development goals—is severely compromised.

Zoning Techniques

The preparation of a zoning ordinance and map is a complex planning and legal task calling for exhaustive studies and sound judgment based on practical experience in the field of planning and planning law. A zoning ordinance must be carefully tailored to the community and its existing land use pattern, or there may be hardships imposed on individual property owners that may result in costly lawsuits and judicial findings of unconstitutionality. The zoning text and map must be prepared so as to bear a just relationship to existing conditions and yet so as to implement the future growth of the community along better lines as envisioned in the land use plan. If challenged in court, the municipality must be able to show sufficient data on which the ordinance is based to meet the legal requirement of reasonableness.

Drafting and subsequent amendment of the zoning ordinance should be the responsibility of the plan commission and planning staff, assisted by an experienced attorney. The drafting should be initiated with a review of the state enabling legislation to determine procedural requirements. The necessary basic data should then be assembled. This should include a current existing land use inventory, along with built form data such as setbacks, lot dimensions, floor area ratios (FAR), impervious surface coverage, and building heights. The necessary land use inventory should be available as a part of the planning data bank maintained by the planning agency. These characteristics can be analyzed using GIS-based parcel databases and 3D mapping, often supplemented by LiDAR or drone imagery. The existence of a computerized, parcel-based land information system will significantly streamline analysis and public communication.

Statistical analyses of the assembled data by an experienced planner is essential to determine representative values of existing zoning metrics and built form characteristics, together with the existing land use data, by delineated analysis areas that may range in size down to individual blocks. A tentative zoning ordinance and attendant zoning district map can then be prepared. The number and type of districts, and the detailed regulations applicable to each district, must be based on careful consideration of the findings of the statistical analyses of the assembled basic data and on the recommendations contained in the adopted land use plan. Issues to be addressed include the number and type of districts required, the sizes, configurations, and locations of the districts as they are to be affixed to the cityscape, and the specific regulations that are to govern each type of district.

The types of districts and attendant regulations may be expected to be quite different for application in areas of existing land use development than for application for areas identified for new development—or redevelopment—in the land use plan. An analysis of the tentative ordinance and map should then be made to identify the extent to which non-conforming uses would be created. Based on that analysis, the tentative ordinance and map may be adjusted and presented for public hearing by the plan commission. The ordinance and maps may then be further adjusted based on the results of the public hearing, and a final recommended ordinance and map approved by the plan commission for presentation to the governing body.

The governing body—common council, village board, town board, or county board—may wish to hold a second hearing on the recommended ordinance. Following such a hearing and any further adjustments to the final zoning ordinance and attendant district map, it may be adopted by the governing body and placed into effect. Depending upon the structure of the cognizant government, responsibility for administration of the ordinance may be lodged with a zoning administrator in the planning department or a building inspector in the public works or engineering department. Regardless of the administrative structure, zoning enforcement and interpretation should be coordinated across planning, legal, and permitting functions, with robust digital tracking systems and opportunities for public feedback.

ZONING DISTRICT MAP

A zoning district map constitutes the means by which the regulations set forth in the text of the zoning ordinance are actually related to the land. While the text of the zoning ordinance sets forth the regulations to be imposed within each zoning district, the indication of where these regulations apply is given by the zoning map. Because it must relate the zoning regulations to the natural and man-made features of the urban and rural landscape in a reasonable manner, the zoning district map is a key element of the zoning ordinance. A good base map is a prerequisite for the preparation of a zoning district map.

The zoning district map should show lake shore, stream, wetland, and floodland boundaries; corporate limit lines; public street and railway rights-of-way lines; public ownership boundaries; and all real property lines. In U.S. Public Land Survey System states, the zoning district map should show the U.S. Public Land Survey township, range, section and quarter-section lines, and identification numbers. Proposed streets, as shown on the adopted official map of the community, should also be shown. The scale of the zoning district map may vary from 1:600 to 1:4,800, depending on the area being mapped and the desired level of detail.

Zoning district boundaries should be readily reproducible upon the ground; that is, they should follow U.S. Public Land Survey lines, real property boundary lines, and center lines of streets. In some cases, the location of the district boundaries may have to be shown by dimensions related to other identifiable features. In urban areas, the cadastral map described in Chapter 5 should be used as the base map for the delineation and display of the zoning district boundaries. The desired accurate reproducibility of the mapped district boundary lines on the ground is facilitated by the monumented control survey system described in Chapter 5. An example of a simple zoning district map is provided in Figure 18.2. Note that the map simply and unambiguously identifies the boundaries of the zoning districts. In rural areas, orthophotography may be used as a base map for the delineation of zoning district boundaries as illustrated in Figure 18.3. Today, many communities use GIS-based digital zoning maps that allow for interactive viewing, real-time updates, and integration with other spatial data. These tools improve boundary accuracy and accessibility for both planners and the public.

Zoning districts should be so delineated and applied as to exert a beneficial effect on the predominant existing land use pattern and to guide changes in accordance with an adopted land use plan. Although zoning should carry out the land use plan, this does not mean that the zoning district map should directly reflect the land use plan. The current zoning of an area for a use that may not be expected to come into existence for many years will tend to promote diffused, uneconomical, and inefficient development at best, and urban blight at worst. Zoning districts should, in addition to protecting sound existing development, provide sufficient developable, or redevelopable, land to accommodate the growth and change anticipated for the short-term future—five to ten years—as opposed to the long-term future—twenty to thirty years—envisioned in the land use plan. Areas identified for the development in the long term should be placed in a holding zone until such time as market demand for the planned long-term use develops. An exclusive agricultural use district can provide a good holding zone. At such time as market demand develops, the area concerned may be rezoned in accordance with the long-term use recommended in the land use plan for the area. Thus, zoning can be used to effectively place development "in time" as well as "in space."

Example of a Typical Zoning Map

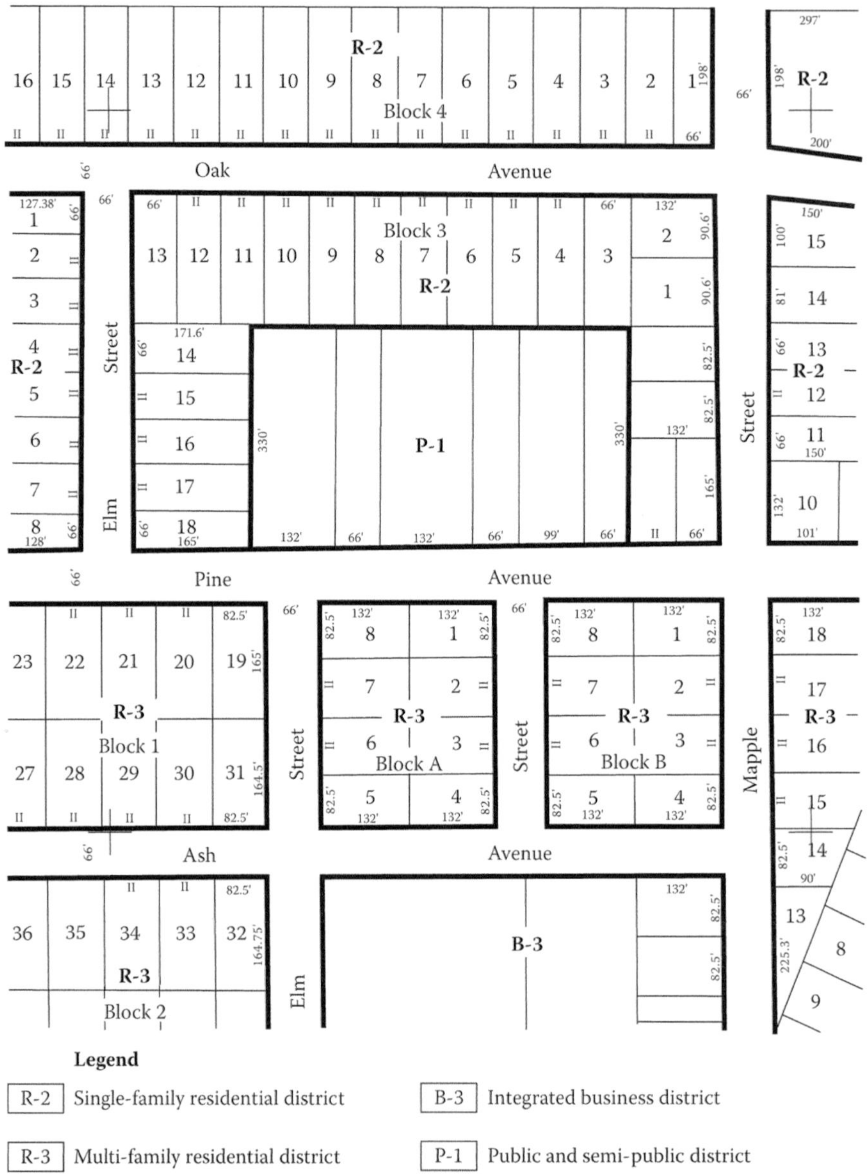

FIGURE 18.2 Illustration of a typical "rectilinear" zoning district map. The base map is provided by a cadastral map such as that described in Chapter 5. Note that the map projection is indicated by the grid tick marks shown. The district boundaries are clearly delineated and can be located on the ground.

Example of Orthophotography Used as a Base for Zoning District Map

Legend

A-1 Agriculture district
B-3 Integrated business district
P-1 Public and semi-public district
R-1 Single family residential district
R-2 Single family residential district

FIGURE 18.3 Illustration of the use of orthophotography as a base for the delineation of zoning districts. Note that the location of some of the zoning district boundaries must be determined by scaled distances. This need and the attendant uncertainty could be avoided by placing dimensions on the orthophotography.

ZONING DISTRICT BOUNDARY DELINEATION RELATED TO RESOURCE PROTECTION

Historically, zoning district boundaries were largely "rectilinear" and directly related to real property boundary lines. Zoning district boundaries could, therefore, be readily and accurately delineated on a cadastral base map. Moreover, the zoning district boundaries were fixed and stable as delineated on the cadastral base maps. The introduction of floodwater- and natural-resource-related zoning districts has complicated the delineation of zoning district boundaries for two reasons: the limits of the areas to be protected are irregular, being related to natural features and elevations rather than artificial property boundaries, and the limits of the areas to be protected are dynamic and subject to change with such factors as weather, land use changes, flood-mitigation infrastructure, climate-driven hydrologic variation, and incremental urban development. To address these challenges, modern delineation relies on GIS-based systems that integrate LiDAR topography, hydrologic and ecological modeling, and periodically updated regulatory datasets to ensure accuracy and adaptability.

Floodland Zoning

An example of the zoning district delineation problems presented by natural resource protection zoning is the problems entailed in floodland zoning. Flooding may be defined as the inundation of the floodplains of a watershed—that is, of the relatively wide, low-lying, flat to gently sloping areas contiguous to and usually lying on both sides of rivers, streams, and watercourses—by stream water moving out of and away from the stream channels. Flooding is a natural and certain process that is unpredictable only in the sense that the exact time of occurrence of a flood of a given magnitude cannot be predetermined, although the average recurrence interval—or annual probability of occurrence—of such a flood is amenable to analysis. How much of a floodplain will be occupied by any given flood depends upon the severity of the flood and, more particularly, upon the peak elevation—or stage—of the floodwaters. Thus, an infinite number of outer limits of natural floodlands may be delineated, each related to a specified recurrence interval as determined by an engineering analysis. The severity of the flood used to delineate the regulatory floodplains for zoning purposes is usually a flood having a recurrence interval of 100 years, that is, a flood having a one percent probability of occurrence in any given year. Based on such an analysis, floodlands may be delineated with sufficient accuracy for regulatory purposes on large-scale topographic maps as continuous irregular areas lying along the streams and watercourses.

The definition of flood-prone areas requires watershed-wide analyses of the riverine areas and flow characteristics of natural streams. The engineering techniques involved require detailed inventories of the physical characteristics of the watershed and riverine areas concerned and the formulation and application of complex mathematical simulation models for flood-related analyses. The analytical techniques involve the conversion of rainfall into stormwater runoff and the conversion of such runoff into peak rates and volumes of flood flows and stages along the streams. The frequencies, as well as the magnitudes of the flows, must be determined, a difficult task since the flows are the result of random meteorological occurrences and site-specific topographic and land use conditions. The analyses required involve application of the sciences of hydrology and hydraulics.

Hydrology may be defined as the study of the physical behavior of water as a resource from its occurrence as precipitation to its entry into streams, lakes, and the groundwater resources and its return to the atmosphere by evapotranspiration. The application of hydrology to the delineation of floodplains focuses specifically on such factors as the frequency and intensity of rainfall, soil and soil moisture conditions, land use, and the volume and timing of the stormwater runoff that reaches receiving surface water bodies. The location, extent, and type of urban development greatly affect the variables concerned due to the changes in imperviousness, reduction in infiltration capacity, and changes in natural storage and in-flow paths and times that accompany the conversion of land from rural to urban uses. Hydraulics may be defined as the study of the physical behavior of water as it flows within natural stream channels and associated floodlands, under and over bridges, culverts, and dams, through lakes and impoundments, and through conduits, such as artificial channels and conduits such as sewers. The application of hydraulics to the delineation of floodlands focuses specifically on such factors as the length, slope, and flow resistance of overland flow paths, receiving streams, and watercourses and on the configuration and capacity of natural and artificial stormwater storage and conveyance facilities.

Figure 9.2 of Chapter 9 illustrates a delineated floodplain along a stream that has a watershed area of about 200 square miles. The hydrologic and hydraulic studies required to produce this map resulted in the preparation of the flood stage profile shown in Figure 18.4. The regulatory flood elevations are provided by the hydraulic

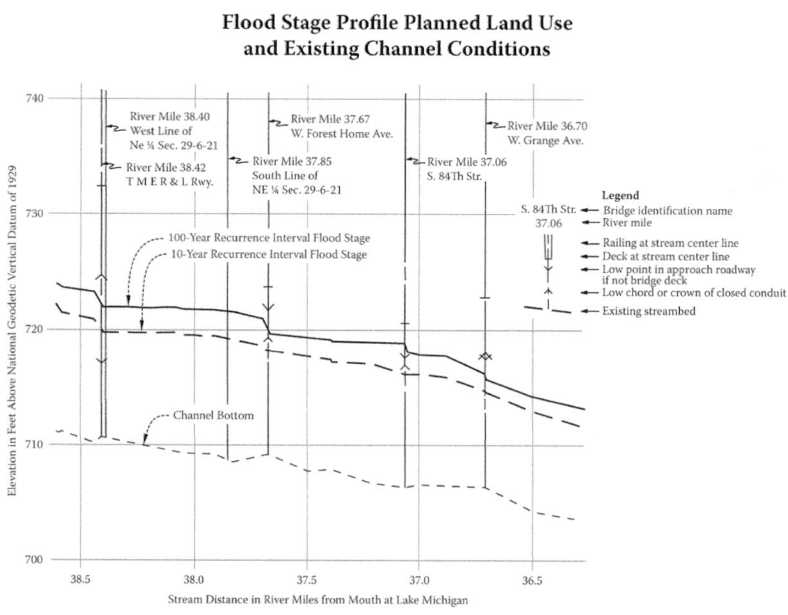

FIGURE 18.4 Illustration of a typical stream flood stage profile. This profile was used to prepare the flood hazard map illustrated in Figure 9.2.

grade line shown on the profile, and those elevations are used to delineate the floodplain boundaries on a large-scale topographic map. Since the two maps should share a common projection and datum, as described in Chapter 5, the resulting irregular floodplain boundary can be readily transferred to a matching cadastral map by analog or digital overlay, and thereby to the zoning district map. The floodplains and stages should be based on existing and planned land use and stormwater management conditions in the watershed.

The floodplain delineation should be sufficiently accurate to avoid the need to determine the existence of a flood hazard by carrying computed flood stage elevations to specific properties by field survey. In this respect, it should be noted that flood stage elevations, however developed, represent approximations and are marked by inherent uncertainties. Although in practice expressed as firm values, flood stage elevations in reality represent uncertain values that lie within a range. That range is not expressed in presenting the findings of the hydrologic and hydraulic analyses. Some regulatory agencies recognize the uncertainty inherent in computed regulatory flood stage values by requiring that buildings proposed to be constructed in and adjacent to floodway fringe areas have the first floor located two feet above the regulatory flood stage. Therefore, in delineating floodland zoning districts, the necessary flood stage values should be recognized as rough approximations, and needless concern with a precision in the flood hazard area determination greater than that provided by mapped boundaries is usually unwarranted. Nevertheless, most floodland zoning ordinances provide that the flood stage elevations as transferred to the site concerned control, and not the position of the mapped floodland boundary.

In the case illustrated by Figures 9.2 and 18.4, good practice would dictate that the entire delineated floodplain should be placed in a floodplain zoning district that precludes the location of any flood-damage-prone development within the district. In the example given, had the stream reach concerned been historically so zoned, the construction of the houses located within the flood hazard area shown would have been avoided. In fact, those houses have, based on the flood hazard mapping, been removed through a flood control program, and the area returned to open parkway use.

In relatively small watersheds—measured in hundreds of square miles—and along the relatively small streams concerned, the use of a single floodplain zoning district that precludes all flood-damage-prone development represents the best practice. Some regulatory agencies, however, promulgate the use of two floodland zoning districts—a floodway district and a floodway fringe area zoning district. The former is defined as encompassing the portion of the total floodplain occupied by moving floodwaters during a major flood event. The delineation of this district entails the exercise of great judgment by the engineer responsible for the required hydraulic analyses. The attendant zoning regulations preclude the location in the district of flood-damage-prone development or development that would interfere with the movement of the floodwaters. The latter district is defined as that part of the total floodplain utilized for floodwater storage during a major flood event. The attendant zoning regulations typically permit the filling of these floodway fringe areas for development. In relatively large watersheds—measured in thousands of square miles—and along the relatively large rivers concerned, such filling and the attendant loss of storage may have insignificant effects upon flood flows and stages. In relatively small watersheds and along the relatively small streams concerned, this practice becomes self-defeating, for the losses in storage may result in significant increases in flood flows and stages and may significantly change the boundaries of the delineated flood hazard area concerned. If this practice is to be followed in relatively small watersheds, the zoning regulations should require that analyses of the effects on flood flows and stages of the proposed fill in the floodway fringe district be analyzed, and, as may be necessary, compensatory storage provided in the stream reach concerned.

It should also be noted that the flood profile and attendant flood stage elevations are dynamic and subject to change, resulting from such factors as changes in the land use pattern of the tributary watershed, in the floodwater storage capacity in the channel and floodway capacities, and in the waterway openings of bridges and culverts. A comprehensive watershed plan should provide the basis for flood hazard area delineation and attendant floodland zoning. Administration of the watershed plan should result in the management of the inevitable changes in the developed characteristics of the watershed so that the regulatory flood flows and stages are not significantly affected.

Today, digital floodplain mapping technologies—such as FEMA's Flood Map Service Center (MSC) and geospatial data portals—allow for interactive viewing and integration with GIS platforms, facilitating real-time analysis, overlays with property boundaries, and improved public accessibility to flood risk information.

Wetland Zoning

Wetland zoning provides another example of the zoning district delineation problems presented by natural resource protection zoning. The term "wetlands" is defined in Chapter 9, and Figure 9.3 illustrates the delineation of wetland areas on orthophotography. The wetland boundaries should be delineated on the orthophotography by field survey, the survey being conducted by an experienced ecologist. If the orthophotography used to map the wetlands is at the same scale and utilizes the same map projection and datum as the cadastral map used as a base map for the preparation of the zoning district map, as recommended in Chapter 5, the irregular mapped wetland boundaries can be directly, efficiently, and accurately transferred to the zoning district map by manual or digital overlay.

Wetland boundaries, like floodway and floodplain boundaries, are dynamic, with the size of the wetlands varying with weather conditions. This dynamic characteristic is illustrated in Figure 18.5. Ideally, the wetland boundary as delineated in the field and shown on the zoning district map, will fall within the normal range indicated in Figure 18.5. Land subdivision control and zoning ordinances may, nevertheless, contain provisions for the refinement of the delineated wetland boundary by additional field surveys. Such refinement should not be necessary and should not significantly change the location of a properly delineated wetland boundary that does indeed fall within the normal range as intended. If, however, the wetland boundaries were delineated on the basis of photo-interpretation alone, then such refinement and field staking will be necessary. Such field staking and refined delineation will be facilitated by the survey control system described in Chapter 5.

Today, digital wetland inventories and mapping tools—such as the U.S. Fish and Wildlife Service's National Wetlands Inventory (NWI), high-resolution LiDAR, and state-maintained GIS data layers—allow planners to integrate real-time wetland data into zoning maps. These tools support more precise, consistent, and publicly

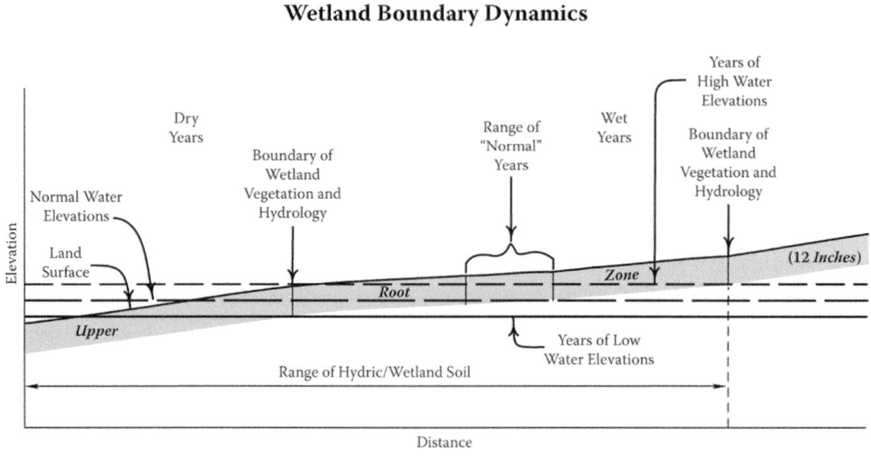

FIGURE 18.5 Illustration of the dynamics of wetland boundaries and the issues involved in the delineation of such boundaries for zoning purposes. The field delineation should place the boundary within the normal range shown. The irregular boundary so delineated can be accurately transferred to the zoning district map if the survey control and mapping system described in Chapter 5 is available.

Source: SEWRPC.

accessible delineations, reducing dependency on manual photo-interpretation and facilitating easier updates in response to ecological change.

ZONING ORDINANCE REGULATIONS

An example of a zoning ordinance—that is, more correctly, of that part of a zoning ordinance consisting of the regulations that must accompany a zoning district map—is provided in Appendix B. These regulations are not to be interpreted as a so-called "model" zoning ordinance but simply as an example of the regulatory portions of the type of zoning ordinances often in use by small and medium-sized urban communities. The ordinance was selected as a means of introduction to such ordinances as well as for its simplicity and brevity. In comparison, a more sophisticated ordinance for even a medium-sized community may occupy over 150 pages of text. It should be noted that the chosen example lacks a number of features of more sophisticated ordinances, such as a performance standards section dealing with air pollutant emissions; fire and explosive hazards; glare, heat, noise, and odor emissions; radioactivity and electrical disturbances; vibration; architectural control standards; and site erosion control. Modern zoning ordinances increasingly incorporate form-based codes, performance-based regulations, and techniques such as transit-oriented development, traditional neighborhood design, and transfer of development rights. Neotraditionalist planning concepts promote mixed-use development. Although not specifically addressed in the ordinance provided, its flexibility provides that mixed-use development can be achieved through the use of the planned development district provided. The example ordinance is specifically based on Wisconsin law and practice. Although the laws and practices governing zoning will differ from state to state, the ordinance should, nevertheless, provide a good example of some of the more common issues and problems as well as practices.

Sections 1.0 and 2.0 of the ordinance present general provisions, including the intent to adopt the ordinance under the police powers of the community. Section 3.0 establishes the zoning districts, references the zoning map, and sets forth the basic use, height, and area regulations applicable within each district. Section 4.0 deals with conditional uses, and Section 5.0 deals with traffic and parking. Section 6.0 deals with exceptions to—or permitted modifications of—the district regulations. Section 7.0 deals with signs, and Section 8.0 deals with non-conforming uses, a troublesome concept. Section 9.0 deals with the important issue of the board of zoning appeals. Section 10.0 deals with amendments, and Section 11.0 sets forth pertinent definitions. Some comments concerning specific requirements of the ordinance follow.

Sections 2.7 and 2.8 of the ordinance deal, in effect, with enforcement of the ordinance. It should be noted that the enforcement provision includes, in addition to fines and imprisonment for violations, the issuance of court injunctions to remove violations. Court injunctions may be sought by the governing body of the municipality, the plan commission, or the zoning administrator. Court injunctions may also be sought by an aggrieved property owner damaged by an ordinance violation. This is an important provision intended to avoid situations in which the payment of provided fines does not, in effect, become special permit fees if the violations are continued. The administration of the ordinance is through the issuance of zoning permits by the zoning administrator as provided in Sections 2.2 and 2.3.

Section 3.0 of the ordinance, reflecting its simplicity, provides for the creation and use of only 13 zoning districts, including single-family, two-family, and multi-family residential districts; neighborhood, community, and central business districts; light and heavy industrial districts; and special districts such as conservancy and floodplain districts. The ordinance is an exclusive use ordinance under which only the specified uses listed are permitted in each district. The ordinance contains a conservancy district, intended for use in protecting environmentally sensitive natural resources, including wetlands, woodlands, wildlife habitat, and environmental corridor areas. Some ordinances may provide separate upland and lowland conservancy districts. The ordinance utilizes a single floodplain district that is intended to exclude flood-damage-prone uses from the entire regulatory flood district. As already noted in this chapter, some ordinances utilize two flood districts, a floodway district and a floodway fringe district, the former essentially excluding all development, the latter permitting filling and development. Such floodway fringe zoning districts may, like other districts, specify permitted uses. Some ordinances may, however, structure the floodway fringe zoning district as an "overlay" district. In such overlay districts, the "rectilinear" zoning districts remain in place and are overlaid by the floodway fringe zoning district that requires filling for development. The permitted uses are those specified in the underlying districts.

The ordinance, which is intended for use in a small urban community, contains an agricultural district that might better be titled an agricultural holding zone or agricultural transition district. Areas of the community for which a market demand for the future use proposed in the land use plan has not yet developed are intended to be placed in this district until the market demand develops, thereby assuring the integrity of the area for the ultimate proposed urban use. The format of the use regulations as provided in the ordinance could be improved by specifically listing the accessory and conditional uses permitted in each of the districts.

Residential density in the ordinance is specified indirectly, but simply, in the form of permitted minimum unit lot areas and widths, supplemented by minimum yard requirements. Such yard requirements are illustrated in Figure 18.6. Modern ordinances often explore density regulation through form-based codes, incentives for affordable housing, and context-sensitive zoning tools.

Height regulations in the districts are expressed in terms of the maximum number of feet permitted. Alternative approaches may express height limitations in terms of the permitted number of stories and, in some instances, as multiples of fronting street widths or as building bulk regulations. Height regulations may be accompanied by required increases in building setbacks from property boundaries with increases in height. Airport runway approach and clear zone zoning, it should be noted, are special forms of height regulation.

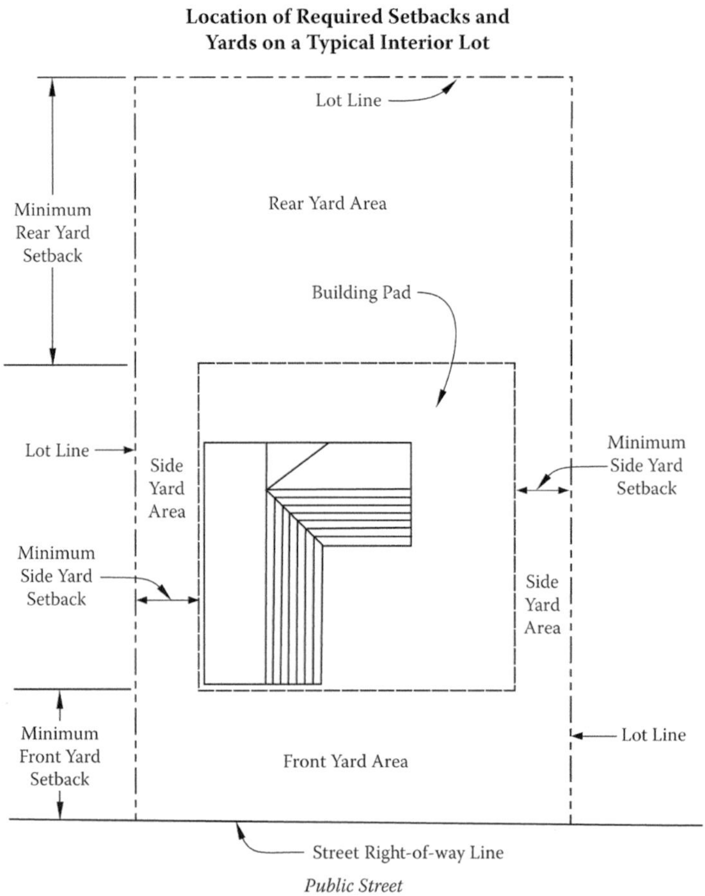

FIGURE 18.6 Illustration of the location of required building setbacks and yards as typically specified in zoning ordinances.

Format for Table Summarizing Zoning Regulations

Use	District	Map Symbol	Height		Min. Front Yard Feet	Min. Side Yard Feet	Min. Rear Yard Feet	Min. Lot Area per Du Sq. Ft.	Min. Lot Width Feet	Min. Floor Area Sq. Ft.	Required Off Street Parking	Conditional Uses
			Stories	Feet								
Single-family Residential, and Garage	R-2	Yellow	2½	35	25	10	40	10,000	70	1,200	None	Elem. Schools Utility Structures Gov't and Cultural Uses

FIGURE 18.7 Illustration of the format of a table that can be used to summarize the regulations contained in a zoning ordinance. Such a table facilitates administration of the ordinance and helps concerned citizens to better understand the provisions of the ordinance.

Section 4.0 of the ordinance deals with conditional uses. Conditional uses are defined as uses of a special nature that are precluded as permitted principal uses in a zoning district but may be authorized under specified conditions. The plan commission should authorize all conditional uses and specify the conditions to be met. Conditional uses may include, for example, utilities such as sewage and water pumping stations, electric power transformer stations, public and parochial elementary schools, and churches in residential areas. Imposed conditions may include landscaping, architectural design, lighting, fencing, and hours of operation.

The basic provisions of a zoning ordinance may be summarized in a tabular form, such as shown in Figure 18.7. Such a table can facilitate administration of the ordinance. More importantly, such a table can be used in public educational efforts about the ordinance and can be helpful in promoting a proposed ordinance for public hearing.

It should be noted that Section 4.5 permits planned unit developments, including portions or all of neighborhood units, as conditional uses. In addition to the residential uses permitted, all neighborhood business uses are permitted. This permits neotraditional neighborhoods, which include mixed residential and commercial uses, to be developed under the ordinance.

The traffic and parking requirements included in the ordinance are not intended in any way to constitute models but are simply examples. Parking requirements particularly require careful consideration in order to meet or solve either inadequate or excessive requirements, considering peak period and normal period demand.

Exceptions may be defined as departures from the requirements of the ordinance permitted by the terms of the ordinance itself. Examples might include the intrusion of open stairs and landings of fire escapes into required yard areas or the exception from height requirements of architectural projections such as spires, belfries, parapet walls, and chimneys.

Board of Appeals

Section 9.0 of the ordinance provides for the establishment of a board of appeals. Such a board may consist of five members appointed by the mayor, village president, or town board chairman and confirmed by the common council or by village board or town board. An alternate member may be appointed who would act only when a regular member is absent or declines to vote because of a conflict of interest. The board is an administrative body with quasi-judicial powers with respect to the enforcement of the zoning ordinance. The board has the following functions: to hear and decide appeals where it is alleged there is error in any decision of the building inspector in enforcing the zoning ordinance; to hear and decide appeals for variances; and to hear and decide appeals for interpretations of the zoning regulations and locations of zoning district boundaries.

It is important that complete and accurate minutes of the proceedings of the board and a record of all actions taken by the board be carefully kept by the secretary of the board. The vote of each member of the board upon each question, the board's findings of fact, and the reasons for the board's determinations should be recorded. These minutes are important because if a decision of the board is appealed to a court of law, the court will not rehear the case but will base its decision on a review of the minutes concerned. In recent years, many municipalities have added digital systems for meeting minutes and case documentation to improve transparency and facilitate electronic submission of appeals and variance applications. Online public access to board decisions, GIS-linked appeal maps, and digital archives is increasingly common and encouraged.

COMMON ZONING PROBLEMS

Zoning ordinances, while essential for plan implementation, often face recurring challenges. Common problems include overzoning, where excessive land is designated for commercial or industrial use beyond foreseeable needs, and underzoning, which fails to allocate sufficient land for essential uses such as affordable housing. Strip and spot zoning can undermine community plans by creating fragmented, inefficient land use patterns. Non-conforming uses present ongoing management difficulties when older developments persist under new regulations, while contract zoning raises legal and ethical concerns when public decisions appear tied to private agreements. Modern zoning practice emphasizes transparency, consistency, and periodic review to ensure that ordinances remain aligned with comprehensive plans, community goals, and evolving land use needs.

FURTHER READING

Elliott, Sclar, et al. *Zoning: A Guide for 21st-Century Planning*. CRC Press, 2020.

Goodman, William I., et al. "Principles and Practices of Urban Planning," Chapter 15, *Zoning*, International City Managers Association, 1968.

19 The Official Map

What is needed from the public's point of view is a police power control that can bar any incompatible development both within and adjacent to the right-of-way, pending its acquisition. The community should also be authorized to control any interim speculation in land. Substantial advantages should derive from a regulation of this kind. Not only will acquisition costs be kept down, but the purpose of the highway will not be defeated by the growth of conflicting development. On a broader level, interim control over land use can help integrate the highway with the general community plan.

Daniel R. Mandelker, Professor
Washington University School of Law, St. Louis, Paper presented at annual meeting of the Highway Research Board, National Academy of Sciences, 1962

INTRODUCTION

Contemporary urban growth, with its evolving patterns of development, mobility, and climate challenges, has created a need to extend urban street systems and to widen, realign, and reconstruct existing trafficways. The attendant growing demand for land for arterial streets and highways, and for collector and land access streets, is accompanied by a demand for land for utility and drainage rights-of-way and for parks and parkways. The reservation of rights-of-way for these purposes is required not only to meet the transportation, stormwater management, and recreational needs of rapidly growing and changing urban areas but also to help give shape and form to urban development.

To adequately meet the growing need to reserve land for public facilities, not only will sound plan preparation at all levels of government be required, but practical plan implementation as well. An interval must necessarily exist between the time a given improvement is incorporated into a long-range plan and the time of actual construction. This time lag is inherent in the public planning process. It is during this time lag that means must be found to effectively reserve land for planned projects and thereby to ensure the integrity of the plan.

The "official map" is one of the oldest plan implementation devices at the disposal of municipalities. Like the use of the zoning ordinance, the use of the official map is an exercise of the police power of a municipality. It is also one of the most effective and efficient devices that can be brought to bear on the problem of reserving land for future public use. Yet it is probably the least understood and the least used of all local plan implementation devices. Zoning ordinances, land subdivision regulations, building and housing codes, and capital improvement programs are all better understood and more widely applied by municipalities.

DOI: 10.1201/9781003500384-19

Historically, the limited availability of accurate geospatial data made official mapping technically and financially challenging. Today, however, high-resolution Geographic Information System (GIS) data, modern cadastral systems, and real-time mapping tools have substantially reduced those barriers. Official maps can now be maintained digitally and updated frequently, allowing for easier integration with zoning maps, capital improvement plans, and transportation modeling systems. Furthermore, official mapping can be paired with right-of-way protection ordinances, development moratoria, or land acquisition incentives to proactively manage urban growth and safeguard infrastructure corridors.

BRIEF HISTORY

The concept and actual use of the official map in the United States dates back to colonial times, when the proprietary founders of well-planned colonial cities, such as Philadelphia, Annapolis, and Williamsburg, caused plans for these cities to be prepared and the necessary streets and public commons preserved for public use by prohibition of the construction of buildings, fences, and other structures in the dedicated areas. As private land ownership expanded beyond proprietary founders, colonial—and later state—legislation authorized appointed commissioners to plat town sites, including streets and public commons, and to obtain deeds of trust from private landowners, through which those owners formally consented to the dedication of streets and open spaces. Washington, D.C., among other cities, was planned, mapped, and initially developed in this manner.

These methods proved too cumbersome with advancing urbanization and were supplanted by early-nineteenth-century state legislation under which municipalities were authorized to reserve rights-of-way for streets in advance of construction and to prohibit any building in the beds of these future streets. No escape from these regulations was provided in the early legislation; no compensation was payable for the removal of a structure built in the bed of a street in violation of the law. The first use of the official map under such legislation and in conjunction with planning for the orderly expansion of an existing urban area was made in New York City in 1806, with other cities soon following this lead. Baltimore completed an official map in 1817, Brooklyn in 1818, and additional cities in Pennsylvania, including Pittsburgh, soon thereafter. The legality of these early official maps was tested and, in spite of the severity of their application as compared to modern practice, upheld in the New York and Pennsylvania courts.

With the birth of the modern city planning movement, modern official map enabling legislation has been adopted by over one-half of the states. The prototype of all modern official map acts was adopted by the State of New York in 1926; subsequently, Maryland, Michigan, Minnesota, New Hampshire, Utah, and Wisconsin, among others, enacted enabling legislation based to a considerable extent on the New York act. The legislation remains essentially sound, placing at the disposal of the municipalities a powerful plan implementation device. In some states this device is also made available, often in modified form, to counties as well as cities, villages, and towns, and to the state departments of transportation.

DEFINITIONS

The laws and practices governing official mapping will differ from state to state. Wisconsin law and practice is used in this text as a sound example of the application of official mapping to plan implementation. The Wisconsin Statutes provide that the common council of any city may establish an official map for the precise designation of right-of-way lines and site boundaries of streets, highways, parks, parkways, and playgrounds. The city may also include on its official map the locations of railway rights-of-way, public transit facility rights-of-way, and those waterways that have been included in a comprehensive surface water drainage plan. Such a map has the force of law and is deemed to be final and conclusive with respect to the location and width of both existing and proposed streets, highways, and parkways; the location and extent of parks and playgrounds; and the location and width of existing and proposed railway rights-of-way shown on the map. The statutes further provide that the official map may be extended to include areas beyond the corporate limits of the city but within its extraterritorial planning approval jurisdiction. This supports coordination across urbanizing areas. The official map enabling legislation in Wisconsin is a subsection of the state planning enabling legislation, and as such is made applicable to villages and towns as well as to cities.

The term "official map" is often misapplied or misused. The common confusion involved stems in part from the fact that local governments may have several maps used for different purposes that are designated as official documents and, therefore, loosely referred to as "official maps." Some of this confusion may also be due to the extensive use of maps in comprehensive community plans. Maps are often used in such plans to present long-range land use and arterial street and highway system plans and when so used are often inappropriately called "official maps." The term "official map" is also sometimes incorrectly applied to the map delineating zoning districts that accompanies a zoning ordinance and is used to administer such an ordinance.

The term "official map" as used in this text applies only to the legally sanctioned map intended to be used as a means of implementing those elements of a municipality's comprehensive plan dealing with streets, highways, parkways, parks, playgrounds, transitways, and drainageways. The basic purpose of an official map is to prohibit the construction of buildings or structures and their associated improvements on land that has been designated for current or future public use. The official map must be adopted by the governing body of the local unit of government concerned, and only after such adoption does it assume its legal force. Good practice would dictate that a certified copy of the resolution adopting the map appear on the face of the map. If this practice is followed, this certificate would constitute a unique identifying feature of an official map.

FUNCTIONS

The primary function of an official map is to implement a municipality's plan of streets and highways in a manner similar to that in which the zoning ordinance and map should implement the municipality's land use plan. The official map permits the community to protect future or expanded rights-of-way from premature development by prohibiting construction within those designated areas. While the cost savings from avoiding building removal can be substantial, even more important is the official map's role in preserving the integrity of the community's long-range transportation system. The official map may also be used to protect the beds of planned transitways.

Another major function of the official map is to support the community's long-range plan for parks, parkways, and open spaces. Because parks and parkways frequently include environmentally valuable natural features, such as scenic and historic places, watercourses and drainageways, floodplains, and wetlands, these features can be protected through an official map. Inclusion of such features as proposed parkland on the official map lends strong legal status to the planned facility and protects the land within the indicated taking lines for public use. This is particularly valuable in areas where natural hazard mitigation overlaps with open space preservation. Similarly, the official map can implement a community's stormwater management plan by reserving land for future drainageways identified in a surface water drainage plan. In addition to potential cost savings, this function supports flood risk reduction and water quality goals.

A related but important benefit of official mapping is that it clearly locates and records existing street lines that form public property boundaries. This helps stabilize local property records and supports accurate surveying and planning. Because the official map shows legally established lines for public improvements, it can serve as a base map for cadastral and planning purposes, expediting a wide range of municipal engineering and development tasks.

RELATIONSHIP TO THE COMPREHENSIVE PLAN

As described in Chapter 10, the term *comprehensive plan* refers to an extensively developed plan for the physical development of a community, including proposals for future land use; transportation; sanitary sewerage, water supply, and stormwater management; and such community facilities as parks, public buildings, and schools. The comprehensive plan is carried out through a series of plan implementation devices, including the official map. The official map allows the municipality to express its intent to reserve land for public purposes without commitment to actual acquisition. Thus, the official map functions as a refinement of the comprehensive plan, reflecting certain aspects in a more precise, accurate, and legally binding manner.

As envisioned by the Wisconsin Statutes, the preparation of an official map involves the initial preparation by the city, village, or town engineer of a precise cadastral base map showing all existing streets, highways, parkways, parks, and playgrounds, and the adoption of this base map by the governing body as the official map. State law further envisions that, after periodic public hearings and plan commission review, the adopted map may be amended to add new proposed public facilities such as future streets, parks, transit corridors, and stormwater management areas. In this way, the official map provides a dynamic mechanism for implementing key elements of the comprehensive plan through the police power. Street and park reservations can be based not on immediate needs alone, as must be the case when such areas are acquired by the exercise of the power of eminent domain, but on future needs as well.

In addition to ensuring that land needed for future public purposes will be available at the price of unimproved land, the adoption of an official map has certain other consequences that tend to give direction and pattern to future development and that can be used to more fully carry out the comprehensive plan. The Wisconsin Statutes provide that where an official map has been established, no public sewer or other municipal street, utility, or improvement may be constructed in any street, highway, or parkway until such street, highway, or parkway has been duly placed on the official map. Similarly, no permit for the erection of any building may be issued unless a street, highway, or parkway giving access to such proposed building has been placed on the official map. Both of these provisions are particularly valuable in controlling development in the outlying rural-urban fringe areas, assuring that such development occurs in conformance with an integrated development plan.

Although the official map is usually applied only to proposed arterial streets and parkways, a strong case can be made for its application in undeveloped or partially developed areas to proposed collector and land access streets as well. If the local community has prepared detailed neighborhood unit development plans, as described in Chapter 13, the collector and land access streets and the neighborhood park and parkway sites shown on such plans can be delineated and placed on the official map. Such mapping helps coordinate fragmented ownership patterns and ensures cohesive neighborhood design, complementing subdivision control and modern infrastructure planning tools such as GIS-based parcel management systems.

EFFECTIVENESS

The official map is particularly effective as a street and highway plan implementation device, although other plan implementation devices such as building setback requirements in zoning ordinances, special building setback line ordinances along major streets, building setback lines on recorded subdivision plats, and private deed restrictions can all be used to reserve land for the future widening of existing streets. However, these alternatives are generally limited in scope and effectiveness. Subdivision control ordinances can be used to protect future streets and highways, but can do so only indirectly and cannot be used to prevent the erection of buildings in the beds of future streets when the erection of such buildings takes place without platted land subdivision. The official map is the only arterial street and highway system plan implementation device that operates on an areawide basis in advance of land development and can thereby effectively assure the integrated development of the street and highway system.

The high order of effectiveness of the official map as a street and highway plan implementation device is attributable to certain characteristics of the official map. Unlike land subdivision control, which operates on a plat-by-plat basis, the official map can operate over a wide planning area well in advance of requests for development. The proper application of the official map necessitates the preparation of precise, or definitive, plans beyond the comprehensive planning stage, and thereby ensures that the broad objectives expressed in the comprehensive plan are reduced to more specific and attainable ones. The official map is a useful device to achieve public acceptance of long-range public facility planning goals by providing clear legal notice of the municipality's intentions. It thereby avoids the altogether too common situation of real property being acquired, and development being undertaken, without knowledge of, or regard for, the long-range plan, and thereby does much to avoid local public resistance when plan implementation becomes imminent.

The effectiveness of the official map as a plan implementation device applies equally well to any application wherein it is essential to preserve an integrated network of public rights-of-way. The location of major stormwater drainage channels and sanitary trunk sewer lines in parkways permits the application of the official map to the implementation of sewerage, drainage, and flood control plans. Because of the legal problems involved, the official map is somewhat less effective as a reservation device for park and open space requirements when such parks and open spaces cover large blocks of land rather than relatively narrow rights-of-way.

A certain practical and desirable degree of flexibility is given to the official map under the Wisconsin Statutes, which provide that changes or additions to the official map made by duly reviewed and approved land subdivision plats do not require the public hearings or governing body actions normally required for the adoption of such changes or additions. This means that the approval and recording of a land subdivision plat automatically amends the official map. If the survey control network described in Chapter 5 is in place, and if the official map and new subdivision plats are based upon that network and therefore have a common map projection and horizontal datum, new subdivision plats can be readily incorporated into the official map by analogue or digital overlay.

Ideally, the official map should be a means of implementing, or carrying out, a community's comprehensive plan. As such, the official map should be prepared within the context of such a plan. However, in the absence of detailed long-range planning, smaller communities often rely on the official map, zoning, and subdivision ordinances to guide development more directly. In such situations, the official map may quite properly combine the expression of the community's long-range street and highway objectives with the implementation device necessary to help achieve these objectives. Thus, in some situations, the official map may serve as both the long-range street and highway system plan and as the primary implementation device for that plan.

LEGAL AND ADMINISTRATIVE CONSIDERATIONS

The Wisconsin Official Map Act includes all of the basic elements of modern official map enabling legislation. As already noted, villages and towns are, by reference in the act, given the same planning powers as cities and, therefore, may prepare, adopt, and administer official maps under the basic enabling legislation. Other state legislation grants modified official map powers to counties and to the state department of transportation. As already noted, the state enabling legislation envisions the adoption of an official map in two stages. In the initial stage, the municipal governing body prepares and adopts a map showing the existing streets, highways, parks, and parkways within the municipality "as laid out, adopted, and established by law." This latter phrase, in effect, defines the initial stage map as the cadastral map described in Chapter 5. The municipality may also include existing railway rights-of-way, public transit facility rights-of-way, and waterways on this initial map. This initial adoption is permitted without public hearing. However, due to the significant impact on private property rights, best practice recommends holding a public hearing even at this early stage.

Also as already noted, in the second stage of map preparation, the governing body may add to the initially adopted official map proposed streets, highways, parkways, parks, playgrounds, transitways, and drainageways, both inside the municipality and within its extraterritorial plat approval jurisdiction. The municipality may also show on the map proposed widenings, narrowings, extensions, or closures of such facilities. The proposed additions must be referred to the plan commission for review and comment, and a public hearing must be held on the proposed additions prior to adoption by the governing body.

The state legislation provides that, following adoption of the second stage of the map, no building permit shall be issued for any building or structure proposed to be erected in the bed of any existing or proposed street, highway, park, parkway, railway right-of-way, public transit facility right-of-way, or drainage shown on the official map unless it can be shown that the property is not yielding a fair return and the owner will be substantially damaged by having to place the proposed building or structure outside the mapped area. It is in this respect that modern official map acts differ from early statutes that provided no variance or appeals process. Under the modern acts, the property owner may file an appeal for relief with the board of appeals—the local quasi-judicial body established to administer the zoning ordinance—which then holds a public hearing and decides the matter upon the merits of the case. Also as already noted, the law also provides that no public sewer or other municipal street utility or improvement shall be constructed in any street, highway, or parkway until it is duly placed on the official map and that no building shall be erected unless it has access to a street, highway, or parkway recorded on the official map. Finally, the state legislation requires that, upon adoption of an official map, the city clerk record with the register of deeds of the county concerned a certificate declaring that the city has established an official map, thereby providing constructive notice to all affected property owners of the municipality's intent.

ENGINEERING CONSIDERATIONS

The comprehensive plan is a general plan, certain elements of which are often presented in the form of small-scale, generalized plan maps. The official map is intended to reflect and refine certain aspects of the comprehensive plan and must, therefore, be capable of more precise and accurate interpretation. Uncertainty over this need for precision and accuracy may be one of the reasons why this powerful plan implementation device has not been more widely utilized. There are two aspects to this uncertainty: one relates to the quality of the available base mapping, the other to the level, or detail, of planning and engineering required to permit a proposed right-of-way line or site boundary to be accurately added to the official map.

Base Mapping Considerations

Maps of the scale and accuracy required for municipal planning and engineering and, therefore, for official mapping are lacking in many communities. Existing municipal maps often may be no more than compilations of paper records, lacking as a basis a map projection, defined horizontal and vertical datums, and a monumented survey control network. These deficiencies, along with the historically high cost of establishing survey control networks and producing quality maps, have historically hindered the broader application of the official map.

Today, improved mapping technologies and widely available survey data have made base mapping less of a barrier. Urban and urbanizing cities, villages, and towns should, as a part of the basic data bank assembled for municipal planning and engineering, have available for use a set of large-scale topographic and cadastral maps. The maps should be prepared to the standards set forth in Chapter 5, the topographic and cadastral maps being based on a common map projection, defined horizontal and vertical datums, and a monumented survey control network. The topographic and cadastral maps should meet specified accuracy standards. The cadastral maps, supplemented by selected planimetric data transferred from the companion topographic maps—the supplemental data consisting primarily of the existing building outlines—provide a proper base map for the preparation of an official map. Indeed, the cadastral map can be adopted as the initial stage of an official map as envisioned by Wisconsin law. As already noted, that initial stage must show the existing streets, highways, parks, and parkways as "laid out, adopted, and established by law"—which is precisely what the cadastral map does. This statutory requirement clearly implies that a properly prepared official map must be based on a good real property boundary line base map reflecting existing conditions. Following this initial adoption, planned streets, highways, parks, and parkways, among other planned rights-of-way and site boundaries, may be added to the map.

If the specifications governing the preparation of the large-scale topographic maps, for example, require the maps to meet National Map Accuracy Standards, 90 percent of all well-defined planimetric features—such as buildings—would have to be located on the map within 1/30 inch of their true position, and 90 percent of all contours and elevations determined from contour lines would have to be accurate to within one-half of the map contour interval. Thus, if the map is at a scale of one inch equals 100 feet, buildings shown on the map may be expected to be located within 3.3 feet of their true position. This level of accuracy, as a practical matter, is adequate for official mapping.

Facility Design Considerations

Uncertainty about the level and detail of the engineering required to permit the placement of "taking" lines on an official map may also have contributed to the under-utilization of this plan implementation device. The placement on an official map of right-of-way lines to accommodate the future widening of existing or the future construction of major new facilities such as streets, highways, and parkways must be preceded by adequate design studies. The preliminary engineering studies for such major new facilities should be conducted by the level and agency of government responsible for the construction of the facility concerned: state, county, or local. The needed rights-of-way may be officially mapped by the level of government responsible for construction, or, in a cooperative effort, or by the municipality within which the major transportation facilities are to be located. With respect to major new arterial facilities—including freeways, expressways, parkways, busways, and light rail transitways—the design studies must specify centerline locations and proposed right-of-way widths, including right-of-way configurations for complex intersections. For these major types of facilities, the design studies must carefully consider vertical as well as horizontal locations, alignments, and configurations and will usually entail the conduct of full-scale preliminary engineering investigations. Such investigations are usually conducted for a specific travel corridor in which the major transportation facility identified in the transportation element of the comprehensive plan is to be located. The needed right-of-way should be officially mapped as soon as the necessary preliminary engineering has been completed and before development in the corridor substantially increases the cost of right-of-way acquisition and construction or precludes construction.

The necessary design studies for most widened urban arterials and for new collector and land access streets, for drainageways and attendant minor parkways, and for new neighborhood parks and school sites are most efficiently and effectively conducted as integral parts of plan preparations for neighborhood units and major activity centers. Set within the context of comprehensive plans, the detailed unit development plans so produced have also been known historically as platting layouts or major land subdivision designs.

The official mapping of proposed widening lines for existing state and county trunk highways may be done by the agency responsible for the maintenance and operation of the facility or, again, by a cooperative effort of the municipalities within which the facility is located. The determination of widening lines can often be based on less extensive preliminary engineering studies than those required for the location of new facilities, requiring only the determination of a desirable cross-section for the widened facility and consideration of intersection treatments.

As already noted, the placement of proposed new urban arterial, collector, and land access streets, the placement of widening lines for urban arterials, and the placement of proposed parks, parkways, and drainageways on official maps should be based on detailed neighborhood unit and major activity center development plans prepared by experienced site planners, adopted by the plan commission and by the governing body of the community concerned. It is important to note that the large-scale topographic and cadastral maps prepared as recommended in Chapter 5, together with soils, floodland, and wetland data collected for comprehensive planning purposes, should provide the basic data required for the conduct of the planning and engineering studies needed to determine locations for most municipal facilities in preparation for the official mapping of those facilities.

MAP FORMAT

In states covered by the U.S. Public Land Survey System, and for all but the smallest communities, it is suggested that the base maps be compiled by U.S. Public Land Survey System section or one-quarter section, and that an individual sheet be used for each section or quarter-section mapped, as shown in Figure 19.1. Smaller communities, however, may be able to prepare an official map at a workable scale

Typical Official Map by One-Quarter Section

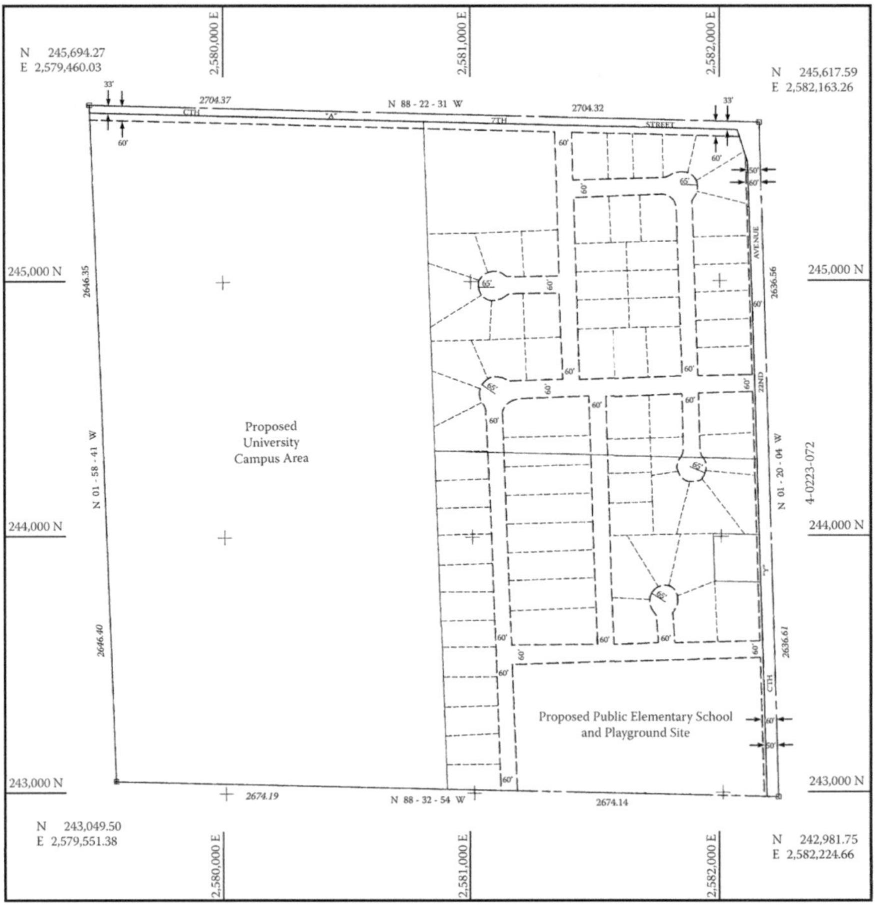

FIGURE 19.1 Illustration of a portion of an official map for a larger community. The U.S. Public Land Survey system one-quarter section sheet format was prepared at an original scale of 1 inch equals 100 feet. Existing street widths and widenings and proposed street widths are given by annotated figures on the map; all other dimensions are determined by scaled measurement on the map.

Example of Official Map for an Entire Community

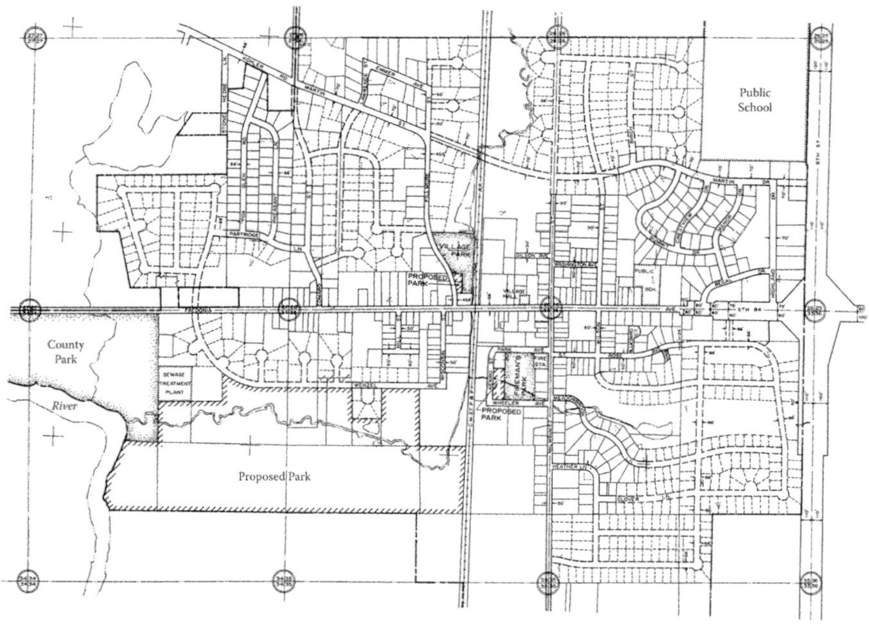

FIGURE 19.2 Illustration of an official map for an entire community. The small size of this community permits use of a single-sheet format at an original scale of 1 inch equals 400 feet. Existing street widths and widenings are given by annotated figures on the map; all other dimensions are determined by scaled measurement on the map.

utilizing a single map sheet, as shown in Figure 19.2. In any case, the scale to be used for the mapping must be adapted to the specific needs of each individual community. Commonly used scales for urban official mapping are one inch equals 100 feet and one inch equals 50 feet. These scales facilitate a reduction on a 10:1 ratio for compiling a wall map of the full community at a final scale of one inch equals 1,000 feet or one inch equals 500 feet by an analog or digital mosaic process, and at a 2:1 ratio for producing neighborhood unit maps used in planning and utility engineering.

TECHNOLOGICAL ADVANCES IN MAPPING

Recent advancements in geospatial technology have significantly enhanced the preparation and maintenance of official maps. Geographic information systems (GIS), high-resolution satellite imagery, LiDAR data, and drone-based aerial surveys now allow for more precise, efficient, and up-to-date mapping of cadastral and topographic features. These technologies enable municipalities to overlay multiple data layers, integrate planning datasets, and update official maps in near real-time, improving accuracy in plan implementation and facilitating better coordination across planning, engineering, and public works departments.

FURTHER READING

Kucirek, J. C. & Buescher, J. H. "Wisconsin's Official Map Law," *Wisconsin Law Review*, Volume 1957.

20 Capital Improvement Programming

The great proof of madness is the disproportion of one's designs to one's means.

Napoleon I
Emperor of the French
1769–1821

INTRODUCTION

Land subdivision control, zoning, and official mapping are the three most basic and most important means available for the implementation of municipal comprehensive plans. All represent exercises of the police power, meaning the authority to regulate land use in support of public health, safety, and general welfare without direct compensation. In contrast, capital improvement programming presents a means of plan implementation through the power of public investment, taxation, and appropriation.

Land subdivision control, zoning, and official mapping are functions that should properly be within the purview of the municipal plan commission and planning staff. Capital improvement programming, however, is a complex task requiring expert knowledge of public finance. Moreover, it is a task that should involve all of the major departments of a city: public works and engineering, health and sanitation, police, fire, library, and, depending upon the structure of the local government, schools. The primary responsibility for the capital improvement programming function may, in some communities, rest with the plan commission or planning department, and indeed, some texts view capital improvement budgeting as performing some of the functions of a comprehensive plan. The responsibility for the capital improvement programming function may, in other communities, and equally appropriately, rest with the board of public works or public works department. This is often considered logical, since most of the improvements included in such a program consist of public works. The position taken in this text, however, is that the responsibility for the preparation of a capital improvement program should rest with the chief executive officer of a municipality—the mayor or city manager, village president, or town chairman—assisted by the chief financial officer. Ideally, the process should be guided by an advisory committee created for this purpose. The committee should comprise the chief executive officer, who should act as chairman, and all of the major municipal department heads, including the planning director. The committee should be staffed by the chief financial officer, for the program involves complex analyses of the financial resources and commitments of the municipality, and expert knowledge of the various ways of funding capital improvement projects, as well as operating budgets, is necessary. The

DOI: 10.1201/9781003500384-20

committee should be responsible for formulating the capital improvement program and presenting it to the governing body for consideration and adoption. The plan commission, through the planning director as a member of the advisory committee, should have an important role in the capital improvement programming process. The plan commission should be responsible for presenting to the committee for consideration needed capital improvements identified in the various elements of the adopted comprehensive plan.

DEFINITION AND CONTEXT

A capital improvement program may be defined as a time schedule for the construction of major physical improvements. From a planning viewpoint, the capital improvement program provides a vital link between planned physical improvements included in the comprehensive plan and the actual construction of those improvements. Such construction provides the public infrastructure facilities necessary to support planned land use development and redevelopment.

The capital improvement program is intended to deal with major and relatively costly physical improvements such as streets; bridges; sanitary trunk sewers and attendant pumping stations and treatment works; water supply transmission, storage, and treatment facilities; stormwater management and flood control works; airport and harbor facilities; public buildings of all types; and major parks and parkways. In some communities, major mass transit improvements may be included. Major rehabilitation or replacement of existing facilities is also classified as a capital improvement to be included in the program together with the construction of new facilities. Depending upon the size of the community, equipment—including fire engines, public works construction equipment, street sweepers, solid waste collection vehicles, snow plows, and other motor vehicles—may, or may not, be classified as capital improvements to be included in the program. In this respect, the term “capital improvement,” representing the major physical improvements to be included in the capital improvement program, should not be confused with the term “capital expenditure” as used by municipal accountants. This term more broadly includes all outlays that add to the physical assets of a municipality and is used in accounting for property control.

The capital improvement program must be understood within the context of an orderly public works development process. That process typically takes place in three successive stages: system planning, preliminary engineering, and final engineering and construction. Provisions must, of course, be made for the subsequent operation and maintenance of the constructed works.

Most public works facilities constitute a part of an integrated system, and each new or rehabilitated facility must function properly as an integral part of the system concerned, such as transportation, sanitary sewerage, water supply, stormwater management and flood control, parks and parkways, libraries, and schools. The necessary system planning should be accomplished as a part of the comprehensive planning process by the planning staff, or, for some systems, cooperatively by the planning and engineering staffs. The system planning will often involve the application of mathematical models to simulate the performance of the existing system and proposed improvements to the system. System plans for sanitary sewerage, water supply, and stormwater management are often termed facility plans. If developed outside of the comprehensive planning process, facility planning may require heroic assumptions to be made concerning future population, household, and employment levels and future land use patterns in the service areas concerned, as well as about such considerations as surface water and groundwater quality objectives. The system planning is intended to identify the general locations of the facilities composing the system and to provide definitive recommendations for the needed future capacity of each link or component of the overall system, together with preliminary estimates of capital and operating costs and a system-level environmental assessment.

In the second stage of the public works development process—preliminary engineering—selected facility improvements recommended in the system plan are refined through more detailed engineering studies. The studies may consider alternative horizontal and vertical alignments and locations for the facility concerned, develop details of the means for providing the needed capacity in the facility, provide a more detailed environmental assessment, and provide more detailed cost estimates.

In the third stage of the public works development process—final engineering and construction—final contract drawings and attendant contract and specification documents are prepared for the proposed facility, together with detailed costs. An accompanying environmental impact statement may be required. Each one of the three stages of the public works development process will involve its own particular means for accommodating necessary public review, and each will require its own official approvals. Failure to achieve the necessary official approvals at a given stage may require reconsideration of the findings and recommendations of the previous stage in the process.

The public works development process is intricately related to the capital improvement programming process. The latter determines the timing for the initiation of preliminary engineering and the final engineering and construction stages of major improvements. The initiation and completion of each stage of the public works development process provides information for the proper review and re-evaluation of the capital improvement program.

BENEFITS OF CAPITAL IMPROVEMENT PROGRAMMING

A number of benefits may be expected to accrue to the community from capital improvement programming. A capital improvement program should assist in the implementation of the comprehensive plan, particularly of those elements of the plan calling for major facility construction or reconstruction. It is through the capital improvement that the decision is made as to when a major facility improvement identified in the comprehensive plan is to be constructed. The timely provision of the facility will in turn influence the timing and pattern of land use development as envisioned in the comprehensive plan. Modern programs also link project timing to sustainability, resilience, and equity goals to ensure that public investments benefit all community sectors.

Capital improvement programming calls attention to deficiencies in the public infrastructure systems and helps promote needed actions to correct those deficiencies.

Capital improvement programming, however, also helps avoid the premature construction of costly facilities as well as promote the timely construction of needed facilities. Capital improvement programming helps the community avoid succumbing to pressures brought by special interest groups for projects that serve the private as opposed to the public interest. In contemporary practice, many cities now use data-driven prioritization frameworks, geographic information systems (GIS), and asset management software to identify and schedule improvements based on measurable needs and performance metrics.

Capital improvement programming will help produce needed cooperation and coordination between the various municipal departments concerned with physical development. Such cooperation and coordination is important, for example, in ensuring that the timing of street repaving projects follow, and not precede, needed underground utility reconstruction or replacement projects. Capital improvement programming will help produce needed cooperation and coordination not only between the various municipal departments, but also between the various levels and agencies of government concerned with physical development, assuring coordination between, for example, state trunk highway reconstruction and municipal improvement projects. Digital capital planning platforms and shared project dashboards now enable more effective interdepartmental coordination and real-time updates.

Importantly, capital improvement programming assures the provision of funding for needed projects in the most financially advantageous manner possible and serves to stabilize the local property tax rate. Capital improvement programming can also assist in evening out the workload of the municipal engineering and public works department staffs, helping maintain a competent construction, administration, and inspection staff without frequent and costly layoffs and subsequent re-hirings and retraining. Newer approaches also explore innovative funding mechanisms such as green bonds, federal infrastructure grants, and public–private partnerships to achieve long-term fiscal sustainability.

While the primary objective of capital improvement programming is to provide an orderly means for funding and constructing needed major improvements over time, and thereby assist in the implementation of certain elements of the comprehensive plan, there are some practical limitations to such programming. Future economic and financial conditions that determine the amount of funding that may be available for capital improvements are never wholly predictable. Emergencies that may require immediate reconstruction of public works facilities are also never predictable. The enactment of new federal or state laws and regulations may also have unanticipated impacts on a capital improvement program. The capital improvement program must, therefore, be flexible. The needed flexibility is facilitated by limiting the time span of the program to five years, each year adopting only the first year of the five-year program as the annual capital improvement budget, while reviewing and revising, as may be found necessary, the remaining four years of the program, and then adding a new fifth year. Many communities now use "rolling" capital improvement programs that are updated annually, supported by long-range financial forecasts, and informed by public participation through digital platforms and workshops.

ELEMENTS OF A CAPITAL IMPROVEMENT PROGRAM

Conceptually, a capital improvement program may be regarded as comprising three parts: the capital improvement schedule, the financial analysis, and the capital improvement budget. The capital improvement schedule is a list of the improvements needed over the next five years, arranged in order of priority. Contemporary programs often include project descriptions, responsible departments, estimated costs, geographic location, and anticipated funding sources. Projects are commonly ranked using a weighted scoring system based on criteria such as urgency, equity, infrastructure condition, and alignment with strategic goals. The financial analysis determines the funding that may be expected to be available over the next five years for capital improvements and identifies the available sources of that funding. Typical funding sources include general obligation bonds, revenue bonds, state or federal grants, developer impact fees, tax increment financing (TIF), and special assessments. Financial forecasting tools are often used to assess debt capacity, track cash flows, and evaluate the long-term affordability of projects. The capital improvement budget consists of the first year of the five-year program and is intended for adoption by the governing body as an integral part of its annual budget. It is important to note that the capital improvement budget is not the annual municipal budget but only one component of that budget that must address other important funding needs, including, importantly, ongoing operating costs. The operating impacts of each capital project—such as staffing, maintenance, and utilities—are now routinely estimated and incorporated into long-term financial plans. In adopting the comprehensive annual budget, the governing body may change the recommended capital improvement budget. Increased public transparency and engagement in budget processes are now common, supported by online dashboards and participatory budgeting initiatives.

PROCEDURE

Prepare List of Projects

The capital improvement programming process is typically initiated by the chief executive of the community—often in coordination with a capital planning advisory committee—issuing a request to all operating departments to submit proposed major public improvements to the committee. The submittals constitute, in effect, an inventory of all capital improvement projects proposed by the various municipal departments. The submittals should be accompanied by pertinent information, including for each proposed project a description of the project, including its nature and location; a statement of the need for and priority of the project; the availability of right-of-way or site area for the project and an estimate of the acquisition cost of any needed land; the estimated capital cost, including preconstruction engineering costs and land acquisition, and construction costs by project year; the estimated continuing operation and maintenance cost; anticipated project initiation and completion dates; estimated potential revenue production, if any; attendant permit-ted abandonment of other facilities; status as an initial or continuing project submittal; potential for future expansion, including adequacy of right-of-way or site; estimated physical and economic life of facility; and suggested means of financing, including TIF and special assessments. In modern practice, many communities also include alignment with adopted strategic goals, equity impact assessments, and GHG emissions implications. Benefit-cost, or cost-effectiveness, analyses of the alternatives to a proposed project may be provided to assist in project evaluation. It is important that the cost estimate be complete by including, for example, the need for and cost of assistance in the construction layout and inspection by the present municipal workforce or the cost of furniture attendant to construction of a new building. The requested information is intended to assist the advisory committee in evaluating the merits of the proposed projects and establishing a relative order of priority for construction.

Conduct Financial Analysis

The chief financial officer of the municipality should be responsible for the assembly and analysis of the financial data required to establish the funding from the various potential sources available to support capital improvement projects. The amounts of all potential future revenues, including property, income, and sales tax revenues; special assessments; license and permit fees; service charges; fines; general obligation and special assessment bonds; federal and state loans and grants; and miscellaneous sources must all be considered together with the municipality's general operating budget and debt service requirements to arrive at an estimate of the revenue stream that can be reasonably assumed to be available annually for capital improvements. Data on all outstanding debt must be assembled so that a future debt retirement schedule can be prepared and the amounts needed annually for debt service determined. The data should be presented by total debt service, principal repayment required, and interest due. Constitutional or statutory debt limits should be indicated. New debt service required and permitted for proposed capital improvements can then be determined. Since it must consider the operating budget needs, this complex task, in effect, constitutes the preparation of a comprehensive financial plan for the municipality. Modern systems often use financial forecasting software and integrated capital planning platforms to support this task.

Prepare Capital Improvement Program

The list of submitted proposed projects and the findings of the financial analysis provide the basic data from which the capital improvement program is formulated. The list of projects must be reviewed and evaluated to establish relative priorities, the estimated available funding must be distributed among the prioritized projects, and a final list of projects must be selected for the program based on consideration of the project priorities, project costs, and available funding. It is now common to involve public engagement processes—such as community surveys or public comment periods—during this stage to increase transparency and support equity in project selection.

Project Evaluation

The list of projects typically may be expected to contain many more proposed improvements than the available funding can support. While the final selection of projects will, in the last analysis, be determined by the amount and type of funds available, the list of projects must be first evaluated on the basis of relative need or desirability. Most important in this evaluation should be identified inadequacies in the performance of the existing infrastructure system and attendant dissatisfaction with the level of service provided. The projects are first assessed on the basis of the inadequacies intended to be corrected within each system of facilities concerned. Thus, for example, proposed street improvement projects may be assessed on the basis of the severity of the levels of traffic congestion on the various segments of the arterial street and highway system or assessed on the condition of the pavements on the various segments of the total street and highway system. Similarly, proposed sanitary sewer improvement projects might be assessed on the basis of the potential for sewage backups into basements or on the severity of clear water infiltration and inflow conditions. The priorities are then reassessed between the various systems and then balanced by considering relative urgencies as influenced by the magnitude of the inadequacies, the requirements of federal, state, and local laws, ordinances and regulations, right-of-way or site availability, and judgments concerning the relationship to the promotion of the general public health, safety, and welfare. Proposed corrections of inadequacies presenting danger to human life or a threat to public health and safety should be given preference over other improvements. Projects to protect public and private property from damage or destruction and projects to conserve resources should be given preference over projects facilitating new development. Some modern evaluation frameworks also consider equity, climate resilience, and environmental justice as critical dimensions of priority. Criteria helpful in evaluating proposed projects and determining relative priorities include protection of life, maintenance of the public health, protection of property, conservation of resources and environmental protection, replacement of obsolete and inefficient facilities, reduction in operating and maintenance costs and energy consumption, promotion of public convenience and comfort, effect on potential future development, and consistency with the adopted comprehensive plan. The relative importance attached to these and other criteria may be expected to differ between communities. As noted in Chapter 2, the value system prevalent in a community will determine the importance attached to public improvements and the importance of the various types of benefits provided by those improvements.

When the relative need or priority of the proposed projects in the list has been agreed upon, and the method of financing and amount of funding available determined, a final list of projects that will constitute the capital improvement program can be determined. The projects may be grouped by four categories on the basis of priority: essential, desirable, acceptable, and deferrable. Projects listed in the essential category would have the highest priority for construction. This category would include projects intended to correct conditions dangerous to human life or threat to public health and safety, to protect public or private property from damage or destruction, and to conserve resources.

Projects listed in the desirable category would include primarily projects that would serve to implement adopted elements of the comprehensive plan and thereby contribute to the sound development of the community. Projects listed in the acceptable category would include projects not in conflict with the comprehensive plan but not absolutely required. Projects listed in the deferred category would represent projects of questionable need and timing, which are recommended for indefinite postponement or elimination from the program. The assignment of priority rankings to the listed projects involves tempered judgment and careful consideration of expediencies, public policies, and public attitudes, and the ranking should be done in a collegial fashion by the advisory committee.

Presentation

The capital improvement program should be presented in well-arranged tabular form, listing projects in the preferred order of construction and year scheduled. Estimated capital costs together with estimated attendant increases in municipal operating and maintenance costs should be provided for each listed project. Proposed methods of financing and estimated financing charges and increases in debt service commitment should also be given for each project listed. The tables should be accompanied by text briefly describing each project, the reasons for the rank ordering of priority, and the urgency of need. The first year of the five-year program should explicitly recommend a capital budget to provide the required expenditures. The total financial need for each of the other four years should also be given and should fall within the financial limitations of the community. Contemporary best practice also includes interactive dashboards or GIS-based visualizations for broader community accessibility.

At least six summary tables should be prepared. The first should present the assembled list of projects. Potential column headings might include the submitting department; the type of project; if a rehabilitation or replacement project, the original construction date and typical useful life; capacity expressed in units, as may be appropriate for the type of facility concerned, such as vehicles per day, gallons per day, miles of operation, employees, and persons; current level of use expressed in the same units as the comparable capacity; capital cost; attendant increase in municipal operating cost; revenue, if any; and suggested source of funding. A final column should indicate whether the project is in conformance or in conflict with the adopted comprehensive plan.

The second table should present historical and projected data on operating costs, capital, and debt service costs—expenditures—by major function such as fire protection, police protection, sewerage, water supply, street, and highways, parks, general government, and perhaps schools. At least 10 years of historical and 5 years of projected data should be presented. The third table should present historical and projected data on revenue by sources such as real property taxes, income taxes, sales taxes, fees, fines, and service charges, special assessments, and state and federal loans and grants. Again, at least 10 years of historical and 5 years of projected data should be presented.

The fourth table should compare historical and projected revenues and expenditures to arrive at estimates of the total funding that may be available annually for capital improvements. This estimate is used to prepare the fifth table—the actual capital improvement program—listing the recommended projects in priority order, together with the capital cost, estimated attendant operating and maintenance cost that may increase or decrease the attendant operating budgets, the sources of funding, and the project timing dates—initiation and completion—by stages such as preliminary engineering, final engineering, and construction. A summary table should be provided showing the effects of the program upon the community with respect to tax requirements.

Adoption and Revision

The recommended capital improvement program is submitted to the governing body, which should hold a public hearing on the program. Based on the information provided by the hearing and in consideration of the relationship between the recommended capital improvement budget and companion necessary operating budgets, the governing body may eliminate some projects, reduce others, shift the order of the projects, or even add new projects before adopting the first year of the recommended program as the capital portion of the total budget for the ensuing year.

Following adoption of the annual capital budget by the governing body, the program for the remaining four years should be reviewed by the committee, any needed new projects added to the list of projects, projects no longer justified eliminated, projects shifted in priority as may be found necessary, an additional year added to replace the year adopted, and the revised list of projects again scheduled over the full five-year period. The process is repeated annually so that a well-conceived five-year program is always available and ready for use, but with only one year being actually committed at a time. As the process is repeated during each succeeding year, proposed projects are typically subject to repeated re-evaluations before actual approval for construction. This repetitive re-evaluation provides additional assurance of the need for, and quality of, the projects in the list, another benefit of capital improvement programming.

21 New Features of Urban Planning

The field of urban planning continues to evolve in response to new challenges, innovations, and societal expectations. While traditional tools such as comprehensive planning, zoning, subdivision regulations, and capital improvement programming remain foundational, emerging priorities now require cities to integrate sustainability, resilience, and technology into the urban fabric in new ways. This chapter highlights three key dimensions of modern urban planning—sustainable cities, resilient cities, and smart cities—and presents representative case studies of how these concepts are being applied.

SUSTAINABLE CITY

The concept of a sustainable city reflects the growing recognition that urban development must meet the needs of the present without compromising the ability of future generations to meet their own needs. Sustainability in urban planning emphasizes reducing environmental impacts, conserving resources, and creating systems that are both efficient and equitable. Planners working toward sustainability seek to integrate land use, energy, water, transportation, and waste systems in ways that promote long-term environmental health and community well-being.

Urban-Scale Electrification and Decarbonization

Sustainable cities are increasingly defined by their ability to transition away from fossil fuel dependency toward clean energy systems. Urban-scale electrification involves the widespread replacement of combustion-based systems—particularly in buildings, vehicles, and industry—with electricity sourced from renewable energy. This transition includes the implementation of electric vehicle charging infrastructure, electrified public transit fleets, and the decarbonization of heating and cooling systems through electric heat pumps and district energy networks.

Planners play a vital role in supporting electrification efforts through land use decisions that accommodate distributed energy resources, utility coordination, and energy zoning overlays. Decarbonization efforts are often guided by climate action plans or greenhouse gas inventories developed at the municipal level, setting benchmarks for carbon neutrality and enabling funding prioritization.

Sustainable Integrated Systems

Modern sustainability strategies emphasize the interconnectedness of urban systems—energy, water, food, transportation, and waste—and the need to optimize them in unison. Integrated systems planning seeks synergies between these domains,

DOI: 10.1201/9781003500384-21

such as using anaerobic digestion of organic waste for biogas production, capturing stormwater for non-potable reuse, or aligning land use and transit investment to reduce auto dependency.

Tools such as life-cycle assessments, material flow analysis, and circular economy frameworks assist planners in evaluating the full environmental impact of urban infrastructure and identifying opportunities for system-wide efficiency. Regional collaboration across jurisdictions is often essential to ensure alignment between infrastructure investments and long-term sustainability goals.

Policy and Regulatory for Sustainable Cities

Policy and regulatory instruments remain critical to the advancement of sustainability in urban planning. Zoning ordinances may be amended to permit mixed-use development, incentivize green buildings, or require energy-efficient construction. Benchmarking and disclosure ordinances for building performance, clean construction requirements, and green infrastructure mandates are becoming increasingly common at the city and county levels.

Planners also support the development of local sustainability standards—such as climate action plans, net-zero codes, or resiliency scorecards—that embed environmental performance into permitting processes. These regulatory tools are evolving rapidly to reflect ambitious state and national climate targets.

RESILIENT CITY

A resilient city is one that is equipped to anticipate, absorb, adapt to, and recover from a wide range of shocks and stresses—whether environmental, economic, or social. In the face of mounting climate risks, urban resilience has become an essential component of planning practice. Planners aim to strengthen the capacity of infrastructure, ecosystems, institutions, and communities to withstand disruptions, minimize damage, and maintain essential functions during and after emergencies.

Climate Change Impacts

Urban areas face increasing exposure to the physical risks of climate change, including sea level rise, extreme heat, flooding, drought, and wildfire. These hazards pose threats to lives, property, infrastructure, and municipal budgets. Effective planning for resilience requires integrating climate projections into land use decisions, infrastructure design, and emergency preparedness.

Tools such as vulnerability assessments and hazard mitigation plans help identify critical assets at risk and prioritize interventions. Planners are increasingly involved in coordinating these assessments across departments and integrating findings into long-range planning documents and building codes.

Infrastructure Designs and Retrofits

Resilience planning requires new thinking in the design and retrofit of urban infrastructure. Green infrastructure—such as rain gardens, bioswales, permeable pavements, and urban tree canopies—provides low-cost, decentralized strategies for managing stormwater and mitigating urban heat. Coastal cities are investing in elevated roadways, seawalls, levees, and other gray infrastructure to protect against storm surge and sea level rise.

Retrofit strategies for existing neighborhoods and facilities often prioritize backup energy systems, flood-proofing of critical facilities, and community shelter hubs. Planners collaborate with engineers, architects, and emergency managers to ensure these interventions align with community needs and available funding mechanisms.

Policy and Regulatory for Resilient Cities

The regulatory foundation for urban resilience is expanding rapidly. Federal guidance—such as FEMA's Community Rating System and the HUD-funded Resilience Planning Grants—has encouraged the integration of resilience into local development regulations. At the state and local level, jurisdictions are adopting "resilience overlays," requiring additional review for developments located in vulnerable areas.

Policy innovations include managed retreat strategies, buyout programs, and zoning restrictions in high-risk flood zones. In many communities, resilience metrics are incorporated into capital improvement programming to guide infrastructure investments toward long-term risk reduction.

SMART CITY

The smart city is characterized by the strategic use of digital technologies to enhance urban systems and improve the quality of life. Sensors, networks, data platforms, and intelligent automation allow cities to manage infrastructure more efficiently, respond to problems in real time, and engage residents in new ways. Planners contribute to smart city initiatives by aligning technological solutions with public goals, ensuring that innovation serves inclusive, livable, and well-governed communities.

Technological Innovations

The smart city movement emphasizes the use of advanced technologies to improve urban services, governance, and quality of life. Core technologies include sensors, broadband networks, autonomous systems, and data analytics platforms. These technologies are used to optimize traffic flow, monitor energy usage, manage solid waste collection, and improve public safety.

Smart city strategies often begin with the development of a digital infrastructure plan that enables the deployment of new technologies while ensuring equity in access. Key challenges include cybersecurity, interoperability, and long-term maintenance. Planners are uniquely positioned to ensure that digital technologies serve broader planning objectives and do not exacerbate existing inequities.

Cyber-Physical Systems

Cyber-physical systems integrate computational infrastructure with physical infrastructure to enable real-time control and responsiveness. Examples include adaptive traffic signal control systems, building automation systems that respond to occupant behavior and weather, and sensor-enabled utility grids.

In the planning context, cyber-physical systems offer new ways to monitor infrastructure performance, collect community feedback, and manage emergencies. Smart stormwater systems, for instance, use real-time sensors and remotely controlled valves to prevent localized flooding during major storms.

CASE STUDIES

Case studies from across the United States and internationally provide insights into how cities are implementing sustainable, resilient, and smart planning strategies:

- New York City, NY: The *OneNYC* plan integrates sustainability and resilience goals, including ambitious electrification targets and coastal flood mitigation projects.
- Austin, TX: The *Austin Climate Equity Plan* emphasizes decarbonization alongside housing affordability, integrating environmental justice into city planning.
- Rotterdam, The Netherlands: This low-lying port city has invested heavily in green-blue infrastructure, such as water plazas, to manage increasing flood risks.
- Barcelona, Spain: Known for its "Superblocks" initiative, Barcelona has redesigned entire neighborhoods to prioritize pedestrians, reduce emissions, and enhance public space.
- San Diego, CA: A leader in smart city deployment, San Diego uses sensors and connected devices to improve energy efficiency and urban mobility.

These examples demonstrate how urban planning continues to evolve to meet the demands of a changing world. Integrating sustainability, resilience, and smart technology not only improves the livability of cities but also enhances their long-term viability in the face of accelerating environmental, economic, and social change.

Appendix A: Example Land Subdivision Control Ordinance Regulations

SECTION 1.00 INTRODUCTION

1.01 Title

This Ordinance shall be known as the "Land Division Control Ordinance of the City of __________," or as "Chapter __________ of the City of __________ Code of Ordinances."

1.02 Statutory Authorization

These regulations are adopted under the authority granted by Section 236.45 of the Wisconsin Statutes.

1.03 Purpose

The purpose of this Ordinance is to regulate and control all land divisions within the corporate limits of the *City* of __________ and within the extraterritorial plat approval jurisdiction of the City in order to promote and protect the public health, safety, aesthetics, and general welfare of the community. More particularly, and without limitation, it is the purpose of this Ordinance to:

A. Implement the City's comprehensive plan and components thereof and facilitate enforcement of community development standards as set forth in the zoning code, building code, and official map.
B. Promote the Wise Use, development, conservation, and protection of the soil, water, wetland, woodland, and wildlife resources in the *City and its area of extraterritorial plat approval jurisdiction*, and to achieve a balanced relationship between land use and development and the supporting and sustaining natural resource base.
C. Further the Orderly Layout and appropriate use of land.
D. Avoid the Harmful Effects of premature division or development of land.
E. Lessen Congestion in the streets and highways.
F. Provide for Proper Ingress to and egress from development sites.
G. Secure Safety from fire, flooding, water pollution, and other hazards and minimize expenditures for flood relief and flood control projects.
H. Prevent and Control Erosion, sedimentation, and other pollution of surface and subsurface waters.

I. Preserve Natural Vegetation and cover and protect the natural beauty of the *City*.
J. Provide Adequate Light and Air.
K. Prevent the Overcrowding of land and avoid undue concentration of population.
L. Facilitate the Division of land into smaller parcels.
M. Facilitate and Ensure the adequate provision of transportation, water, sewerage, stormwater management, schools, parks, playgrounds, and other public facilities and services.
N. Ensure Adequate Legal Description and proper survey monumentation of divided land.
O. Provide Adequate, Affordable Housing.
P. Restrict Building in areas of unsuitable soils.
Q. Provide for the Administration and enforcement of this Ordinance.

1.04 Abrogation and Greater Restrictions

It is not the intent of this Ordinance to repeal, abrogate, annul, impair, or interfere with any existing easements, covenants, agreements, rules, regulations, permits, or approvals previously adopted or issued pursuant to law. However, where this Ordinance imposes greater restrictions, and such restrictions do not contravene rights vested under law, the provisions of this Ordinance shall govern.

1.05 Interpretation

The provisions of this Ordinance shall be interpreted to be minimum requirements and shall be liberally construed in favor of the *City*, and shall not be deemed a limitation or repeal of any other power granted by the Wisconsin Statutes.

1.06 Severability

If any section, provision, or portion of this Ordinance is adjudged unconstitutional or invalid by a court of competent jurisdiction, the remainder of this Ordinance shall not be affected thereby.

1.07 Repeal

All other ordinances or parts of ordinances of the *City* inconsistent or conflicting with this Ordinance, to the extent of the inconsistency only, are hereby repealed.

1.08 Disclaimer of Liability

The *City* does not guarantee, warrant, or represent that only those areas delineated as floodplains on plats and certified survey maps will be subject to periodic inundation, nor does the *City* guarantee, warrant, or represent that the soils shown to be unsuited for a given land use from tests required by the Ordinance are the only unsuited soils

and facilities. These documents shall be subject to review as to form by the *City* Attorney at the subdivider's expense.

B. The Association shall be established by the owner or applicant and shall be operating prior to the sale of any lots or units in the subdivision or condominium.
C. Membership in the association shall be mandatory for all purchasers of lots or units therein and their successors and assigns.
D. The Association shall be responsible for maintenance and insurance of common areas and facilities.
E. A Land Stewardship Plan for any common open space to be retained in a natural state shall be included in the submittal of association documents.
F. The Members of the association shall share equitably the costs of maintaining, insuring, and operating common areas and facilities.
G. The Association shall have or hire adequate staff to administer, maintain, and operate common areas and facilities.
H. The Subdivider shall arrange with the *City* Assessor a method of assessment of any common areas and facilities, which will allocate to each lot, parcel, or unit within the land division or condominium a share of the total assessment for such common areas and facilities.
I. The *City* may require that it receive written notice of any proposed transfer of common areas or facilities by the association or the assumption of maintenance of common areas or facilities. Such notice shall be given to all members of the association and to the *City* at least 30 days prior to such transfer.
J. In the Event that the association established to own and maintain common areas and facilities, or any successor organization thereto, fails to properly maintain all or any portion of the aforesaid common areas or facilities, the *City* may serve written notice upon such association setting forth the manner in which the association has failed to maintain the aforesaid common areas and facilities. Such notice shall set forth the nature of corrections required and the time within which the corrections shall be made. Upon failure to comply within the time specified, the association, or any successor association, shall be considered in violation of this Ordinance, in which case the *City* shall have the right to enter the premises and take the needed corrective actions. The costs of corrective actions by the *City* shall be assessed against the properties that have the right of enjoyment of the common areas and facilities.

2.07 Improvements

Before approval of any final plat located within the corporate limits of the *City*, the subdivider shall install street and other improvements as hereinafter provided. In the alternative, if such improvements are not installed at the time the final plat is submitted for approval, the subdivider shall, before the recording of the plat, enter into a development agreement with the *City* agreeing to install the required improvements, and shall file with said agreement a bond or letter of credit with good and sufficient surety meeting the approval of the *City* Attorney or a certified check in the amount equal to the estimated cost of the improvements. Said estimate shall be made by the *City* Engineer, as a guarantee that such improvements will be completed by the subdivider or his or her subcontractors not later than one year from the date the plat is recorded and as a further guarantee that all obligations to subcontractors for work on the subdivision are satisfied.

A. Contracts and contract specifications for the construction of street and utility improvements within public street rights-of-way, as well as contractors and subcontractors providing such work, shall be subject to approval of the *City* Engineer.
B. Governmental Units to which these bond and contract provisions apply may file, in lieu of said contract and bond, a letter from officers authorized to act on their behalf agreeing to comply with the provisions of this section.
C. Survey Monuments. Before final approval of any plat within the corporate limits of the *City*, the subdivider shall cause survey monuments to be installed as required by and placed in accordance with the requirements of Section 236.15 of the Wisconsin Statutes, and as may be required by the *City* Engineer. The City Engineer may waive the placing of monuments, as provided in Section 236.15(1)(h) of the Wisconsin Statutes, for a reasonable time, not to exceed one year, on condition that the subdivider provide a letter of credit, certified check, or surety bond equal to the estimated cost of installing the monuments to ensure the placing of such monuments within the time required by statute. Additional time may be granted upon show of cause.
D. *Plats Outside Corporate Limits. Before final approval by the City of any plat located outside the corporate limits of the City but within the extraterritorial plat approval jurisdiction of the City, the subdivider shall give evidence that he or she has complied with all street and utility improvement requirements of the Town in which of the land being platted is located.*

2.08 Development Agreement

Before or as a condition of receiving final approval from the *Common Council* of any final plat, condominium plat, or certified survey map for which public improvements are required by this Ordinance; or for which public improvements, dedications, or fees are being deferred under this Ordinance; or for which phasing approval is being granted under Section 9.02 of this Ordinance, the subdivider shall sign and file with the *Common Council* a development agreement. The development agreement shall be approved as to form by the *City* Attorney, and shall be approved by the *Common Council* prior to approval of the final plat, condominium plat, or certified survey map.

2.09 Exceptions and Modifications

Where, in the judgment of the *City* Plan Commission, it would be inappropriate to apply literally the provisions of Sections 7.00 and 8.00 of this Ordinance because exceptional or undue hardship would result, the Plan Commission may waive or modify any requirement to the extent deemed just and proper. Such relief shall be granted without detriment to the public good, without impairing the intent and purpose of this Ordinance or the desirable general development of the community in accordance with an adopted comprehensive plan or component thereof. No exception or modification shall be granted unless the Plan Commission finds that all the following facts and conditions exist and so indicates in the minutes of its proceedings:

A. Exceptional Circumstances: There are exceptional, extraordinary, or unusual circumstances or conditions where a literal enforcement of the requirements of this Ordinance would result in severe hardship. Such hardships should not apply generally to other properties or be of such a recurrent nature as to suggest that this Ordinance should be changed.
B. Preservation of Property Rights: That such exception or modification is necessary for the preservation and enjoyment of substantial property rights possessed by other properties in the same vicinity.
C. Absence of Detriment. That the exception or modification will not create substantial detriment to adjacent property and will not materially impair or be contrary to the purpose and spirit of this Ordinance or the public interest.
D. A Simple Majority Vote of the full membership of the Plan Commission shall be required to grant any exception or modification of this Ordinance, and the reasons shall be entered into the minutes of the Commission.

2.10 Violations

No person, firm, or corporation shall build upon, divide, convey, record, or place monuments on any land in violation of this Ordinance or the Wisconsin Statutes. No person, firm, or corporation shall be issued a building permit by the *City* authorizing the building on, or improvement of, any subdivision, minor land division, replat, or condominium within the jurisdiction of this Ordinance not of record as of the effective date of this Ordinance, until the provisions and requirements of this Ordinance have been fully met. The *City* may institute appropriate action or proceedings to enjoin violations of this Ordinance.

2.11 Penalties and Remedies

Any person, firm, or corporation who fails to comply with the provisions of this Ordinance shall, upon conviction thereof, forfeit not less than $100 plus any additional applicable costs incurred by the *City* for each offense, and the penalty for default of payment of such forfeiture and costs shall be imprisonment in the County Jail until payment thereof, but not exceeding six months. Each day a violation exists or continues shall constitute a separate offense. Violations and concomitant penalties shall include the following:

A. Recordation improperly made carries penalties as provided in Section 236.30 of the Wisconsin Statutes.
B. Conveyance of lots in unrecorded plats carries penalties as provided for in Section 236.31 of the Wisconsin Statutes.

C. Monuments disturbed or not placed carries penalties as provided for in Section 236.32 of the Wisconsin Statutes.
D. An Assessor's Plat made under Section 70.27 of the Wisconsin Statutes may be ordered as a remedy by the *City*, at the expense of the subdivider, when a subdivision is created by successive divisions.

2.12 Appeals

Any person aggrieved by an objection to a plat or a failure to approve a plat may appeal such objection or failure to approve, as provided in Sections 236.13(5) and 62.23 (7)(e) of the Wisconsin Statutes, within 30 days of notification of the rejection of the plat. Where failure to approve is based on an unsatisfied objection, the agency making the objection shall be made a party to the action. The court shall direct that the plat be approved if it finds that the action of the approving or objecting agency is arbitrary, unreasonable, or discriminatory.

SECTION 3.00 LAND DIVISION PROCEDURES

3.01 Pre-Application Staff Conference

It is recommended that, prior to the filing of an application for the approval of a preliminary plat, condominium plat, or certified survey map, the subdivider consult with the Plan Commission and/or its staff in order to obtain their advice and assistance. It is recommended that a conceptual plan of the proposed subdivision, condominium, or certified survey map be brought by the applicant to the meeting, but such conceptual plan is not required. This consultation is neither formal nor mandatory, but is intended to inform the subdivider of the purpose and objectives of these regulations, the comprehensive plan, and duly adopted plan implementation devices of the *City* and to otherwise assist the subdivider in planning the development. In so doing, both the subdivider and Plan Commission may reach mutual conclusions regarding the general program and objectives of the proposed development and its possible effects on the neighborhood and community. The subdivider will gain a better understanding of the subsequent required procedures.

3.02 Preliminary Plat Review within the City

Before submitting a final plat for approval, the subdivider shall prepare a preliminary plat and complete an application and review checklist. The preliminary plat shall be prepared in accordance with this Ordinance and the subdivider shall file an adequate number of copies of the plat for distribution in accordance with this Section; the completed application and checklist; and the preliminary plat review fee with the *City* Clerk at least 60 days prior to the meeting of the *Common Council* at which action is desired.

A. The City Clerk shall, within two normal working days after filing, transmit:
   1. Four copies to the County Planning Agency;
   2. Two copies to the Director of Plat Review, Wisconsin Department of Administration;

3. Additional copies to the Director of Plat Review, Wisconsin Department of Administration, for re-transmission as follows:
   a. Two copies to the Wisconsin Department of Transportation (WisDOT) if the subdivision abuts or adjoins a state trunk highway or a connecting highway;
   b. Two copies to the Wisconsin Department of Commerce if the subdivision is not served by a public sewer and provision for such service has not been made; and
   c. Two copies to the Wisconsin Department of Natural Resources (WDNR) if lands included in the plat lie within 500 feet of the ordinary high water mark of any navigable stream, lake, or other navigable body of water, or if any shoreland areas are located within the plat.

B. In Lieu of the Procedure Set Forth Above, the subdivider may, pursuant to Section 236.12(6) of the Wisconsin Statutes, submit the original of the preliminary plat directly to the plat review section of the Wisconsin Department of Administration, who will prepare and forward copies of the plat at the subdivider's expense to the objecting agencies. When the subdivider elects to use this alternative procedure, it shall be the responsibility of the subdivider to submit to the *City* Clerk the additional copies required for the reviews required below.

C. The *City* Clerk shall also transmit, within two normal working days after filing, eight copies of the preliminary plat to the City Plan Commission and one copy each to the *City* Engineer, Director of Public Works, *City* Planner, Fire Chief, Parks Director, and Land Information Officer for review and recommendations concerning matters within their jurisdiction. The recommendations of City officials shall be transmitted to the Plan Commission within 20 days from the date the plat is received. The preliminary plat shall then be reviewed by the Plan Commission for conformance with this Ordinance, and all other *City* ordinances, rules, regulations, and the comprehensive plan and components thereof.

D. The *City* Clerk shall also transmit, within two normal working days after filing, one copy each of the preliminary plat to the Southeastern Wisconsin Regional Planning Commission, affected public and private utility companies, and the affected school district or districts for their review and recommendation concerning matters within their jurisdiction. Their recommendations shall be transmitted to the Plan Commission within 20 days from the date the plat is received.

3.03 Approval of a Preliminary Plat Located within the City

A. The Objecting Agencies shall, within 20 days of the date of receiving their copies of the preliminary plat, notify the subdivider and all other approving and objecting agencies of any objections. If there are no objections, they shall so certify on the face of the copy of the plat and shall return that copy to the City Clerk. If an objecting agency fails to act within 20 days, it shall be deemed to have no objection to the plat.

B. The City Plan Commission shall promptly review the preliminary plat, after objections and comments have been received by the objecting and reviewing agencies and officials, for conformance with this Ordinance and all applicable laws, ordinances, and comprehensive plans and components of such plans. The Plan Commission shall comment and recommend action on the preliminary plat to the Common Council.
C. The Common Council shall, within 90 days of the date of filing of the preliminary plat with the City Clerk, approve, approve conditionally, or reject such plat. One copy of the plat shall thereupon be returned to the subdivider with the date and action endorsed thereon; and if approved conditionally or rejected, a letter setting forth the conditions of approval or the reasons for rejection shall accompany the plat. One copy each of the plat and letter shall be placed in the Plan Commission's permanent file.
D. Failure of the Common Council to act within 90 days shall constitute an approval of the plat as filed, unless the review period is extended by mutual consent.
E. Approval or conditional approval of a preliminary plat shall not constitute automatic approval of the final plat, except that if the final plat is submitted within 24 months after the last required approval of the preliminary plat and conforms substantially to the preliminary plat, including any conditions of that approval, and to local plans and ordinances, the final plat shall be entitled to approval as provided in Section 236.11(1)(b) of the Wisconsin Statutes.

3.04 Final Plat Review within the City

A final plat shall be prepared in accordance with this Ordinance and the subdivider shall file an adequate number of copies of the plat for distribution in accordance with this Section; the completed application; and the final plat review fee with the City Clerk at least 25 days prior to the meeting of the Common Council at which action is desired.

A. The *City* Clerk shall, within two normal working days after filing, transmit:
   1. Four copies to the County Planning Agency;
   2. Two copies to the Director of Plat Review, Wisconsin Department of Administration;
   3. Additional copies to the Director of Plat Review, Wisconsin Department of Administration, for re-transmission as follows:
      a. Two copies to the Wisconsin Department of Transportation (WisDOT) if the subdivision abuts or adjoins a state trunk highway or a connecting highway;
      b. Two copies to the Wisconsin Department of Commerce if the subdivision is not served by a public sewer and provision for such service has not been made; and
      c. Two copies to the Wisconsin Department of Natural Resources (WDNR) if lands included in the plat lie within 500 feet of the ordinary

high water mark of any navigable stream, lake, or other navigable body of water, or if any shoreland areas are located within the plat.

B. In Lieu of the Procedure Set Forth Above, the subdivider may, pursuant to Section 236.12(6) of the Wisconsin Statutes, submit the original of the final plat directly to the plat review section of the Wisconsin Department of Administration, who will prepare and forward copies of the plat at the subdivider's expense to the objecting agencies. When the subdivider elects to use this alternative procedure, it shall be the responsibility of the subdivider to submit to the *City* Clerk the additional copies required for the reviews required below.

C. The *City* Clerk shall also transmit, within two normal working days after filing, eight copies of the final plat to the *City* Plan Commission and one copy to each of the affected public or private utilities.

D. The City Plan Commission shall examine the final plat as to its conformance with the approved preliminary plat; conditions of approval of the preliminary plat; this Ordinance and all ordinances, rules, regulations, comprehensive plans or components thereof which may affect it; and shall recommend approval or rejection of the plat to the *Common Council.*

E. Partial Platting. The final plat may, if permitted by the Plan Commission, constitute only that portion of the approved preliminary plat that the subdivider proposes to record at that time; however, it is required that each phase be final platted and designated as a phase of the approved preliminary plat.

3.05 Approval of a Final Plat Located within the City

The objecting agencies, shall, within 20 days of the date of receiving their copies of the final plat, notify the subdivider and all other approving and objecting agencies of any objections. If there are no objections, they shall so certify on the face of the copy of the plat and shall return that copy to the City Plan Commission. If an objecting agency fails to act within 20 days, it shall be deemed to have no objection to the plat.

A. Submission. If the final plat is not submitted within 24 months of the last required approval of the preliminary plat, the *Common Council* may refuse to approve the final plat.

B. The *City* Plan Commission shall, within 45 days of the date of filing of the final plat with the *City* Clerk, recommend approval or rejection of the plat and shall transmit the final plat and application along with its recommendation to the *Common Council.*

C. Notification. The *City* Plan Commission shall, when it determines to recommend approval or rejection of a plat to the *Common Council*, give at least 10 days prior written notice of its recommendation to the clerk of any municipality within 1,000 feet of the plat.

D. The *Common Council* shall, in accordance with Section 236.12 of the Wisconsin Statutes, within 60 days of the date of filing the original final plat with the City Clerk, approve or reject such plat. The *Common Council* may act on the plat at the same meeting at which the Plan Commission makes its recommendation. One copy of the plat shall thereupon be returned to the subdivider with the date and action endorsed thereon. If the plat is rejected, the reasons shall be stated in the minutes of the meeting and a written statement of the reasons forwarded to the subdivider. One copy each of the plat and letter shall be placed in the *City* Clerk's permanent file. The *Common Council* shall not inscribe its approval on the final plat unless the *City* Clerk certifies on the face of the plat that the copies were forwarded to objecting agencies as required herein, with the date they were forwarded, and that no objections have been filed within 20 days or, if filed, that they have been met.

E. Failure of the *Common Council* to act within 60 days, the time having not been extended and no unsatisfied objections having been filed, and all fees payable by the subdivider having been paid, shall constitute approval of the final plat.

F. Recordation. After the final plat has been approved by the *Common Council* and required improvements either installed or a contract and sureties insuring their installation is filed, the *City* Clerk shall cause the certificate inscribed upon the plat attesting to such approval to be duly executed and the plat returned to the subdivider for recording with the County Register of Deeds. The Register of Deeds shall not record the plat unless it is offered for recording within 30 days after the date of the last approval and within 24 months after the first approval, as required in Section 236.25(2)(b) of the Wisconsin Statutes.

G. Copies. The subdivider shall file 10 copies of the recorded final plat with the City Clerk. The Clerk shall distribute copies of the plat to the City Engineer, Building Inspector, Assessor, Land Information Officer, Planning Director, and other affected City and County departments for their files.

3.06 Plats within the Extraterritorial Plat Approval Jurisdiction of the City

A. When the Land to be Subdivided lies within the extra-territorial plat approval jurisdiction of the *City*, the subdivider shall proceed as specified in Sections 3.01 through 3.05 except:

1. If the subdivider elects to initially submit the proposed plat to a local municipality, the plat shall, as a matter of courtesy, be first submitted to the Town Clerk concerned. The Town Clerk shall then assume the responsibility for transmitting the plat to the objecting agencies and other approving authorities, including the *City*.
2. Approving agencies include the *Common Council*, Town Board, and County Planning Agency; and the subdivider must comply with the land subdivision ordinances of the *City*, Town, and County.

B. The Subdivider may proceed with the installation of such improvements and under such regulations as the Town Board of the Town within whose limits the plat lies may require. Wherever connection to any *City* utility is desired, permission for such connection shall be approved by the *Common Council*.

C. All Improvement Requirements specified by the Town Board or any special improvement district in matters over which they have jurisdiction shall be met before the final plat is filed.

3.07 Minor Land Division

When it is proposed to divide land into more than one, but less than five, parcels or building sites, inclusive of the original remnant parcel, any one of which is five acres or less in area, by a division or by successive divisions of any part of the original parcel within a five-year period; or when it is proposed to divide a block, lot, or outlot within a recorded subdivision plat into more than one, but less than five, parcels or building sites, inclusive of the original remnant parcel, without changing the exterior boundaries of the subdivision plat, or the exterior boundaries of blocks within the subdivision plat, and the division does not result in a subdivision, the subdivider may effect the division by use of a certified survey map. The subdivider shall prepare the certified survey map in accordance with this Ordinance and shall file sufficient copies of the map and the completed application with the *City* Clerk at least 10 days prior to the meeting of the Plan Commission at which action is desired. The Plan Commission may for good reason, such reason being set forth in the minutes of the meeting concerned, accept for review and approval certified survey maps that consist of a single parcel.

A. A Pre-Application Staff Conference similar to the consultation suggested in Section 3.01 of this Ordinance is recommended.
B. The *City Clerk* shall, within two normal working days after filing, transmit the copies of the map and letter of application to the *City* Plan Commission.
C. The *City* Plan Commission shall transmit a copy of the map to all affected *City* boards and commissions for their review and recommendations concerning matters within their jurisdiction. Copies may also be transmitted to the County Planning Agency and to the Southeastern Wisconsin Regional Planning Commission for review and comment. Their recommendations shall be transmitted to the *City* Plan Commission within 20 days from the date the map is received. The map shall be reviewed by the Plan Commission for conformance to this Ordinance, and all other ordinances, rules, regulations, and comprehensive plans and components thereof as may be applicable.
D. The *City* Plan Commission shall, within 45 days from the date of filing of the map, recommend approval, conditional approval or rejection of the map, and shall transmit the map along with its recommendations to the *Common Council.*
E. The *Common Council* shall approve, approve conditionally and thereby require resubmission of a corrected map, or reject such map within 60 days from the date of filing of the map unless the time is extended by agreement with the subdivider. If the map is rejected, the reason shall be stated in the minutes of the meeting and a written statement forwarded to the subdivider. If the map is approved, the *Common Council* shall cause the *City* Clerk to so certify on the face of the original map.

F. Recordation. After the certified survey map has been approved by the *Common Council*, the *City* Clerk shall cause the certification inscribed upon the map attesting to such approval to be duly executed and the map returned to the subdivider for recording with the County Register of Deeds. The Register of Deeds shall not record the map unless it is offered for recording within 30 days after the date of the last approval and within 12 months after the first approval.
G. Copies. The subdivider shall file 10 copies of the recorded certified survey map with the *City* Clerk. The Clerk shall distribute copies of the map to the *City* Engineer, *City* Planner, Building Inspector, Assessor, Land Information Officer, and other affected *City* and County officials for their files.

3.08 Replats

A. When It Is Proposed to replat a recorded subdivision, or part thereof, so as to vacate or alter areas within a plat dedicated to the public, or to change the boundaries of a recorded subdivision, or part thereof, the subdivider or person wishing to replat shall vacate or alter the recorded plat as provided in Sections 236.40 through 236.44 of the Wisconsin Statutes. If the replat is proposing to change the boundaries of a recorded subdivision, or part thereof, the subdivider or person wishing to replat shall then proceed as specified in Sections 3.01 through 3.06 of this Ordinance.
B. The *City* Clerk shall schedule a public hearing before the Plan Commission when a preliminary plat of a replat of lands within the *City* is filed, and shall cause notices of the proposed replat and public hearing to be published and mailed to the owners of record of all properties within the limits of the exterior boundaries of the proposed replat and to the owners of all properties within 200 feet of the exterior boundaries of the proposed replat.

3.09 Condominium Plats

A condominium plat prepared by a land surveyor registered in Wisconsin is required for all condominium plats or any amendments or expansions thereof. Such plat shall comply in all respects with the requirements of Section 703.11 of the Wisconsin Statutes and shall be reviewed and approved or denied in the same manner as a subdivision plat as set forth in Sections 3.01 through 3.06 of this Ordinance. Such plat shall comply with the design standards, improvements, and all other requirements of this Ordinance that would otherwise apply to subdivision plats, including, but not limited to, those set forth in Sections 7.00 and 8.00 of this Ordinance.

SECTION 4.00 PRELIMINARY PLAT

4.01 General Requirements

A preliminary plat shall be required for all subdivisions and condominiums and shall be based upon a survey by a registered land surveyor and the plat prepared on tracing cloth or paper of good quality at a scale no smaller than one inch equals 100 feet and shall show correctly on its face the following information:

A. Title or Name under which the proposed subdivision is to be recorded. Such title shall not be the same or similar to a previously approved and recorded plat, unless it is an addition to a previously recorded plat and is so stated on the plat.
B. Location of proposed subdivision by quarter section, township, range, county, and state.
C. Date, graphic scale, and north arrow.
D. Names and addresses of the owner, subdivider, and land surveyor preparing the plat.
E. The Entire Area Contiguous to the proposed plat owned or controlled by the subdivider shall be included on the preliminary plat even though only a portion of said area is proposed for immediate development. The Plan Commission may waive this requirement where it is unnecessary to fulfill the purposes and intent of this Ordinance and undue hardship would result from strict application thereof.

4.02 Site Analysis Information

In the absence of an adopted neighborhood unit development plan, the following site analysis information shall be inventoried and mapped at a scale no smaller than one inch equals 100 feet in sufficient detail, with brief descriptions if necessary, to allow for the proper evaluation of a preliminary plat. The site analysis map and accompanying descriptions shall be included with the submittal of the preliminary plat. The map shall include:

A. Topographic Features, with two-foot intervals for slopes less than 12 percent and at no more than five-foot intervals for slopes 12 percent and greater. Elevations shall be marked on such contours, referenced to National Geodetic Vertical Datum (NGVD) of 1929. Any rock outcrops, slopes of 12 percent or greater, ridge lines, and hilltops shall be noted.
B. Hydrologic Characteristics, including lakes, ponds, rivers, streams, creeks, drainage ditches, wetlands, floodplains, shoreland areas, and surface drainage patterns. The boundaries of wetlands shall be as delineated and mapped by the Southeastern Wisconsin Regional Planning Commission. The boundaries of the 100-year recurrence interval floodplain, as determined by the Federal Emergency Management Agency or the Southeastern Wisconsin Regional Planning Commission, shall be shown. Where such floodplain data are not available, the floodplain boundaries and related stages shall be determined by a registered professional engineer retained by the subdivider and the engineer's report providing the required data shall be subject to review and approved by the *City* Engineer.
C. Delineations of Natural Resource Areas, including the boundaries of primary and secondary environmental corridors and isolated natural resource areas

as identified by the Southeastern Wisconsin Regional Planning Commission, and the location and type of any rare or endangered species habitat.

D. Soil Types, as shown on the soil survey maps prepared by the U.S. Soil Conservation Service (now known as the Natural Resources Conservation Service).
E. Existing Vegetation, including the boundaries and characteristics of woodlands, hedgerows, and prairies. Predominant species of hedgerows and woodlands shall be identified. Unless located within an area proposed to be maintained in open space, specimen trees shall be located and identified by species, size, and health.
F. Historic, Cultural, and Archaeological Features, with a brief description of the historic character of buildings, structures, ruins, and burial sites.
G. Scenic Vistas, both into the proposed subdivision from adjacent roads and public areas and views from within the proposed subdivision.
H. The Location and Classification of existing streets and highways within or adjacent to the proposed subdivision and desirable or undesirable entry points into the subdivision.
I. Existing Land Uses within the proposed subdivision and within 200 feet therefrom, including cultivated and non-cultivated fields, paved areas, buildings, structures, and all encumbrances, such as easements or covenants.
J. Public Parks and Open Space Areas within or adjacent to the proposed subdivision, and potential open space connections between the proposed subdivision and adjacent lands.
K. Existing and Proposed Zoning on and adjacent to the proposed subdivision.

4.03 Plat Data

All preliminary plats shall show the following:

A. Length and Bearing of the exterior boundaries of the proposed subdivision referenced to two corners established in the U.S. Public Land Survey and the total acreage encompassed thereby. The lengths of lines shall be given to the nearest 0.01 foot and bearings to the nearest one second of arc. The arc length, chord length, radius length, and bearing shall be given for all curved lines.
B. Topographic Features, including existing contours, with two-foot intervals for slopes less than 12 percent and at no more than five-foot intervals for slopes 12 percent and greater. Elevations shall be marked on such contours, referenced to NGVD of 1929.
C. Boundaries of the 100-year recurrence interval floodplain and related regulatory stages, as determined by the Federal Emergency Management Agency or the Southeastern Wisconsin Regional Planning Commission. Where such data are not available, the floodplain boundaries and related stages shall be determined by a registered professional engineer retained by the subdivider and the engineer's report providing the required data shall be submitted with the plat for review and approval by the *City* Engineer.

D. Location and Water Elevations at the date of the survey of all lakes, ponds, rivers, streams, creeks, and drainage ditches within the plat and within 200 feet of the exterior boundaries of the plat. Approximate high and low water elevations and the ordinary high water mark referenced to NGVD 1929 shall also be shown. The status of navigability of the lakes, ponds, rivers, streams, creeks, and drainage ditches shall be indicated based upon a determination by the Wisconsin Department of Natural Resources.
E. Lake and Stream Meander Lines proposed to be established.
F. Boundaries of Primary and Secondary Environmental Corridors and isolated natural resource areas, as delineated and mapped by the Southeastern Wisconsin Regional Planning Commission. The boundaries of wetlands shall also be shown. The wetland boundaries shall be determined on the basis of a field survey made to identify, delineate, and map those boundaries; and the name of the person, agency, or firm identifying, delineating, and mapping the boundaries shall be provided together with the date of the field survey concerned.
G. The Location of Woodlands as mapped by the Southeastern Wisconsin Regional Planning Commission and existing vegetation to be retained within the proposed subdivision.
H. Location, right-of-way width, and names of all existing and proposed streets, highways, alleys, or other public ways, pedestrian and bicycle ways, utility rights-of-way, active and abandoned railway rights-of-way, vision corner easements, and other easements within or adjacent to the plat.
I. Type, Width, and Elevation of any existing street pavements within or adjacent to the plat, together with any legally established centerline elevations, referenced to mean NGVD (1929).
J. Approximate Radii of all curved lines within the exterior boundaries of the plat.
K. Location and Names of any adjoining subdivisions, parks, cemeteries, public lands, and watercourses, including impoundments. The owners of record of abutting unplatted lands shall also be shown.
L. All Existing Structures, together with an identification of the type of structure, such as residence, garage, barn, or shed; the distances of such structures from existing and proposed property lines, wells, watercourses, and drainage ditches; and existing property boundary lines in the area adjacent to the exterior boundaries of the proposed plat and within 100 feet thereof. The proposed use of existing structures to be retained shall be noted. All wells within the exterior boundaries of the plat, and within 50 feet of the exterior boundaries of the plat, shall be shown.
M. Locations of all civil division boundary lines and U.S. Public Land Survey system section and one-quarter section lines within the plat and within 100 feet of the exterior boundaries of the plat.
N. Approximate Dimensions of all lots, the minimum lot area required by the zoning district in which the plat is located, and proposed lot and block numbers.

O. Building or Setback Lines that are proposed to be more restrictive than the regulations of the zoning district in which the plat is located.
P. Location, Approximate Dimensions, and Area of any sites to be reserved or dedicated for parks, playgrounds, drainageways, open space preservation, or other public use.
Q. Location, Approximate Dimensions, and Area of any proposed common areas or facilities.
R. Location, Approximate Dimensions, and Area of any sites that are to be used for multi-family housing, shopping centers, church sites, or other non-public uses not requiring lotting.
S. Location, Size, and Invert Elevation of any existing sanitary or storm sewers, culverts and drain pipes, the location of manholes, catch basins, hydrants, electric and communication facilities, whether overhead or underground, and the location and size of any existing water and gas mains within or adjacent to the plat. If no sewers or water mains are located on or immediately adjacent to the proposed subdivision, the nearest such sewers or water mains that might be extended to serve the proposed subdivision shall be indicated by their direction and distance from the plat, and by their size and invert elevations. All elevations shall be referenced to NGVD (1929).
T. Any Proposed Lake and Stream Access, and the width of the proposed access, to be provided within the exterior boundaries of the plat.
U. Any Proposed Lake and Stream Improvement or relocation, and notice of application for approval by the Wisconsin Department of Natural Resources, when applicable.
V. The Approximate Location of any existing onsite sewage treatment and disposal facilities.
W. Any Additional Information requested by the *City* Plan Commission.

4.04 Street Plans and Profiles

The *City* Engineer or Plan Commission may require that the subdivider provide street plans and profiles showing the existing ground surface, proposed and established street grades, including extensions for a reasonable distance beyond the limits of the proposed subdivision when requested. All elevations shall be based on NGVD (1929), and plans and profiles shall meet the approval of the *City* Engineer.

4.05 Soil Borings and Tests

A. The Plan Commission, upon recommendation of the City Engineer, may, in order to determine the suitability of specific areas for the construction of buildings and supporting roadways, require that soil borings and tests be made to ascertain subsurface soil conditions and depths to bedrock and to the groundwater table. The number of such borings and tests shall be adequate to portray for the intended purpose the character of the soil and the depths to bedrock and groundwater from the undisturbed surface.

B. Where a Subdivision is not to be served by public sanitary sewer, soil borings and tests shall be made to determine the suitability of the site for the use of onsite sewage treatment and disposal systems. Such borings and tests shall meet the requirements of Chapters Comm 83 and 85 of the Wisconsin Administrative Code. The location of the borings shall be shown on the preliminary plat and the findings, with respect to the suitability of the site for the use of onsite sewage treatment and disposal systems, shall be set forth in a separate report submitted with the plat.

4.06 Soil and Water Conservation

The Plan Commission, upon the recommendation of the *City* Engineer, after determining from a review of the preliminary plat that the soil, slope, vegetation, and drainage characteristics of the site are such as to require substantial cutting, clearing, grading, and other earthmoving operations in the development of the subdivision or otherwise entail a severe erosion hazard, may require the subdivider to provide soil erosion and sedimentation control plans and specifications. Such plans shall generally follow the guidelines and standards promulgated by the County Land Conservation Committee, and shall be in accordance with the requirements set forth in the City Erosion Control and Stormwater Management Ordinance.

4.07 Covenants and Condominium or Homeowners Association Documents

A. A Draft Copy of any proposed protective covenants whereby the subdivider intends to regulate land use in the proposed subdivision shall accompany the preliminary plat. The proposed covenants shall be subject to review and approval by the City Attorney as to form.

B. A Draft Copy of any proposed condominium or homeowners association declarations, covenants, or other documents shall accompany the preliminary plat. These documents shall include the information specified in Section 2.06. The proposed documents shall be subject to review and approval by the City Attorney as to form.

4.08 Surveyor's Certificate

The surveyor preparing the preliminary plat shall certify on the face of the plat that it is a correct representation of the exterior boundaries of the proposed plat and of all existing land divisions and features within and adjacent thereto; and that the surveyor has fully complied with the provisions of this Ordinance and of Chapter 236 of the Wisconsin Statutes.

SECTION 5.00 FINAL PLAT

5.01 General Requirements

A final plat prepared by a registered land surveyor shall be required for all subdivisions and condominiums. It shall comply in all respects with the requirements of Chapter 236 of the Wisconsin Statutes.

5.02 Final Plat Data

The Plat shall show correctly on its face, in addition to the information required by Section 236.20 of the Wisconsin Statutes, the following:

A. Length and Bearing of the centerline of all streets. The lengths shall be given to the nearest 0.01 foot and bearings to the nearest one second of arc. The arc, chord, and radius lengths and the chord bearings, together with the bearings of the radii at the ends of the arcs and chords, shall be given for all curved streets.
B. Street Width along the line of any obliquely intersecting street to the nearest 0.01 foot.
C. Active and Abandoned Railway rights-of-way within and abutting the exterior boundaries of the plat.
D. Building or Setback Lines required by the *City* Plan Commission or other approving or objecting agency that are more restrictive than the regulations of the zoning district in which the plat is located, or that are proposed by the subdivider and are to be included in recorded private covenants.
E. Easements for any Public sanitary sewers, water supply mains, stormwater management facilities, drainageways, or access ways.
F. All Lands Reserved for future public acquisition or reserved for the common use of property owners within the plat. If property reserved for common use is located within the subdivision or condominium, the information required by Section 2.06 shall be submitted with the Final Plat, together with any associated deed or plat restrictions required by the Plan Commission.
G. Boundaries of the 100-year recurrence interval floodplain and related regulatory stages as determined by the Federal Emergency Management Agency or the Southeastern Wisconsin Regional Planning Commission. Where such data are not available, the floodplain boundaries and related stages shall be determined by a registered professional engineer retained by the subdivider and the engineer's report providing the required data shall be submitted with the plat for review and approval by the *City* Engineer.
H. Location and Right-of-Way of existing and proposed bicycle and pedestrian ways and utility rights-of-way.
I. Notations or Any Restrictions required by the *City* Plan Commission or other approving or objecting agency relative to access control along any public ways within or adjacent to the plat; the provision and use of planting strips; or provisions for the protection of any existing wetlands

or other environmentally significant lands within the exterior boundaries of the plat.

5.03 Deed Restrictions

The *City* may require that deed restrictions be filed with the final plat. When required, such restrictions shall be recorded with the final plat.

5.04 Survey Accuracy

The *City* Engineer shall examine all final plats within the City and may make, or cause to be made by a registered land surveyor under the supervision or direction of the *City* Engineer, field checks for the accuracy and closure of survey, proper kind and location of monuments, and liability and completeness of the drawing. In addition:

A. The Maximum Error of Closure before adjustment of the survey of the exterior boundary of the subdivision shall not exceed, in horizontal distance or position, the ratio of one part in 10,000, nor in azimuth, of four seconds of arc per interior angle. If field measurements exceed this maximum, new field measurements shall be made until a satisfactory closure is obtained. When a satisfactory closure of the field measurements has been obtained, the survey of the exterior boundary shall be adjusted to form a closed geometric figure.
B. All Street, Block, and Lot Dimensions shall be computed as closed geometric figures based upon the control provided by the closed exterior boundary survey. If field checks disclose an error for any interior line of the plat greater than the ratio of one part in 5,000, or an error in measured angle greater than one minute of arc for any angle where the shorter side forming the angle is 300 feet or longer, necessary corrections shall be made. Where the shorter side of a measured angle is less than 300 feet in length, the error shall not exceed the value of one minute multiplied by the quotient of 300 divided by the length of the shorter side; however, such error shall not in any case exceed five minutes of arc.
C. The *Common Council* shall receive the results of the City Engineer's examination prior to approving the final plat. The City Engineer may, however, in accordance with Section 2.07C of this Ordinance, waive the placing of monuments for a reasonable time, not to exceed one year, on condition that the subdivider provide a letter of credit, certified check, or surety bond equal to the estimated cost of installing the monuments, to ensure the placing of such monuments within the time required by Statute. In that case, the *City* Engineer's examination required under this section and any related field checks shall be made after the required monuments have been installed. The letter of credit, certified check, or surety bond concerned shall not be released until the *City* Engineer is satisfied with the accuracy of the land surveying concerned.

5.05 Surveying and Monumenting

All final plats shall meet all surveying and monumenting requirements of Section 236.15 of the Wisconsin Statutes.

5.06 State Plane Coordinate System

Where the plat is located within a one-quarter section, the corners of which have been located, monumented, and placed on the State Plane Coordinate System through high-order horizontal control surveys conducted to standards established by the Southeastern Wisconsin Regional Planning Commission, the plat shall be tied directly to two adjacent section or quarter-section corners defining a quarter-section line so located, monumented, and placed on the State Plane Coordinate System. The grid bearing and distance of each tie shall be determined by field measurements. The Wisconsin State Plane Coordinates, together with a description of the monuments marking the section or quarter-section corners to which the plat is tied shall be shown on the plat. All distances and bearings shall be referenced to the Wisconsin State Plane Coordinate System, South Zone, based on the North American Datum of 1927, and shall be adjusted to the control survey network established to the standards promulgated by the Southeastern Wisconsin Regional Planning Commission for the area concerned. Where the field measurements differ from the control survey data by more than one part in 10,000, in the alternative to adjusting the field measured distances and bearings of the ties to the control survey network, the surveyor shall show both the measured field distances and bearings and the recorded and published control survey distances and bearings concerned. Under this alternative, the discrepancies shall be brought to the attention of the custodian of the control survey data for the area concerned by the surveyor. All distances shall be recorded to the nearest 0.01 foot and all bearings to the nearest one second of arc. The grid bearing and distance of the tie shall be determined by a closed survey meeting the error of closure herein specified for the survey of the exterior boundaries of the subdivision.

5.07 Certificates

All final plats shall provide all the certificates required by Section 236.21 of the Wisconsin Statutes; and, in addition, the surveyor shall certify that he or she has fully complied with all the provisions of this Ordinance.

5.08 Filing and Recording

A. The Final Plat shall be submitted for recording in accordance with Section 3.05F of this Ordinance.
B. The County Register of Deeds shall record the plat as provided by Section 236.25 of the Wisconsin Statutes.
C. The Subdivider shall file a copy of the final plat with the City Clerk, as provided by Section 236.27 of the Wisconsin Statutes.

SECTION 6.00 CERTIFIED SURVEY MAP

6.01 General Requirements

A certified survey map prepared by a registered land surveyor shall be required for all minor land divisions. It shall comply in all respects with the requirements of Section 236.34 of the Wisconsin Statutes. The minor land division shall comply with the design standards and improvement requirements set forth in Sections 7.00 and 8.00 of this Ordinance.

A preliminary map or sketch map may be submitted by the subdivider to the *City* Plan Commission or its staff for review and comment prior to the submission of a proposed certified survey map for review and approval.

6.02 Required Information

The map shall show correctly on its face, in addition to the information required by Section 236.34 of the Wisconsin Statutes, the following:

A. Inset Map of the area concerned showing the location of the proposed certified survey map in relation to the U.S. Public Land Survey section and quarter-section lines and abutting and nearby public streets and highways.
B. Date, graphic scale, and north point.
C. Name and addresses of the owner, subdivider, and land surveyor preparing the plat.
D. All Existing Structures, together with an identification of the type of structure, such as residence, garage, barn, or shed; the distances of such structures from existing and proposed property lines, wells, watercourses, and drainage ditches; and existing property boundary lines in the area adjacent to the exterior boundaries of the proposed certified survey map and within 100 feet thereof. The proposed use of existing structures to be retained shall be noted. All wells within the exterior boundaries of the proposed certified survey map, and within 50 feet of the exterior boundaries of the map, shall be shown.
E. Location, Approximate Dimensions, and Area of any sites to be reserved or dedicated for parks, playgrounds, drainageways, open space preservation, or other public use.
F. Building or Setback Lines required by the *City* Plan Commission, or other approving or objecting agency, that are more restrictive than the regulations of the zoning district in which the certified survey map is located, or which are proposed by the subdivider and are to be included in recorded private covenants.
G. Location and Names of any adjoining streets, highways, subdivisions, parks, cemeteries, public lands, and watercourses, including impoundments. The owners of record of abutting unplatted lands shall also be shown.
H. Length and Bearing of the centerline of all streets. The lengths shall be given to the nearest 0.01 foot and the bearings to the nearest one second of arc. The arc, chord, and radius lengths, and the chord bearing, together with

the bearings of the radii of the ends of the arcs and chords, shall be given for all curved lines.

I. Street Width along the line of any obliquely intersecting street line to the nearest 0.01 foot.
J. Active and Abandoned Railway rights-of-way within and abutting the exterior boundaries of the proposed certified survey map and the location and right-of-way of existing and proposed bicycle and pedestrian ways.
K. Notations or Any Restrictions required by the *City* Plan Commission or other approving or objecting agency relative to access control along any public ways within or adjacent to the proposed certified survey map; the provision and use of planting strips; or provisions for the protection of any existing wetlands or other environmentally significant lands within the exterior boundaries of the proposed certified survey map.
L. Easements for any Public sanitary sewers, water supply mains, stormwater management facilities, drainageways, or access ways.

6.03 Additional Information

The Plan Commission may require that the following additional information be provided when necessary for the proper review and consideration of the proposed land division:

A. Topographic Features, including existing and/or proposed contours, with two-foot intervals for slopes less than 12 percent and at no more than five-foot intervals for slopes 12 percent and greater. Elevations shall be marked on such contours, referenced to NGVD of 1929. The requirement to provide topographic data may be waived if the parcel or parcels proposed to be created are fully developed.
B. Soil Types as shown on the soil survey maps prepared by the U.S. Soil Conservation Service (now known as the Natural Resources Conservation Service).
C. The Square Footage and elevation of the first floor of all buildings proposed to remain on the site or sites included in the certified survey map.
D. The *City* Plan Commission, upon recommendation of the *City* Engineer, may, in order to determine the suitability of the site concerned for the construction of buildings and supporting roadways, require that soil borings and tests be made to ascertain subsurface soil conditions and depths to bedrock and to the groundwater table. The number of such borings and tests shall be adequate to portray for the intended purpose the character of the soil and the depths to bedrock and groundwater from the undisturbed surface.
E. Where the Site is Not to be Served by public sanitary sewer, soil borings and tests shall be made to determine the suitability of the site for the use of onsite sewage treatment and disposal systems. Such borings and tests shall meet the requirements of Chapters Comm 83 and 85 of the Wisconsin Administrative Code. The location of the borings shall be shown on the map and the findings, with respect to suitability for the use of onsite sewage

treatment and disposal systems, set forth in a separate report submitted with the proposed certified survey map.

F. Boundaries of Primary and Secondary Environmental Corridors and isolated natural resource areas, as delineated and mapped by the Southeastern Wisconsin Regional Planning Commission. The boundaries of wetlands shall also be shown. The wetland boundaries shall be determined on the basis of a field survey made to identify, delineate, and map those boundaries; and the name of the person, agency, or firm identifying, delineating, and mapping the wetland boundaries shall be provided together with the date of the field survey concerned.

G. Boundaries of the 100-year recurrence interval floodplain and related regulatory stages, as determined by the Federal Emergency Management Agency or the Southeastern Wisconsin Regional Planning Commission. Where such data are not available, the floodplain boundaries and related stages shall be determined by a registered professional engineer retained by the subdivider, and the engineer's report providing the required data shall be submitted for review and approval by the *City* Engineer.

H. The Location of Woodlands, as mapped by the Southeastern Wisconsin Regional Planning Commission, within the proposed certified survey map.

I. The Approximate Location of existing and proposed onsite sewage treatment and disposal facilities.

J. Historic, Cultural, and Archaeological Features, with a brief description of the historic character of buildings, structures, ruins, and burial sites.

K. Location and Water Elevations at the date of the survey of all lakes, ponds, rivers, streams, creeks, and drainage ditches within the proposed certified survey map and within 200 feet of the exterior boundaries of the map. Approximate high and low water elevations and the ordinary high water mark referenced to NGVD 1929 shall also be shown. The status of navigability of the lakes, ponds, rivers, streams, creeks, and drainage ditches shall be indicated based upon a determination of the Wisconsin Department of Natural Resources.

L. The *City* Plan Commission may require that the entire area contiguous to the land encompassed within the proposed certified survey map and owned or controlled by the subdivider be included in the certified survey map even though only a portion of said area is proposed for immediate development. The *City* Plan Commission may also require the submission of a sketch plan, drawn to scale, showing the entire contiguous holdings owned or controlled by the subdivider and identifying proposed future development of the parcel, including general street and parcel locations.

6.04 State Plane Coordinate System

Where the map is located within a one-quarter section, the corners of which have been located, monumented, and placed on the State Plane Coordinate System through high-order horizontal control surveys conducted to standards established by the Southeastern Wisconsin Regional Planning Commission, the map shall be tied directly to two adjacent section or quarter-section corners defining a quarter-section line so located, monumented, and placed on the State Plane Coordinate System. The grid bearing and distance of each tie shall be determined by field measurements. The Wisconsin State Plane Coordinates, together with a description of the monuments marking the section or quarter-section corners to which the map is tied shall be shown on the map. All distances and bearings shall be referenced to the Wisconsin State Plane Coordinate System, South Zone, based upon the North American Datum of 1927, and shall be adjusted to the control survey network established to the standards promulgated by the Southeastern Wisconsin Regional Planning Commission for the area concerned. Where the field measurements differ from the control survey data by more than one part in 10,000, in the alternative to adjusting the field measured distances and bearings of the ties to the control survey network, the surveyor shall show both the measured field distances and bearings and the recorded and published control survey distances and bearings concerned. Under this alternative, the discrepancies shall be brought to the attention of the custodian of the control survey data for the area concerned by the surveyor. All distances shall be recorded to the nearest 0.01 foot and all bearings to the nearest one second of arc. The grid bearing and distance of the tie shall be determined by a closed survey meeting the error of closure herein specified for the survey of the exterior boundaries of the certified survey map.

6.05 Certificates

All certified survey maps shall provide all of the certificates required for final plats by Section 236.21 of the Wisconsin Statutes. The *Common Council* shall certify its approval on the face of the map. In addition, the surveyor shall certify that he or she has fully complied with all of the provisions of this Ordinance.

6.06 Recording

After the certified survey map has been duly approved by the *Common Council*, the *City* Clerk shall cause the certificate to be inscribed upon the map attesting to such approval and the map recorded as provided for under Section 3.07F of this Ordinance.

SECTION 7.00 DESIGN STANDARDS

7.01 Street Arrangement

A. General Requirements. In any new land division or condominium, the street layout shall conform to the arrangement, width, type, and location indicated on the adopted County jurisdictional highway system plan, the adopted *City* official map, or the adopted *City* comprehensive plan or plan component. In areas for which such plans have not been completed, or are of insufficient detail, the street layout shall recognize the functional classification of the various types of streets and shall be developed and located in proper relation to existing and proposed streets, to the topography, to such natural features as streams and existing trees, to public convenience and safety, to the proposed use of the land to be served by such streets, and to the most advantageous development of adjoining areas. The land division or condominium shall be designed so as to provide each lot with satisfactory frontage on a public street.

B. Arterial Streets shall be arranged so as to provide ready access to centers of employment, centers of governmental activity, community shopping areas, community recreation, and points beyond the boundaries of the community. They shall also be properly integrated with and related to the existing and planned system of arterial streets and highways and shall be, insofar as practicable, continuous and in alignment with existing or planned streets with which they are to connect.

C. Collector Streets shall be arranged so as to provide ready collection of traffic from residential areas and conveyance of this traffic to the arterial street and highway system and shall be properly related to the arterial streets to which they connect.

D. Land Access Streets shall be arranged to conform to the topography, to discourage use by through traffic, to permit the design of efficient storm and sanitary sewerage systems, and to require the minimum street area necessary to provide safe and convenient access to abutting property.

E. Proposed Streets shall extend to the boundary lines of the lot, parcel, or tract being subdivided or developed unless prevented by topography or other physical conditions or unless, in the opinion of the Plan Commission, such extension is not necessary or desirable for the coordination of the layout of the land division or condominium or for the advantageous development of adjacent lands.

F. Arterial Street Protection. Whenever an existing or planned arterial street is located adjacent to or within a proposed land division or condominium, adequate protection of residential lots, limitation of access to the arterial street, and separation of through and local traffic shall be provided through the use of alleys, frontage streets, or cul-de-sac or loop streets. A restricted non-access easement along any property line abutting an arterial street may be required. Permanent screening or landscape plantings may be required in any restricted non-access area.

G. Development Control or Reserve Strips shall not be allowed on any plat or certified survey map to control access to streets or alleys, except where control of such strips is placed with the *City* under conditions approved by the Plan Commission.

H. Access shall be provided in commercial and industrial districts for off-street loading and service areas.

I. Street Names shall be approved by the *City* and shall not duplicate or be similar to existing street names elsewhere in the *City*. Existing street names shall be continued into the land division or condominium wherever possible. Where an adopted *City* or County street address system plan exists, the street names shall be assigned in accordance with the recommendations of such plan or plans.

7.02 Limited Access Highway and Railroad Right-of-Way Treatment

Whenever a proposed land division or condominium contains or is adjacent to a limited access highway or railroad right-of-way the design shall provide the following treatment:

A. Non-Access Easement and Planting Area: When lots within a proposed land division or condominium back upon the right-of-way of an existing or planned limited access highway or railroad, a non-access easement and planting area at least 50 feet in depth shall be provided adjacent to the highway or railroad right-of-way. The minimum lot depth required by the *City* zoning ordinance shall be increased by 50 feet to accommodate the non-access easement and planting area. This non-access easement and planting area shall be a part of all lots and shall have the following restriction lettered on the face of the plat or certified survey map: "This area is reserved for the planting of trees and shrubs. No access shall be permitted across this area. The building of structures, except public or private utility structures and fences, is prohibited hereon."
B. Plats Located in Commercial and Industrial Zoning Districts shall provide, on each side of a limited access highway or railroad right-of-way, streets approximately parallel to such highway or railroad. A distance of not less than 150 feet shall be provided to allow for the appropriate use of the land between such streets and the highway or railroad.
C. Streets Parallel to a Limited Access Highway or railroad right-of-way, when intersecting an arterial or collector street that crosses said highway shall be located at a minimum distance of 250 feet from said street or railroad right-of-way. Such distance, where desirable and practicable, shall be determined with due consideration of the minimum distance required for the future separation of grades by means of desirable approach gradients.
D. Land Access Streets immediately adjacent to arterial streets and railroad rights-of-way shall be avoided in residential areas.

7.03 Street, Bicycle, and Pedestrian Way Design Standards

A. Minimum Width. The minimum right-of-way and roadway width of all proposed streets and alleys shall be as specified by the comprehensive plan or component thereof, official map, or County jurisdictional highway system plan. If no width is specified therein, the minimum right-of-way and roadway widths for arterial, collector, and land access streets shall be as shown on Table A.1 for streets having an urban cross-section. If the Plan Commission determines that a permanent rural cross-section may be used, the minimum right-of-way and roadway widths set forth in Table A.2 shall apply.
B. Cross-Sections for collector and land access streets having an urban cross-section shall be as shown on Table A.1. If the Plan Commission determines that a permanent rural cross-section may be used, the cross-sections for

TABLE A1
Required Urban Cross-Sections for Streets and Other Public Ways

Type of Street or Public Way	Right-of-Way Width to Be Dedicated	Roadway, Terrace, Sidewalk, and Related Widths
Arterial Streets	120 feet, or as required by the City Official Map or Comprehensive Plan	As determined by the City Plan Commission
Collector Streets	80 feet	48-foot pavement (face of curb to face of curb) 10-foot terraces 5-foot sidewalks 1-foot sidewalk buffers
Land Access Streets	60 feet	32-foot pavement[a] (face of curb to face of curb) 8-foot terraces 5-foot sidewalks 1-foot sidewalk buffers
Minor Land Access Streets[b]	50 feet	28-foot pavement (face of curb to face of curb) 5-foot terraces 5-foot sidewalks[c] 1-foot sidewalk buffers
Alley	25 feet	20-foot pavement 2.5-foot buffers
Cul-de-Sac	75-foot outside radius	61-foot outside curb radius 37-foot radius for center island[d] 24-foot pavement (face of curb to face of curb) 8-foot terrace 5-foot sidewalk 1-foot sidewalk buffer
Pedestrian and Bicycle Ways	20 feet[e]	10-foot pavement[f] 5-foot buffer

Source: SEWRPC

[a] The 32-foot pavement width is suggested only for use with land access streets serving relatively low density, single-family residential areas. For land access streets serving higher density single- and multi-family residential, commercial, and industrial areas, a minimum width of 36 feet should be provided.

[b] A 50-foot right-of-way and 28-foot pavement width for land access streets would be applicable on relatively short loop and cul-de-sac streets in areas of single-family homes with attached garages and driveways, with adequate area available on each lot for off-street parking and snow storage, and where no bus or truck traffic other than occasional school buses and service or delivery trucks would be expected to operate over the street.

[c] Sidewalks may be eliminated on one side of minor land access streets.

[d] The center island should be tapered with a face-of-curb radius of 37 feet at its widest end, a face-of-curb radius of 20 feet at its narrowest end, and a length of 100 feet.

[e] Recommended right-of-way width for combined bicycle and pedestrian ways separate from street rights-of-way.

[f] The pavement width of pedestrian and bicycle ways in areas of high use may be increased to 12 feet, and the buffers decreased to four feet each.

TABLE A.2
Required Rural Cross-Sections for Streets and Other Public Ways

Type of Street or Public Way	Right-of-Way Width to Be Dedicated	Roadway and Related Widths
Arterial Streets	120 feet, or as required by the City Official Map or Comprehensive Plan	As determined by the City Plan Commission
Collector Streets	80 feet	24-foot pavement 6-foot shoulders[a] 22-foot terraces/drainage swales[b]
Land Access Streets	66 feet	22-foot pavement 4-foot foot shoulders[a] 18-foot terraces/drainage swales[b]
Cul-de-Sac	75-foot outside radius	53-foot outside pavement radius 35-foot radius for center island[c] 18-foot traveled way 22-foot terrace/drainage swale[b]
Pedestrian and Bicycle Ways	20 feet[d]	10-foot pavement[e] 5-foot buffer

Source: SEWRPC

[a] Shoulders may be paved or gravel.

[b] The Plan Commission may require sidewalks to be provided on one or both sides of any street, if the Commission determines that sidewalks will be needed to accommodate anticipated pedestrian traffic. Such sidewalks shall be located at the outside edge of the terrace/drainage swale, with a one-foot wide buffer between the sidewalk and outside edge of the street right-of-way, unless otherwise directed by the Plan Commission upon the recommendation of the City Engineer.

[c] The center island should be tapered with a face-of-curb radius of 35 feet at its widest end, a face-of-curb radius of 20 feet at its narrowest end, and a length of 93 feet.

[d] Recommended right-of-way width for combined bicycle and pedestrian ways separate from street right-of-way.

[e] The pavement width of pedestrian and bicycle ways in areas of high use may be increased to 12 feet, and the buffers decreased to four feet each.

collector and land access streets set forth in Table A.2 shall apply. The cross-sections for arterial streets should be based on detailed engineering studies.

C. Cul-de-Sac Streets designed to have one end permanently closed shall not exceed 1,000 feet in length unless provisions are made for adequate emergency access and water main configuration. Cul-de-sac streets shall terminate in a circular turnabout having a minimum right-of-way radius of 75 feet. Cul-de-sac turnabouts with an urban cross-section shall have a minimum outside face-of-curb radius of 61 feet; and have a tapered landscaped island with a face-of-curb radius of 37 feet at its widest end, a face-of-curb radius of 20 feet at its narrowest end, and a length of 100 feet.

Cul-de-sac turnabouts with a rural cross-section shall have a minimum outside pavement radius of 53 feet; and have a tapered landscaped island with a face-of-curb radius of 35 feet at its widest end, a face-of-curb radius of 20 feet at its narrowest end, and a length of 93 feet.

D. Temporary Termination of streets intended to be extended at a later date shall be accomplished with the construction of a temporary “T”-shaped turnabout contained within the street right-of-way.

E. Bicycle and Pedestrian Ways with a right-of-way width of not less than 20 feet may be required where deemed necessary by the Plan Commission to provide adequate bicycle and pedestrian circulation or access to schools, shopping centers, churches, or transportation facilities. Bicycle and pedestrian ways in wooded and wetland areas shall be so designed and constructed as to minimize the removal of trees, shrubs, and other vegetation, and to preserve the natural beauty of the area.

F. Grades

1. Street grades shall be established wherever practicable so as to avoid excessive grading, the promiscuous removal of ground cover and tree growth, and general leveling of the topography. All changes in street grades shall be connected by vertical curves of a minimum length equivalent in feet to 30 times the algebraic difference in the rates of grade for arterial streets, and one-half this minimum for all other streets.
2. Unless necessitated by exceptional topography, subject to the approval of the Plan Commission, the maximum centerline grade of any street or public way shall not exceed the following:
   a. Arterial streets: 6 percent.
   b. Collector streets: 8 percent.
   c. Land access streets, alleys and frontage streets: 10 percent.
   d. Bicycle ways: 5 percent; however, steeper grades are acceptable for distances up to 500 feet.
   e. Pedestrian ways: 12 percent. Steps or stairs shall be provided if the grade will exceed 12 percent.
3. The grade of any street shall in no case exceed 12 percent or be less than one-half of 1 percent for streets with an urban cross-section, and 1 percent for streets with a rural cross-section.

G. Crowns. Unless otherwise approved, roadway pavements shall be designed with a centerline crown. Offset crowns or continuous cross-slopes may be utilized upon approval of the *City* Engineer. Alley pavements shall be “V”-shaped, with a centerline gutter for drainage.

H. Radii of Curvature. When a continuous street centerline deflects at any one point by more than 10 degrees, a circular curve shall be introduced having a radius of curvature on said centerline of not less than the following:

1. Arterial streets and highways	500 feet
2. Collector streets	300 feet
3. Land access streets	100 feet

A tangent at least 100 feet in length shall be provided between reverse curves on arterial and collector streets.

I. Elevations of Arterial Streets shall be set so that they will not be overtopped by a 50-year recurrence interval flood.

J. Bridges and Culverts. All new and replacement bridges and culverts over navigable waterways, including pedestrian and other minor bridges, shall be designed so as to accommodate the peak rate of discharge of a 100-year recurrence interval flood event without raising the peak stage, either upstream or downstream, more than 0.01 foot above the peak stage for the 100-year recurrence interval flood, as established by the Southeastern Wisconsin Regional Planning Commission or the Federal Emergency Management Agency. Larger permissible flood stage increases may be acceptable for reaches having topographic or land use conditions that could accommodate the increased stages without creating additional flood damage potential upstream or downstream of the proposed structure, providing that flood easements or other appropriate legal measures have been secured from all property owners affected by the excess stage increases. Such bridges and culverts shall be so designed and constructed as to facilitate the passage of ice flows and other debris.

K. Half-Streets. Where an existing dedicated or platted half-street is adjacent to the proposed land division or condominium plat, the other half of the street shall be dedicated by the subdivider. The platting of new half-streets shall not be permitted.

7.04 Street Intersections

A. Right Angle. Streets shall intersect each other at as nearly right angles as topography and other limiting factors of good design permit.

B. The Maximum Number of streets converging at one intersection shall not exceed two.

C. The Number of Intersections along arterial streets and highways shall be held to a minimum. Wherever practicable, the distance between such intersections shall not be less than 1,200 feet.

D. Continuation of Land Access and Collector Streets. Land access and collector streets shall not necessarily continue across arterial streets; but if the centerlines of such streets approach the arterial streets from opposite sides within 300 feet of each other, measured along the centerline of the arterial or collector streets, then the location of the collector and/or land access streets shall be so adjusted so that a single intersection is formed.

E. Corner Curves. Property lines at intersections of arterial streets and at intersections of collector and arterial streets shall be rounded to an arc with a minimum radius of 15 feet, or a greater radius if required by the *City* Engineer.

F. Vision Clearance Easements shall be provided at street intersections as may be required by the *City* zoning ordinance and by any approving or objecting authority concerned.

7.05 Blocks

A. General Requirements. The widths, lengths, and shapes of requirements; the need for convenient bicycle, pedestrian, and motor vehicle access; traffic safety; and the limitations and opportunities of topography.
B. The Length of Blocks in residential areas shall not as a general rule be less than 600 feet nor more than 1,200 feet in length unless otherwise dictated by exceptional topography or other limiting factors of good design.
C. The Width of Blocks shall be sufficient to provide for two tiers of lots of appropriate depth except where otherwise required to separate residential development from arterial streets and railroad rights-of-way. The width of lots or parcels reserved or laid out for commercial or industrial use shall be adequate to provide for off-street parking and loading required by the contemplated use and the *City* zoning ordinance.

7.06 Lots

A. General Requirements. The size, shape, and orientation of lots shall be appropriate for the location of the land division and for the type of development and use contemplated. The lots should be designed to provide an aesthetically pleasing building site, and a proper architectural setting for the buildings contemplated. Lot lines shall follow municipal boundary lines rather than cross them.
B. Side Lot Lines shall be at right angles to straight street lines or radial to curved street lines on which the lots face, unless a non-conventional lot layout is approved by the Plan Commission.
C. Double Frontage Lots shall be prohibited except where necessary to provide separation of residential development from arterial streets or to overcome specific disadvantages of topography and orientation.
D. Public Street Frontage. Every lot shall front or abut for a distance of at least 30 feet on a public street.
E. The Area and Dimensions of Lots shall conform to the requirements of the *City* zoning ordinance. Lots shall contain sufficient area to permit compliance with all required setbacks, including those set forth in the *City* zoning ordinance and those that may be required to meet the requirements of Chapter Trans 233 of the Wisconsin Administrative Code. Buildable lots that will not be served by a public sanitary sewerage system shall be of sufficient size to permit the use of a private onsite wastewater treatment system designed in accordance with Chapter Comm 83 of the Wisconsin Administrative Code.
F. Re-division of Lots. Wherever a lot, parcel, or tract is subdivided into lots or parcels that are more than twice the minimum lot area required in the zoning district in which the lot or parcel is located, the Plan Commission may require that such lots or parcels be arranged and dimensioned to allow re-division into smaller lots or parcels that will meet the provisions of this Ordinance and the zoning ordinance.

G. Depth. Lots shall have a minimum average depth of 100 feet. Excessive depth in relation to width shall be avoided and a proportion of two to one (2:1) shall be considered a desirable ratio, unless a deeper lot is needed to protect natural resources. The depth of lots or parcels reserved or laid out for commercial or industrial use shall be adequate to provide for off-street parking and loading areas required by the contemplated use and the City zoning ordinance.
H. The Width of Lots shall conform to the requirements of the *City* Zoning Ordinance, and in no case shall a lot be less than 60 feet in width at the building setback line, unless otherwise provided by the *City* zoning ordinance.
I. Corner Lots, when located in a district that permits a lot width less than 100 feet, shall have an extra width of 10 feet to permit adequate building setbacks from side streets.
J. The Shape of lots shall be approximately rectangular, with the exception of lots located on a curved street or on a cul-de-sac turnabout. Flag lots shall be prohibited, except where necessary to accommodate exceptional topography or to preserve natural resources.
K. Lands Lying Between the Meander Line and the Water's Edge and any otherwise unplattable lands that lie between a proposed land division or condominium and the water's edge shall be included as part of lots, outlots, or public dedications in any plat abutting a lake or stream.
L. Restrictions Prohibiting Development. Whenever a lot appearing on a final plat, condominium plat, or certified survey map is not intended to be buildable, or is intended to be buildable only upon certain conditions, an express restriction to that effect, running with the land and enforceable by the *City*, shall appear on the face of the plat or map.

7.07 Building and Setback Lines

Building setback lines appropriate to the location and type of development contemplated, which are more restrictive than required in the applicable zoning district, may be permitted or required by the Plan Commission and shall be shown on the final plat, condominium plat, or certified survey map. Examples of the application of this provision would include requiring greater setbacks on cul-de-sac lots to achieve the necessary lot width at the setback line, requiring greater setbacks to conform to setbacks of existing adjacent development, requiring greater setbacks to accommodate a coving design, requiring greater setbacks to avoid placing buildings within easements or vision clearance triangles, setting special yard requirements to protect natural resources, or requiring greater setbacks along arterial streets and highways to meet the requirements of Chapter Trans 233 of the Wisconsin Administrative Code.

7.08 Easements

A. Utility Easements. The Plan Commission may require utility easements of widths deemed adequate for the intended purpose. Such easements shall be located as determined by the applicable utility company, but preferably should be located along rear and side lot lines and should be designed to avoid the location of such facilities as electric power transformers in the flow lines of drainage swales and ditches. All lines, pipes, cables and similar equipment shall be installed underground unless the Plan Commission finds that the topography, soils, depth to bedrock, woodlands, wetlands, or other physical barriers would make underground installation impractical, or that the lots to be served by said facilities can be served directly from existing overhead facilities and requiring underground installation would constitute an undue hardship upon the subdivider. Associated equipment and facilities that are appurtenant to underground electric power, communications, and gas facility systems, including but not limited to, substations, pad-mounted transformers, pad-mounted sectionalizing switches, above-grade pedestal-mounted terminal boxes, junction boxes, meter points, and similar equipment may be installed at ground level. A landscape screening plan for such above-ground equipment shall be submitted by the subdivider to the affected utility and the Plan Commission for approval. All utility easements shall be noted on the final plat, condominium plat, or certified survey map followed by reference to the use or uses for which they are intended.

B. Drainage Easements. Where a land division or condominium is traversed by a drainageway or stream, an adequate easement shall be provided as required by the Plan Commission. The location, width, alignment, and improvement of such drainageway or easement shall be subject to the approval of the *City* Engineer; and parallel streets or parkways may be required in connection therewith. Where necessary, storm water drainage shall be maintained by landscaped open channels of adequate size and grade to hydraulically accommodate maximum potential volumes of flow. These design details are subject to review and approval by the *City* Engineer.

7.09 Protection of Natural Resources

Where natural drainage channels, floodplains, wetlands, or other environmentally sensitive areas are encompassed in whole or in part within a proposed land division or condominium, the Plan Commission may require that such areas be dedicated or that restriction be placed on the plat or certified survey map to protect such resources. The Plan Commission may further require that such areas be included in outlots designated on the plat or certified survey map and restricted from development.

7.10 Park, Open Space, and Other Public Sites

A. In the Design of a subdivision or condominium plat or a certified survey map, due consideration shall be given to the dedication or reservation of suitable sites of adequate size for future schools, parks, playgrounds, public access to navigable waters, and other public purposes. Accordingly, each subdivider of land in the *City* shall dedicate park and open space lands designated on the *City* official map or comprehensive plan or component thereof, or, where no park or open space lands are directly involved, pay a public site fee. Proposed school sites shall be reserved by the subdivider for future acquisition by the School Board.

1. Dedication of public parks and open space sites. Whenever a proposed public playground, park, parkway, trail corridor, public open space site, or other public lands designated on the City's official map or comprehensive plan or component thereof is encompassed, in whole or in part, within a proposed land division or condominium, the public lands shall be made a part of the subdivision or condominium plat or certified survey map and shall be dedicated to the public by the subdivider. Should the value of the land to be dedicated be less than the value of the public site fee, the subdivider shall be required to pay the *City* the difference between the value of the land dedicated and the public site fee. Should the value of the land to be dedicated exceed the public site fee, any lands in excess of the value of the public site fee shall be reserved for a period not to exceed three years, unless extended by mutual agreement, for purchase by the *City* at the price agreed upon and set forth in the Development Agreement. If the reserved lands are not acquired within the three-year period, the land will be released from reservation to the owner. Land values shall be determined in accordance with Subsection 7.10B.
2. Reservation of school sites. Whenever a proposed public school site designated on the *City*'s official map or comprehensive plan or component thereof is encompassed, in whole or in part, within a proposed land division or condominium, the proposed school site shall be made a part of the plat and reserved at the time of final plat or certified survey map approval for a period not to exceed three years, unless extended by mutual agreement, for acquisition by the School Board at a price agreed upon and set forth in the Development Agreement.
3. Public site fee option. If a proposed land division or condominium does not encompass a proposed public playground, park, parkway, open space site, or other public lands, the subdivider shall pay a public site fee to be used for the acquisition of public sites to serve the future inhabitants of the proposed subdivision, minor land division, or condominium at the time of application for final plat or certified survey map approval at the rate and in accordance with the procedures set forth in Section 10.06.

B. The Value of Land to be dedicated for park or open space purposes shall be agreed upon by the *City* and the subdivider on the basis of full and fair market value of the land to be dedicated. If the value cannot be agreed upon by the *City* and the developer, an appraisal board consisting of one appraiser selected by the *City* and retained at the *City*'s expense, one appraiser selected by the subdivider and retained at the subdivider's expense, and a third appraiser selected by the other two appraisers and retained at a cost shared equally by the *City* and the subdivider, shall determine the value of the land.

C. Navigable Streams or Lakeshores shall have a public access-way at least 60 feet in width platted to the low water mark at intervals of not more than one-half mile and connecting to existing public streets, unless wider access or greater shoreline intervals are agreed upon by the Wisconsin Department of Administration, the Wisconsin Department of Natural Resources, and the *City*, as required by Section 236.16(3) of the Wisconsin Statutes.

SECTION 8.00 REQUIRED IMPROVEMENTS

8.01 General Requirements

All required improvements shall be constructed in accordance with plans and specifications approved by the *City* Engineer.

8.02 Survey Monuments

The subdivider shall install survey monuments placed in accordance with the requirements of Section 236.15 of the Wisconsin Statutes and as may be required by the *City* Engineer.

8.03 Grading

A. Following the Installation of temporary block corner monuments or other survey control points by the subdivider and establishment of street grades by the *City* Engineer, the subdivider shall grade the full width of the right-of-way of all streets proposed to be dedicated in accordance with plans and specifications approved by the *City* Engineer. The subdivider shall grade the roadbeds in the street rights-of-way to subgrade.
B. Streets and Lots shall be brought to finished grades as specified in a site grading plan approved by the *City* Engineer.

8.04 Surfacing

Following the installation, inspection, and approval by the *City* Engineer of utility and stormwater drainage improvements, the subdivider shall surface all roadways in streets proposed to be dedicated to the public to widths prescribed by this Ordinance, the *City* official map, or comprehensive plan or component thereof. Said surfacing shall be done in accordance with plans and specifications approved by the *City* Engineer. The cost of surfacing in excess of 48 feet in width that is not required to serve the needs of the land division or condominium should be borne by the *City* or other unit or agency of government having jurisdiction over the street.

8.05 Curb and Gutter

A. Following the Installation and the *City*'s inspection and approval of all utility and stormwater drainage improvements, the subdivider shall construct concrete curbs and gutters in accordance with plans and specifications approved by the *City* Engineer. This requirement may be waived where a permanent rural street section has been approved by the Plan Commission.

The cost of installation of all inside curbs and gutters for dual roadway pavements shall be borne by the *City* or the unit or agency of government having jurisdiction.

B. Curb Ramps shall be installed in accordance with the Americans with Disabilities Act and Section 66.0909 of the Wisconsin Statutes, and as approved by the *City* Engineer.

8.06 Rural Street Sections

When permanent rural street sections have been approved by the Plan Commission, the subdivider shall finish grade all shoulders and road ditches, install all necessary culverts at intersections and, if required, surface ditch inverts to prevent erosion and sedimentation in accordance with plans and specifications approved by the *City* Engineer.

8.07 Sidewalks

The subdivider shall construct a concrete sidewalk on one side of all frontage streets and both sides of all other streets within the land division or condominium. The construction of all sidewalks shall be in accordance with plans and specifications approved by the *City* Engineer. Wider than standard sidewalks may be required by the *City* Engineer in the vicinity of schools, commercial areas, and other places of public assembly. The Plan Commission may waive the requirement for sidewalks upon a finding that such walks are not required because of the provision of a separate network of pedestrian ways, low vehicular or pedestrian traffic volumes, or lot arrangement.

8.08 Public Sanitary Sewerage and Private Sewage Disposal Systems

A. The Subdivider Shall Construct sanitary sewers in such a manner as to make adequate sanitary sewerage service available to each lot within the land division or condominium. Where public sanitary sewer facilities are not available, the subdivider shall make provision for adequate private sewage disposal systems as specified by the City, County, and State agencies concerned.

B. The Subdivider Shall Install sewer laterals to the street lot line. If, at the time of final platting, sanitary sewer facilities are not available to the plat, but will become available within a period of five years from the date of plat recording, the subdivider shall install or cause to be installed sanitary sewers and sewer laterals to the street lot line in accordance with this Section and shall cap all laterals as may be specified by the *City* Engineer. The size, type, and installation of all sanitary sewers proposed to be constructed shall be in accordance with the plans and specifications approved by the *City* Engineer.

C. The Subdivider Shall Assume the cost of installing all sanitary sewers, laterals, and appurtenances required to serve the land division or condominium development proposed. If sewers greater than eight inches in diameter are required to accommodate sewage flows originating from outside of the proposed development, the cost of such larger sewers shall be prorated either in proportion to the ratio of the total area of the land division or condominium development to the total tributary drainage area to be served by such larger sewer, or in proportion to the contributing sewage flows, as may be agreed upon between the subdivider and the *City*, and the excess cost either borne by the *City* or assessed against the total tributary drainage area.

8.09 Stormwater Management Facilities

A. The Subdivider Shall Construct stormwater drainage facilities, which may include curbs and gutters, catch basins and inlets, storm sewers, road ditches, open channels, and storage facilities as may be required. All such facilities are to be of adequate size and grade to hydraulically accommodate potential volumes of flow. The type of facilities required and the design criteria shall be determined by the *City* Engineer. Storm drainage facilities shall be so designed as to prevent and control soil erosion and sedimentation and present no hazard to life or property. The size, type, and installation of all stormwater management facilities proposed to be constructed shall be in accordance with the plans and specifications approved by the *City* Engineer.

B. The Subdivider Shall Assume the costs entailed in constructing stormwater conveyances and storage facilities necessary to serve the proposed development and to carry the existing stormwater flows through the proposed development. If larger conveyance and storage facilities are required to accommodate flows originating from outside of the proposed development, or to avoid flooding attendant to increased flows downstream of the proposed development caused not by the development but by pre-existing development upstream, the cost of such facilities shall be prorated in proportion to the contributing rates of flows, and the excess cost shall be borne by the *City* or assessed against the tributary drainage areas concerned.

8.10 Water Supply Facilities

A. The Subdivider Shall Construct water mains in such a manner as to make adequate water service available to each lot within the land division or condominium. If municipal water service is not available, the subdivider shall make provision for adequate private water systems as specified by the *City*, County, and State agencies concerned. The *City* Plan Commission may require the installation of water laterals to the street lot line. The size, type, and installation of all public water mains proposed to be constructed shall be in accordance with plans and specifications approved by the *City* Engineer.

B. The Subdivider Shall Assume the cost of installing all water mains eight inches in diameter or less in size. If water mains greater than eight inches in diameter are required to serve areas outside the proposed development, the excess cost shall be borne by the *City*.

8.11 Other Utilities

A. The Subdivider Shall Cause gas, electrical power, and telephone and other communication facilities to be installed in such a manner as to make adequate service available to each lot in the land division or condominium, in accordance with Section 7.08.
B. Plans Indicating the proposed location of all gas, electrical power, telephone, and other communications distribution and transmission lines required to serve the land division or condominium shall be approved by the *City* Engineer.

8.12 Street Lamps

A. The Subdivider Shall Install public street lamps along all streets proposed to be dedicated. The Plan Commission shall approve the design and location of all street lamps, which shall be compatible with the neighborhood and type of development proposed.
B. In Lieu of or in Addition to the Installation of public street lamps, the *City* Plan Commission may permit the installation of private post lamps on each lot of a land division and at appropriate locations within a condominium. The type and location of such post lamps shall be approved by the *City* Engineer.

8.13 Street Signs

The subdivider shall install at the intersection of all streets proposed to be dedicated a street sign of a design specified by the *City* Engineer.

8.14 Street Trees

A. The Subdivider Shall Plant or provide funding for the planting of at least one tree of a species approved by the Plan Commission of at least two inches in diameter measured at six inches above the top of the root ball at an average spacing of 50 feet along the frontage of all streets proposed to be dedicated. The required trees shall be planted in the area between the sidewalk and curb in accordance with plans and specifications approved by the *City* Engineer.
B. The Requirement for street trees may be waived by the Plan Commission if substantial alternative landscaping, including trees, is to be provided within the land division or condominium in accordance with a landscaping plan approved by the *City* Plan Commission.

8.15 Erosion and Sedimentation Control

A. The Subdivider Shall Prepare an erosion and sedimentation control plan addressing the installation and maintenance of soil erosion and sedimentation control measures. Such plans shall meet the requirements set forth in the *City* Erosion Control and Stormwater Management Ordinances.

B. The Subdivider Shall Plant those grasses, trees, and groundcover of species and size specified by the Plan Commission, upon recommendation of the *City* Engineer, necessary to prevent soil erosion and sedimentation, in accordance with the approved erosion and sedimentation control plan.

C. The Subdivider Shall Install those protection and rehabilitation measures, such as fencing, sloping, seeding, riprap, revetments, jetties, clearing, dredging, snagging, drop structures, brush mats, willow poles, and grade stabilization structures, set forth in the approved erosion and sedimentation control plan.

8.16 Landscaping

A. The Subdivider Shall Install landscaping in accordance with a landscaping plan approved by the *City* Plan Commission. If plantings are not installed prior to approval of a final plat or condominium plat, a landscaping schedule shall be specified in the Development Agreement and appropriate sureties shall be provided.

B. Maintenance of All Landscaping included in an approved landscaping plan shall be the responsibility of the property owner, or, for landscaping installed in common areas, the homeowners or condominium owners association. Provisions for the maintenance of such landscaping shall be included in the homeowners association documents required under Section 2.06.

SECTION 9.00 CONSTRUCTION

9.01 Commencement

No construction or installation of improvements shall commence in a proposed land division or condominium development until a development agreement has been executed, the *City* Engineer has given written authorization to proceed, and a preconstruction meeting of concerned parties, such as the utilities and contractors concerned, has been called by the *City* Engineer.

9.02 Phasing

The *Common Council* may permit the construction and installation of public improvements in phases corresponding to the development phases of a final plat.

9.03 Building Permits

No building permits shall be issued for a structure on any lot not of record on the date of adoption of this Ordinance until all the requirements of this Ordinance have been met.

9.04 Plans

Each of the following plans and accompanying construction specifications shall, except for the landscaping plan, be approved by the *City* Engineer and any other agency having relevant approving authority before commencement of the installation of the relevant improvement. The landscaping plan shall be approved by the *City* Plan Commission.

A. Street Plans and Profiles showing existing and proposed grades, elevations, cross-sections, materials, and other details of required improvements.
B. Sanitary Sewer Plans and profiles showing the locations, grades, sizes, elevations, materials, and other details of required facilities.
C. Plans for Stormwater Management Facilities showing the locations, grades, sizes, elevations, materials, and other details of required facilities, together with the path of drainage to the receiving storm sewer, drainage channel, or watercourse.
D. Water Supply and Distribution Plans and profiles showing the locations, sizes, elevations, materials, and other details of required facilities.
E. Grading Plans showing existing and proposed topographic contours, mass and finished grading plans, proposed top of building foundation and finished yard grade elevations, and such supplemental information as required by the *City* Engineer.
F. Erosion and Sedimentation Control Plans showing those structures necessary to retard the rate of runoff water and those measures and practices that will minimize erosion and sedimentation, in accordance with Section 8.15.
G. Landscaping Plans showing and describing in detail the location, size, and species of any proposed new trees, shrubs, and other vegetation; existing trees, shrubs, and other vegetation proposed to be retained; nonliving durable material such as rocks, sand, gravel, or mulch; and structures such as walls, fences, and entrance signs.
H. Additional Special Plans or information required by the *City* staff, Plan Commission, or *Common Council.*

9.05 Earth Moving

Earth moving, such as grading, topsoil removal, mineral extraction, stream course changing, road cutting, waterway construction or enlargement, removal of stream or lake bed materials, excavation, channeling, clearing, ditching, drain tile laying, dredging, and lagooning, shall be so conducted as to minimize erosion and sedimentation and disturbance of the natural fauna, flora, watercourse, water regimen, and topography.

9.06 Preservation of Existing Vegetation

The subdivider shall make every effort to protect and retain all existing desirable trees, shrubs, grasses, and groundcover not actually lying in public roadways, drainageways, building foundation sites, private driveways, soil absorption waste disposal areas, and bicycle and pedestrian ways. Trees shall be protected and preserved during construction in accordance with the approved landscaping plan and with sound conservation practices, including the preservation of trees by well islands or retaining walls, whenever abutting grades are altered.

9.07 Inspection

The subdivider, prior to commencing any work within the land division or condominium, shall make arrangements with the *City* Engineer to provide for inspection. The *City* Engineer shall inspect and approve all completed work prior to approval of the final plat or release of the sureties.

9.08 Completion of Improvements

All of the improvements required under this Ordinance shall be completed prior to the final approval of a subdivision or condominium plat by the *Common Council*, except that in lieu of completion of construction, a certified check, surety bond, or letter of credit approved by the *City* Attorney may be furnished as provided in Section 2.07.

9.09 As-Built Plans

Within 30 days following completion and acceptance by the *City* Engineer of all improvements, the subdivider shall provide reproducible copies of plans and profiles that accurately show the location, extent, and horizontal and vertical location and alignment of all improvements as actually constructed. Horizontal locations shall be expressed in terms of Wisconsin State Plane Coordinates, North American Datum of 1927 and vertical locations shall be referenced to the NGVD of 1929.

SECTION 10.00 FEES

10.01 General

The subdivider shall pay to the *City* Treasurer all fees as hereinafter required and at the times specified before being entitled to record the Plat or Certified Survey Map concerned.

10.02 Preliminary Plat or Certified Survey Map Review Fee

A. The Subdivider Shall pay a fee as set forth in the *City* fee schedule to the *City* Treasurer at the time of first application for approval of any preliminary plat or certified survey map to assist in defraying the cost of review.

B. A Reapplication Fee as set forth in the *City* fee schedule shall be paid to the *City* Treasurer at the time of reapplication for approval of any preliminary plat or certified survey map that has previously been reviewed.

10.03 Improvement Review Fee

A. The Subdivider Shall pay a fee or present a letter of credit or a bond equal to 1 percent of the cost of the required public improvements as estimated by the *City* Engineer at the time of the submission of improvement plans and specifications to partially cover the cost to the *City* of reviewing such plans and specifications.
B. The Fee May be recomputed, upon demand of the subdivider or *City* Engineer, after completion of improvement construction in accordance with the actual cost of such improvements and the difference, if any, shall be paid by or remitted to the subdivider. Evidence of cost shall be in such detail and form as required by the *City* Engineer.

10.04 Construction Review Fee

The subdivider shall pay a fee equal to the actual cost to the *City* for such inspection as the *City* Engineer deems necessary to assure that the construction of the required improvements is in compliance with the plans, specifications, and ordinances of the *City* or any other governmental authority.

10.05 Final Plat Review Fee

A. The Subdivider Shall pay a fee as set forth in the *City* fee schedule for each lot or parcel within the final plat to the *City* Treasurer at the time of first application for approval of said plat to assist in defraying the cost of review.
B. A Reapplication Fee as set forth in the *City* fee schedule shall be paid to the *City* Treasurer at the time of a reapplication for approval of any final plat that has previously been reviewed.

10.06 Public Site Fee

A. If Required by the Plan Commission under Section 7.10, a fee for the acquisition of public sites to serve the future inhabitants of the proposed land division or condominium shall be paid to the *City* Treasurer at the time of first application for approval of a final plat or certified survey map of said land division or condominium in the amount set forth in the *City* fee schedule.
B. Public Site Fees shall be placed in a nonlapsing separate Service District Fund by the *City* Treasurer to be used only for the acquisition of playground, park, parkway, or other open space site that will serve the proposed land division or condominium. Said fund shall be established on the basis of the service area of existing or proposed park or open space sites.

10.07 Engineering Fee

A. The Subdivider Shall pay a fee equal to the actual cost to the *City* for all engineering work incurred by the *City* in connection with the plat.

B. Engineering Work shall include the preparation of construction plans and standard specifications. The *City* Engineer may permit the subdivider to furnish all, some, or part of the required construction plans and specifications, in which case no engineering fees shall be levied for such plans and specifications.

10.08 Special Legal and Fiscal Review Fees

The subdivider shall pay a fee equal to the cost of any special legal or fiscal analyses that may be undertaken by the *City* in connection with the proposed land division or condominium plat, including the drafting of contracts between the *City* and the subdivider. These fees may also include the cost of obtaining independent professional opinions of engineers, landscape architects, and land planners requested by the *City* Plan Commission in connection with the review of the land division or condominium plat being considered.

10.09 Appeal of Fees

The Subdivider shall have the right to challenge the amount of any fees levied under Sections 10.03, 10.04, 10.07, and 10.08 of this Ordinance by an appeal to the *Common Council*. Upon receipt of such an appeal, the *Common Council*, upon due notice, shall hold a public hearing at which the Subdivider and the *City* officials concerned can present their case. Based on review of relevant records and the testimony presented at the public hearing, the *Common Council* shall make a determination with respect to the fairness of the amount of the fees challenged and shall make a determination to decrease, affirm, or increase the fees concerned.

SECTION 11.00 DEFINITIONS

11.01 General Definitions

For the purposes of this Ordinance, the following definitions shall apply. Words used in the present tense include the future; the singular number includes the plural number; and the plural number includes the singular. The word "shall" is mandatory. Any words not defined in this Section shall be presumed to have their customary definitions as given in standard reference dictionaries.

11.02 Specific Words and Phrases

Advisory Agency. Any agency, other than an objecting agency, to which a plat or certified survey map may be submitted for review and comment. An advisory agency may give advice to the *City* and suggest that certain changes be made to the plat or certified map, or it may suggest that a plat or certified survey map be approved or denied. Suggestions made by an advisory agency are not, however, binding on the *Common Council* or Plan Commission. Examples of advisory agencies include the

Southeastern Wisconsin Regional Planning Commission, school districts, and local utility companies.

Alley. A public way affording secondary access to abutting properties.

Approving Authorities. Each governmental body having authority to approve or reject a preliminary or final plat. Approving authorities are set forth in Section 236.10 of the Wisconsin Statutes.

Arterial Street. A street used, or intended to be used, primarily for fast or heavy through traffic, whose function is to convey traffic between municipalities and activity centers. Arterial streets are designated in the Regional Transportation System Plan prepared and adopted by the Southeastern Wisconsin Regional Planning Commission.

Block. An area of land bounded by streets, or a combination of streets, public parks, cemeteries, railroad rights-of-way, bulkhead lines, shorelines of waterways, and *City*, village, or town boundaries.

Building. Any structure having a roof supported by columns or walls.

Building Line. A line parallel to a lot line and at a specified minimum distance from the lot line to comply with the building setback requirements of the *City* Zoning Ordinance and the requirements of this Ordinance.

Building Setback Line. See Building Line.

Certified Survey Map. A map, prepared in accordance with Section 236.34 of the Wisconsin Statutes and this Ordinance, for the purpose of dividing land into not more than four parcels; or used to document for recording purposes survey and dedication data relating to single parcels.

City Engineer. A registered professional engineer who is a full-time employee of the *City*, or a consulting engineer who provides resident staff services to the *City*, and who is duly appointed by the *Common Council* to the position.

Collector Street. A street used, or intended to be used, to carry traffic from land access streets to the system of arterial streets, including the principal entrance streets to residential developments.

Common Open Space. See Open Space, Common.

Comprehensive Plan. The extensively developed plan, also called a master plan, adopted by the Plan Commission and certified to the *Common Council* pursuant to Section 62.23 of the Wisconsin Statutes, or a Comprehensive Plan adopted by the *Common Council* pursuant to Section 66.1001 of the Wisconsin Statutes. Components of a comprehensive plan include, but are not limited to, a land use, transportation system, park and open space, sanitary sewer, public water supply, and stormwater management system elements, and neighborhood unit development plans. Devices for the implementation of such plans include zoning, official mapping, land division control, and capital improvement programs.

Condominium. A form of ownership combining individual unit ownership with shared use and ownership of common property or facilities, established in accordance with Chapter 703 of the Wisconsin Statutes. Common areas and facilities are owned by all members of the condominium association on a proportional, undivided basis. A condominium is a legal form of ownership, and not a specific building type or style.

Condominium Association. An association, whose members consist of owners of units in a condominium, which administers and maintains the common property and common elements of a condominium.

Condominium Declaration. The instrument by which property becomes subject to Chapter 703 of the Wisconsin Statutes.

Condominium Unit. A part of a condominium intended for any type of independent use, including one or more cubicles of air at one or more levels of space or one or more rooms or enclosed spaces located on one or more floors (or parts thereof) in a building. A unit may include two or more noncontiguous areas.

County Planning Agency. The agency created by the County Board and authorized by Statute to plan land use and to review subdivision plats and certified survey maps.

Covenant. A restriction on the use of land, usually set forth in the deed.

Cul-de-Sac Street. A local street with only one outlet and having an appropriate turn-about for vehicular traffic.

Deed Restriction. A restriction on the use of a property set forth in the deed.

Development Agreement. An agreement entered into by and between the *City* and a subdivider whereby the *City* and subdivider agree as to the design, construction, and installation of required public improvements; the payment for such public improvements; dedication of land; and other matters related to the requirements of this Ordinance. The Development Agreement shall not come into effect unless and until a Letter of Credit or other appropriate surety has been provided to the *City* by the subdivider.

Environmental Corridor. See Primary Environmental Corridor, Secondary Environmental Corridor, and Isolated Natural Resource Area.

Extraterritorial Plat Approval Jurisdiction. The unincorporated area within 1.5 miles of a fourth class *City* or a village and within three miles of all other cities. Where such jurisdictions overlap, the jurisdiction over the overlapping area is divided on a line, all parts of which are equidistant from the boundaries of each municipality, so that not more than one municipality exercises extraterritorial plat approval jurisdiction over any area.

Final Plat. A map prepared in accordance with the requirements of Chapter 236 of the Wisconsin Statutes and this Ordinance for the purpose of creating a subdivision.

Floodplains. Those lands, including the floodplains, floodways, and channels, subject to inundation by the 100-year recurrence interval flood or, where such data are not available, the maximum flood of record.

Frontage. The total dimension of a lot abutting a public street measured along the street line.

Frontage Street. A land access street auxiliary to and located on the side of an arterial street for control of access and for service to the abutting development.

Hedgerow. A row of shrubs or trees planted for enclosure or separation of fields.

Homeowners Association. An association combining individual home ownership with shared use, ownership, maintenance, and responsibility for common property or facilities, including private open space, within a land division.

Isolated Natural Resource Area. An area containing significant remnant natural resources at least five acres in area and at least 200 feet in width, as delineated and mapped by the Southeastern Wisconsin Regional Planning Commission.

Land Access Street. A street used, or intended to be used, primarily for access to abutting properties.

Land Division. A generic term that includes both subdivisions and minor land divisions, as those terms are defined in this Section.

Landscaping. Living plant material, such as grass, groundcover, flowers, shrubs, vines, hedges, and trees; nonliving durable material such as rocks, pebbles, sand, mulch, wood chips or bark; and structures such as walls and fences.

Letter of Credit. An irrevocable written agreement guaranteeing payment for improvements, entered into by a bank, savings and loan, or other financial institution authorized to do business in the State of Wisconsin that has a financial standing acceptable to the *City*, and that secures a subdivider's obligation to pay the cost of designing, constructing, and installing required public improvements and certain other obligations in connection with an approved land division or condominium.

Lot. A parcel of land having frontage on a public street, occupied or intended to be occupied by a principal structure or use and sufficient in size to meet lot width, lot frontage, lot area, setback, yard, parking, and other requirements of the *City* Zoning Ordinance.

Lot, Corner. A lot abutting two or more streets at their intersection, provided that the corner of such intersection shall have an angle of 135 degrees or less.

Lot, Double Frontage. A lot, other than a corner lot, with frontage on more than one street. Double frontage lots shall normally be deemed to have two front yards and two side yards and no rear yard.

Lot, Flag. A lot not fronting on or abutting a public street and where access to the public street system is by a narrow strip of land, easement, or private right-of-way. Flag lots generally are not considered to conform to sound planning principles.

Minor Land Division. A minor land division is any division of land that:

1. Creates more than one, but fewer than five, parcels or building sites, inclusive of the original remnant parcel, any one of which is five acres or less in area, by a division or by successive divisions of any part of the original parcel within a period of five years; or
2. Divides a block, lot, or outlot within a recorded subdivision plat into more than one, but fewer than five, parcels or building sites, inclusive of the original remnant parcel, without changing the exterior boundaries of said plat or the exterior boundaries of blocks within the plat, and the division does not result in a subdivision.

Municipality. An incorporated city or village.

National Map Accuracy Standards. Standards governing the horizontal and vertical accuracy of topographic maps and specifying the means for testing and determining such accuracy, endorsed by all federal agencies having surveying and mapping functions and responsibilities. These standards have been fully reproduced in Appendix D of SEWRPC Technical Report No. 7, Horizontal and Vertical Survey Control in Southeastern Wisconsin.

Navigable Water. Lake Michigan, all natural inland lakes within Wisconsin, and all rivers, streams, ponds, sloughs, flowages, and other waters within the territorial limits of Wisconsin that are navigable under the laws of this State. The Wisconsin Supreme Court has declared navigable all bodies of water with a bed differentiated from adjacent uplands and with levels of flow sufficient to support navigation by a recreational craft of the shallowest draft on an annually recurring basis. The Wisconsin Department of Natural Resources is responsible for determining if a water body or watercourse is navigable.

Objecting Agency. An agency empowered to object to a subdivision plat pursuant to Chapter 236 of the Wisconsin Statutes. The *City* may not approve any plat upon which an objection has been certified until the objection has been satisfied. The objecting agencies include the Wisconsin Department of Administration, the Wisconsin Department of Commerce, the Wisconsin Department of Transportation, and the County Planning Agency.

Official Map. A document prepared and adopted pursuant to Section 62.23(6) of the Wisconsin Statutes that shows the location of existing and planned streets, parkways, parks, playgrounds, railway rights-of-way, waterways, and public transit facilities.

Open Space. Any site, parcel, lot, area, or outlot of land or water that has been designated, dedicated, reserved, or restricted from further development. Open space may be privately or publicly owned, but shall not be part of individual residential lots. Open space shall be substantially free of structures, but may contain recreational facilities approved by the *City*.

Open Space, Common. Privately owned land within a land division or condominium that has been restricted in perpetuity from further development and is set aside for the use and enjoyment by residents of the land division or condominium. Common open space shall be substantially free of structures, but may contain recreational facilities approved by the City.

Open Space, Public. Land within a land division or condominium that has been dedicated to the public for recreational or conservation purposes. Open space lands shall be substantially free of structures, but may contain recreational facilities approved by the *City*.

Ordinary High Water Mark. The point on the bank or shore up to which the presence and action of surface water is so continuous as to leave a distinctive mark such as by erosion, destruction, or prevention of terrestrial vegetation, predominance of aquatic vegetation, or other easily recognized characteristic.

Outlot. A parcel of land, other than a buildable lot or block, so designated on the plat, that is used to convey or reserve parcels of land. Outlots may be created to restrict a lot that is unbuildable due to high groundwater, steep slopes, or other physical constraints, or to create common open space. Outlots may also be parcels of land intended to be re-divided into lots or combined with lots or outlots in adjacent land divisions in the future for the purpose of creating buildable lots. An outlot may also be created if a lot fails to meet requirements for a private onsite wastewater treatment system, but may be buildable if public sewer is extended to the lot or land division.

Section 236.13(6) of the Wisconsin Statutes prohibits using an outlot as a building site unless it complies with all the requirements imposed for buildable lots. The *City* will generally require that any restrictions related to an outlot be included on the face of the plat.

Parcel. A single piece of land separately owned, either publicly or privately, and capable of being conveyed separately.

Plat. A map prepared, as required by Section 2.02 of this Ordinance, for the purpose of recording a subdivision, minor land division, or condominium.

Prairies. Open, generally treeless areas that are dominated by native grasses, as delineated and mapped by the Southeastern Wisconsin Regional Planning Commission.

Preliminary Plat. A map showing the salient features of a proposed subdivision submitted to an approving authority for purposes of preliminary consideration. A preliminary plat precisely describes the location and exterior boundaries of the parcel proposed to be divided, and shows the approximate location of lots and other improvements.

Primary Environmental Corridor. A concentration of significant natural resources at least 400 acres in area, at least two miles in length, and at least 200 feet in width, as delineated and mapped by the Southeastern Wisconsin Regional Planning Commission.

Public Improvement. Any sanitary sewer, storm sewer, open channel, water main, street, park, sidewalk, bicycle or pedestrian way, or other facility for which the *City* may ultimately assume the responsibility for maintenance and operation.

Public Way. Any public street, highway, bicycle or pedestrian way, drainageway, or part thereof.

Replat. The process of changing, or the plat or map that changes, the boundaries of a recorded subdivision plat, certified survey map, or a part thereof. The division of a large block, lot, or outlot within a recorded subdivision plat or certified survey map without changing the exterior boundaries of said block, lot, or outlot is not a replat.

Reserve Strip. Any land that would prohibit or interfere with the orderly extension of streets, bicycle or pedestrian ways, sanitary sewer, water mains, stormwater facilities or other utilities or improvements between two abutting properties.

Secondary Environmental Corridor. A concentration of significant natural resources at least 100 acres in area and at least one mile in length. Where such corridors serve to link primary environmental corridors, no minimum area or length criteria apply. Secondary environmental corridors are delineated and mapped by the Southeastern Wisconsin Regional Planning Commission.

Shorelands. Those lands lying within the following distances: 1,000 feet from the ordinary high water elevation of a navigable lake, pond, or flowages; or 300 feet from the ordinary high water elevation of a navigable stream, or to the landward edge of the floodplain, whichever is greater.

Soil Mapping Unit. Soil type, slope, and erosion factor boundaries as shown on the operational soil survey maps prepared by the U.S. Soil Conservation Service (now known as the Natural Resources Conservation Service).

Subdivider. Any person, firm or corporation, or any agent thereof, dividing or proposing to divide land resulting in a subdivision, minor land division, or replat, or any person who creates a condominium under Chapter 703 of the Wisconsin Statutes.

Subdivision. A division of a lot, parcel, or tract of land by the owner thereof or the owner's agent for the purpose of sale or of building development, including condominium development, where:

1. The act of division creates five or more parcels or building sites, inclusive of the original remnant parcel, any one of which is five acres or less in area, by a division or by successive divisions of any part of the original property within a period of five years; or
2. The act of division creates six or more parcels or building sites, inclusive of the original remnant parcel, of any size by successive divisions of any part of the original property within a period of five years.

Surety Bond. A bond guaranteeing performance of a contract or obligation through forfeiture of the bond if said contract or obligation is unfulfilled by the subdivider.

Tract. A parcel lying in more than one U.S. Public Land Survey section.

Unit. See condominium unit.

Wetland. An area where water is at, near, or above the land surface long enough to be capable of supporting aquatic or hydrophytic vegetation, and that has soils indicative of wet conditions.

Woodlands. Upland areas at least one acre in extent covered by deciduous or coniferous trees as delineated and mapped by the Southeastern Wisconsin Regional Planning Commission.

SECTION 12.00 ADOPTION AND EFFECTIVE DATE

12.01 Plan Commission Recommendation

The *City* Plan Commission recommended the adoption of this Land Division Control Ordinance at a meeting held on the ________ day of ________, 20________.

12.02 Public Hearing

The *C ommon Council* held a public hearing on the proposed Land Division Control Ordinance on the ________ day of ________, 20________.

12.03 Common Council Approval

The *Common Council* of the *City* of ______________________ concurred with the recommendations of the Plan Commission and adopted the Land Division Control Ordinance at a meeting held on the ________ day of ________, 20________.

12.04 Effective Date

This Land Division Control Ordinance shall take effect upon adoption by the *Common Council* and the filing of proof of publication in the office of the *City* Clerk.

Date of Publication ________
Effective Date: ________

____

Mayor

ATTEST:

____

City Clerk

Appendix B: Example Zoning Ordinance Regulations

SECTION 1.0 INTRODUCTION

1.1 Authority

These regulations are adopted under the authority granted by Sections 61.35 and 62.23*(7)* of the Wisconsin Statutes. Therefore, the Village Board of ________, Wisconsin, do ordain as follows:

1.2 Purpose

The purpose of this Ordinance is to promote the health, safety, morals, prosperity, aesthetics, and general welfare of this community.

1.3 Intent

It is the general intent of this Ordinance to regulate and restrict the use of all structures, lands, and waters; regulate and restrict lot coverage, population distribution and density, and the size and location of all structures so as to: lessen congestion to promote the safety and efficiency of the streets and highways; secure safety from fire, flooding, panic and other dangers; provide adequate light, air, sanitation, and drainage; prevent overcrowding; avoid undue population concentration; facilitate the adequate provision of public facilities and utilities; stabilize and protect property values; further the appropriate use of land and conservation of natural resources; preserve and promote the beauty of the community; and implement the community's comprehensive plan or plan components. It is further intended to provide for the administration and enforcement of this Ordinance and to provide penalties for its violation.

1.4 Abrogation and Greater Restrictions

It is not intended by this Ordinance to repeal, abrogate, annul, impair, or interfere with any existing easements, covenants, deed restrictions, agreements, ordinances, rules, regulations, or permits previously adopted or issued pursuant to laws. However, wherever this Ordinance imposes greater restrictions, the provisions of this Ordinance shall govern.

1.5 Interpretation

In their interpretation and application, the provisions of this Ordinance shall be held to be minimum requirements and shall be liberally construed in favor of the *Village* and shall not be deemed a limitation or repeal of any other power granted by the Wisconsin Statutes.

1.6 Severability

If any section, clause, provision, or portion of the Ordinance is adjudged unconstitutional or invalid by a court of competent jurisdiction, the remainder of this Ordinance shall not be affected thereby.

1.7 Repeal

All other ordinances or parts of ordinances of the *Village* inconsistent or conflicting with this Ordinance, to the extent of the inconsistency only, are hereby repealed.

1.8 Title

This Ordinance shall be known as, referred to, or cited as the "ZONING ORDINANCE, *VILLAGE* OF __________, WISCONSIN."

1.9 Effective Date

This Ordinance shall be effective after a public hearing, adoption by the *Village Board of Trustees*, and publication or posting as provided by law.

SECTION 2.0 GENERAL PROVISIONS

2.1 Jurisdiction

The jurisdiction of this Ordinance shall include all lands and waters within the corporate limits of the *Village of* ____.

2.2 Compliance

No structure, land or water shall hereafter be used and no structure or part thereof shall hereafter be located, erected, moved, reconstructed, extended, enlarged, converted, or structurally altered without a *zoning* permit except minor structures and without full compliance with the provisions of this Ordinance and all other applicable local, county, and state regulations.

The Duty of the *Zoning Administrator*, with the aid of the Police Department, shall be to investigate all complaints, give notice of violations, and to enforce the provisions of this Ordinance. The *Zoning Administrator* and his duly appointed deputies may enter at any reasonable time onto any public or private lands or waters to make a zoning inspection.

2.3 Zoning Permit

Applications for a zoning permit shall be made in duplicate to the *Zoning Administrator* on forms furnished by the *Zoning Administrator* and shall include the following where applicable:

Names and Addresses of the applicant, owner of the site, architect, professional engineer, or contractor.

Description of the Subject Site by lot, block, and recorded subdivision or by metes and bounds; address of the subject site; type of structure; existing and proposed operation of use of the structure or site; number of employees; and the zoning district within which the subject site lies.

Plat of Survey prepared by a registered land surveyor showing the location, boundaries, dimensions, elevations, uses, and size of the following: subject site; existing and proposed structures; existing and proposed easements, streets, and other public ways; off-street parking, loading areas, and driveways; existing highway access restrictions; existing and proposed street, side, and rear yards. In addition, the plat of survey shall show the location, elevation, and use of any abutting lands and their structures within *forty (40)* feet of the subject site.

Proposed Sewage Disposal Plan if municipal sewerage service is not available. This plan shall be approved by the *Village* Engineer who shall certify in writing that satisfactory, adequate, and safe sewage disposal is possible on the site as proposed by the plan in accordance with applicable local, county, and state board of health regulations.

Proposed Water Supply Plan if municipal water service is not available. This plan shall be approved by the *Village* Engineer who shall certify in writing that an adequate and safe supply of water will be provided.

Additional Information as may be required by the *Village Plan Commission, Village Engineer, Zoning, Building, Plumbing, or Health Inspectors.* Fee Receipt from the *Village* Treasurer in the amount of *Twenty-five Dollars* ($25).

Zoning Permit shall be granted or denied in writing by the *Zoning Administrator* within thirty (30) days. The permit shall expire within *six (6)* months unless substantial work has commenced. Any permit issued in conflict with the provisions of this Ordinance shall be null and void.

2.4 Site Restrictions

No land shall be used or structure erected where the land is held unsuitable for such use or structure by the *Village Plan Commission* by reason of flooding, concentrated runoff, inadequate drainage, adverse soil or rock formation, unfavorable topography, low percolation rate or bearing strength, erosion susceptibility, or any other feature likely to be harmful to the health, safety, prosperity, aesthetics, and general welfare of this community. The *Village Plan Commission*, in applying the provisions of this section, shall in writing recite the particular facts upon which it bases its conclusion that the land is not suitable for certain uses. The applicant shall have an opportunity to present evidence contesting such unsuitability if he so desires. Thereafter the *Village Plan Commission* may affirm, modify, or withdraw its determination of unsuitability.

All Lots shall abut upon a public street, and each lot shall have a minimum frontage of *thirty (30)* feet.

All Principal Structures shall be located on a lot; and only one principal structure shall be located, erected, or moved onto a lot.

No Zoning Permit shall be issued for a lot that abuts a public street dedicated to only a portion of its proposed width and located on that side thereof from which the required dedication has not been secured.

Private Sewer and Water. In any district where public sewerage service is not available, the width and area of all lots shall be sufficient to permit the use of an on-site sewage disposal system designed in accordance with Section H65 of the Wisconsin Administrative Code. In any district where a public water service or public sewerage service is not available, the lot width and area shall be determined in accordance with Section H65 of the Wisconsin Administrative Code, but for one-family dwellings shall be no less than *one hundred (100)* feet and no less than *20,000* square feet, respectively.

Lots Abutting More Restrictive district boundaries shall pro-vide side and rear yards not less than those required in the more restrictive abutting district. The street yards on the less restrictive district shall be modified for a distance of not more than *sixty (60)* feet from the district boundary line so as to equal the average of the street yards required in both districts.

2.5 Use Restrictions

The following use restrictions and regulations shall apply:

Principal Uses. Only those principal uses specified for a district, their essential services, and the following uses shall be permitted in that district.

Accessory Uses and structures are permitted in any district but not until their principal structure is present or under construction. Residential accessory uses shall not involve the conduct of any business, trade, or industry. Accessory uses include incidental repairs; storage; parking facilities; gardening; servant's, owner's, itinerant agricultural laborer's, and watchman's quarters not for rent; private swimming pools; and private emergency shelters.

Conditional Uses and their accessory uses are considered as special uses requiring review, public hearing, and approval by the *Village Plan Commission* in accordance with Section 4.0. Any development within *five hundred (500)* feet of the existing or proposed rights-of-way of freeways, expressways, interstate and controlled access trafficways and within *fifteen hundred (1,500)* feet of their existing or proposed interchange or turning lane rights-of-way shall be deemed to be conditional uses. Such development shall be specifically reviewed and approved by the *Village Plan Commission* as provided in Section 4.0.

Unclassified or Unspecified Uses may be permitted by the Board of *Zoning Appeals* after the *Village Plan Commission* has made a review and recommendation provided that such uses are similar in character to the principal uses permitted in the district.

Temporary Uses, such as real estate sales field offices or shelters for materials and equipment being use in the construction of permanent structure, may be permitted by the *Board of Zoning Appeals.*

Performance Standards listed in Section 9.0 shall be complied with by all uses in all districts.

2.6 Reduction or Joint Use

No lot, yard, parking area, building area, or other space shall be reduced in area or dimension so as not to meet the provisions of this Ordinance. No part of any lot, yard, parking area, or other space required for a structure or use shall be used for any other structure or use.

2.7 Violations

It shall be unlawful to construct or use any structure, land, or water in violation of any of the provisions of this Ordinance. In case of any violation, the *Board of Trustees*, the *Zoning Administrator*, the *Village Plan Commission*, or any property owner who would be specifically damaged by such violation may institute appropriate action or proceeding to enjoin a violation of this Ordinance.

2.8 Penalties

Any person, firm, or corporation who fails to comply with the provisions of this Ordinance shall, upon conviction thereof, forfeit not less than *Ten Dollars ($10)* nor more than *Two Hundred Dollars ($200)* and costs of prosecution for each violation and in default of payment of such forfeiture and costs shall be imprisoned in the County Jail until payment thereof, but not exceeding *thirty (30)* days. Each day a violation exists or continues shall constitute a separate offense.

SECTION 3.0 ZONING DISTRICTS

3.1 Establishment

For the purpose of this Ordinance, the *Village of* __________ is hereby divided into the following 14 zoning districts:

R-1 Single-Family Residential District
R-2 Single-Family Residential District
R-3 Multi-Family Residential District
B-1 Neighborhood Business District
B-2 Community Business District
B-3 Integrated Business District
B-4 Highway Business District
M-1 Industrial District
M-2 Heavy Industrial District
A-1 Agricultural District

C-1 Conservancy District
F-1 Floodplain District
P-1 Public and Semipublic District
P-2 Planned Unit Development District

Boundaries of These Districts are hereby established as shown on a map entitled "Zoning Map, *Village* of ________, Wisconsin," dated ________, which accompanies and is a part of this Ordinance. Such boundaries shall be construed to follow: corporate limits; U.S. Public Land Survey lines; lot or property lines; centerlines of streets, highways, alleys, easements, and railroad rights-of-way or such lines extended; unless otherwise noted on the Zoning Map.

Vacation of public streets and alleys shall cause the land vacated to be automatically placed in the same district as the abutting side to which the vacated land reverts.

Annexations to or consolidations with the Village subsequent to the effective date of this Ordinance shall be placed in the *A-1 Agricultural District*, unless the annexation ordinance temporarily places the land in another district. Within one (1) year the *Village Plan Commission* shall evaluate and recommend a permanent district classification to the *Village Board.*

3.2 Zoning District Map

A certified copy of the Zoning District Map shall be adopted and approved with the text as part of this Ordinance and shall bear upon its face the attestation of the *Village* President and *Village* Clerk and shall be available to the public in the office of the *Village* Clerk.

Changes thereafter to the districts shall not be effective until entered and attested on this certified copy.

3.3 Residential Districts

R-1 Single-Family Residential District:

Principal Use		One-family dwelling with a garage
Conditional Uses		See Section 4.4
Lot	Width	Minimum *120* ft.
	Area	Minimum *40,000* sq. ft.
Building	Area	Minimum *2,000* sq. ft.
	Height	Maximum *35* ft.
Yards	Street	Minimum *50* ft.
	Rear	Minimum *50* ft.
	Side	Minimum *20* ft.

R-2 Single-Family Residential District:

Principal Use		One-family dwelling with a garage
Conditional Uses		See Sections 4.4 and 4.5
Lot	Width	Minimum *70* ft.

	Area	Minimum *10,000* sq. ft.
Building	Area	Minimum *1,200* sq. ft.
	Height	Maximum *35* ft.
Yards	Street	Minimum *25* ft.
	Rear	Minimum *40* ft.
	Side	Minimum *10* ft.

R-3 Multi-Family Residential District:

Principal Use		Multi-family dwelling with garages
Conditional Uses		See Section 4.4 and 4.5
Lot	Width	Minimum *120* ft.
	Area	Minimum *15,000* sq. ft. with no less than *2,000* sq. ft. per efficiency; *2,500* sq. ft. per one-bedroom unit; *3,000* sq. ft. per two-bedroom unit.
Building	Area	Minimum *600* sq. ft. per family
	Height	Maximum *35* ft.
Yards	Street	Minimum *35* ft.
	Rear	Minimum *50* ft.
	Side	Minimum *20* ft.

3.4 Business Districts

B-1 Neighborhood Business District

Principal Uses. The following uses provided that they shall be retail establishments selling and storing only new merchandise: bakeries, barber shops, bars, beauty shops, business offices, clinics, clothing stores, clubs, cocktail lounges, confectioneries, delicatessens, drug stores, fish markets, florists, fraternities, fruit stores, gift stores, grocery stores, hardware stores, house occupations, hobby shops, lodges, meat markets, optical stores, packaged beverage stores, professional offices, restaurants, self-service and pickup laundry and dry cleaning establishments, soda fountains, sporting goods, super-markets, tobacco stores, and vegetable stores. Existing residences shall comply with all the provisions of the *R-3 Residential District.*

Conditional Uses		See Sections 4.4 and 4.6.
Building	Height	Maximum *35* sq. ft.
Yards	Street	Minimum *25* ft.
	Rear	Minimum *50* ft.
	Side	*None* or if provided a minimum of *10* ft.

B-2 Community Business District

Principal Uses. All uses permitted in the *B-1 Neighborhood Business District* and the following: apartment hotels, appliance stores, caterers, churches, clothing repair shops, crockery stores, department stores, electrical supply, financial institutions, food lockers, furniture stores, furniture upholstery shops, heating supply, hotels, laundry and dry-cleaning establishments employing not over seven persons, liquor stores, music stores, newspaper offices and press rooms, night clubs, office supplies, pawn shops, personal service establishments, pet shops, places of entertainment, photographic supplies, plumbing supplies, printing, private clubs, private schools, publishing, radio broadcasting studios, secondhand stores, signs, television broad-casting studios, trade and contractor's offices, upholsterer's shops, and variety stores. Existing residences shall comply with all the provisions of the *R-3 Residential District.*

Conditional Uses		See Sections 4.4, 4.6, and 4.9.
Building	Height	Maximum *45* ft.
Yards	Street	Minimum *10* ft.
	Rear	Minimum *30* ft.
	Side	*None* or if provided a minimum of *10* ft.

A-1 Integrated Business District:

Principal Uses		None
Conditional Uses		All *B-2 Community Business District* uses. See Sections 4.4, 4.6, and 4.9.
Lot	Frontage	Minimum *200* ft.
	Area	Minimum *2* acres.
Building	Height	Maximum *45* ft.
Yards	Street	Minimum *80* ft.
	Rear	Minimum *40* ft.
	Side	*None* or if provided a minimum of *10* ft.

3.5 Industrial Districts

M-1 Industrial District

Principal Uses. Automotive body repairs; automotive upholstery; cleaning, pressing, and dyeing establishments; commercial bakeries; commercial greenhouses; distributors; farm machinery; food locker plants; laboratories; machine shops; manufacture and bottling of nonalcoholic beverages; painting; printing; publishing; storage and sale of machinery and equipment; trade and contractors' offices; warehousing; and wholesaling. Manufacture, fabrication, packing, packaging, and assembly of products from furs, glass, leather, metals, paper, plaster, plastics, textiles, and wood. Manufacture, fabrication, processing, packaging, and packing of confections; cosmetics; electrical appliances; electronic devices; food except cabbage, fish and fish products, meat and meat products, and pea vining; instruments; jewelry;

pharmaceuticals; tobacco; and toiletries. Existing residences shall comply with all the provisions of the *R-3 Residential District.*

Conditional Uses		See Sections 4.4 and 4.7.
Building	Height	Maximum *45* ft.
Yards	Street	Minimum *25* ft.
	Rear	Minimum *30* ft.
	Side	Minimum *20* ft.

M-2 Heavy Industrial District

Principal Uses. All *M-1 Industrial District* principal uses, freight yards, freight terminals and transhipment depots, inside storage, breweries, and crematories. Existing residences shall comply with all the provisions of the *R-3 Residential District.*

Conditional Uses		See Sections 4.4 and 4.7.
Building	Height	Maximum *60* ft.
Yards	Street	Minimum *10* ft.
	Rear	Minimum *30* ft.
	Side	Minimum *10* ft.

3.6 Agricultural District

A-1 Agricultural District

Principal Uses. Apiculture, dairying, floriculture forestry, general farming, grazing, greenhouses, hatcheries, horticulture, livestock raising, nurseries, orchards, paddocks, pasturage, poultry raising, stables, truck farming, and viticulture Farm dwellings for those resident owners and laborers actually engaged in the principal permitted uses are accessory uses and shall comply with all the provisions of the *R-2 Residential District.*

Conditional Uses		See Sections 4.4 and 4.7.
Farm	Frontage	Minimum *200* ft.
	Area	Minimum *10* acres.
Structure	Height	Maximum *50* ft.
Yards	Street	Minimum *80* ft.
	Rear	Minimum *50* ft.
	Side	Minimum *50* ft.

3.7 Conservancy District

C-1 Conservancy District

Principal Uses. Fishing; hunting; preservation of scenic, historic, and scientific areas; public fish hatcheries; soil and water conservation; sustained yield forestry; stream bank and lake shore protection; water retention; and wildlife preserves.

Conditional Uses. Drainage; water measurement and water control facilities; grazing; accessory structures, such as hunting or fishing lodges; orchards; truck farming; utilities; and wildcrop harvesting. The above uses shall not involve the dumping, filling, cultivation, mineral, soil or peat removal or any other use that would disturb the natural fauna, flora, watercourses, water regimen, or topography.

Structures None permitted except accessory to the principal or conditional uses.

3.8 Floodplain District

F-1 Floodplain District

Principal Uses. Flood overflows, impoundments, parks, sustained yield forestry, fish hatcheries, wildlife preserves, water measurement and water control facilities.

Conditional Uses. All uses permitted in the *A-1 Agricultural District* except residential uses; all uses permitted in the *P-1 Public and Semipublic District* except the caging of animals; warehousing, storage, parking, and loading areas. The above uses shall not include the storage of materials that are buoyant, flammable, explosive, or injurious to human, animal, or plant life nor substantially reduce the flood water storage capacity of the flood plain. See Section 4.4.

Buildings. All buildings shall have their first floors constructed at an elevation no less than two (2) feet above the level of the *100-year recurrence interval flood.*

3.9 Public and Semipublic District

P-1 Public and Semipublic District

Principal Uses. Parks, arboretums, playgrounds, fishing, wading, swimming, beaches, skating, sledding, sustained yield forestry, wildlife preserves, soil and water conservation, water measurement and water control facilities.

Conditional Uses. All structures; see Sections 4.4 and 4.9.

SECTION 4.0 CONDITIONAL USES

4.1 Permit

The *Village Plan Commission* may authorize the *Zoning Administrator* to issue a conditional use permit for conditional uses after review and a public hearing, provided that such conditional uses and structures are in accordance with the purpose and intent of this Ordinance and are found to be not hazardous, harmful, offensive, or otherwise adverse to the environment or the value of the neighborhood or the community.

4.2 Application

Applications for conditional use permits shall be made in duplicate to the *Zoning Administrator* on forms furnished by the *Zoning Administrator* and shall include the following:

Names and Addresses of the applicant, owner of the site, architect, professional engineer, or contractor, and all opposite and abutting property owners of record.

Description of the Subject Site by lot, block, and recorded subdivision or by metes and bounds; address of the subject site; type of structure; proposed operation or use of the structure or site; number of employees; and the zoning district within which the subject site lies.

Plat of Survey prepared by a registered land surveyor showing all of the information required under Section 2.3 for a *Zoning* Permit and, in addition, the following: mean and historic high water lines, on or within *forty (40)* feet of the subject premises, and existing and proposed landscaping.

Additional Information as may be required by the *Village Plan Commission, Village Engineer, Zoning Administrator, Building, Plumbing, or Health Inspectors.*

Fee Receipt from the Village Treasurer in the amount of *Twenty-five Dollars ($25).*

4.3 Review and Approval

The *Village Plan Commission* shall review the site, existing and proposed structures, architectural plans, neighboring uses, parking areas, driveway locations, highway access, traffic generation and circulation, drainage, sewerage and water systems, and the proposed operation.

Any Development within *five hundred (500)* feet of the existing or proposed rights-of-way of freeways, expressways, interstate and controlled access trafficways and within *fifteen hundred (1,500)* feet of their existing or proposed interchange or turning lane rights-of-way shall be specifically reviewed by the highway agency that has jurisdiction over the trafficway. The *Village Plan Commission* shall request such review and await the Highway Agency's recommendations for a period not to exceed *sixty (60)* days before taking final action.

Conditions, such as landscaping, architectural design, type of construction, construction commencement and completion dates, sureties, lighting, fencing, planting screens, operational control, hours of operation, improved traffic circulation, deed restrictions, highway access restrictions, increased yards, or parking requirements, may be required by the *Village Plan Commission* upon its finding that these are necessary to fulfill the purpose and intent of this Ordinance.

Compliance with all other provisions of this Ordinance, such as lot width and area, yards, height, parking, loading, traffic, highway access, and performance standards, shall be required of all conditional uses. Variances shall only be granted as provided in Section 11.0.

4.4 Public and Semipublic Uses

The following public and semipublic uses shall be conditional uses and may be permitted as specified:

Airports, airstrips, and landing fields in the *M-1* and *M-2 Industrial Districts, A-1 Agricultural District*, and *P-1 Public and Semipublic District*, provided the site area is not less than *twenty (20)* acres.

Governmental and Cultural Uses, such as fire and police stations, community centers, libraries, public emergency shelters, parks, playgrounds, and museums, in all residential and business districts; *M-1* and *M-2 Industrial Districts*, and *P-1 Public and Semipublic District.*

Utilities in all districts provided all principal structures and uses are not less than *fifty (50)* feet from any residential district lot line.

Public Passenger Transportation Terminals, such as heliports, bus and rail depots, except airports, airstrips, and landing fields, in all *Business Districts* and *M-1* and *M-2 Industrial Districts* provided all principal structures and uses are not less than *one hundred (100)* feet from any residential district boundary.

Public, Parochial, and Private Elementary and secondary schools and churches in the *R-2* and *R-3 Residential Districts* and *P-1 Public and Semipublic District* provided the lot area is not less than *two (2)* acres and all principal structures and uses are not less than *fifty (50)* feet from any lot line.

Colleges; Universities; Hospitals; sanitariums; religious, charitable, penal and correctional institutions; cemeteries and crematories in the *A-1 Agricultural District* and *P-1 Public and Semipublic District* provided all principal structures and uses are not less than *fifty (50)* feet from any lot line.

4.5 Residential Uses

The following residential and quasi-residential uses shall be conditional uses and may be permitted as specified:

Planned Residential Developments, such as cluster and planned neighborhood unit developments in the *R-2 Residential District* and garden apartments, row housing and group housing in the *R-3 Residential District.* The district regulations may be varied provided that adequate open space shall be provided so that the average intensity and density of land use shall be no greater than that permitted for the district in which it is located. The proper preservation, care, and maintenance by the original and all subsequent owners of the exterior design; all common structures, facilities, utilities, access and open spaces shall be assured by deed restrictions enforceable by the *Village.* The following provisions shall be complied with:

Development		Minimum *10* acres.
Lot	Area	Minimum *of 2/3* of the minimum lot area for the district in which located. Minimum *3,000* sq. ft. for row houses.
	Width	Minimum of *2/3* of the minimum lot width for the district in which located. Minimum *20* sq. ft. for row houses.
Building	Area	Minimum building area for the district in which located.
	Height	Maximum *35* ft.
	Rooms	All living rooms shall have windows opening onto a yard.
Yards	Street	Minimum *20* ft.
	Rear	Minimum *50* ft.
	Side	Minimum *20* ft. from street rights-of-way, exterior property lines of the development, and other buildings.

B-1 Neighborhood Businesses District Uses

Clubs, fraternities, lodges, and meeting places of a noncommercial nature in the *R-3 Residential District* provided all principal structures and uses are not less than *twenty-five (25)* feet from any lot line.

Rest homes, nursing homes, homes for the aged, clinics, and children's nurseries in the *R-2* or *R-3 Residential Districts* provided all principal structures and uses are not less than *fifty (50)* feet from any lot line.

Home occupations and professional offices in the *R-2* or *R-3 Residential Districts.*

4.6 Highway-Oriented Uses

The following commercial uses shall be conditional uses and may be permitted as specified:

Drive-In Theaters in the B-4 Business District provided that a planting screen at least *twenty-five (25)* feet wide is created along any side abutting a residential district and no access is permitted to or within *one thousand (1,000)* feet of an arterial street.

Drive-In Establishments serving food or beverages for consumption outside the structure in the *B-4 Business District.*

Motels in the *B-4 Business District.*

Funeral Homes in the *B-2, B-3,* and *B-4 Business Districts* provided all principal structures and uses are not less than *twenty-five (25)* feet from any lot line.

Drive-In Banks in the *B-2, B-3,* and *B-4 Business Districts.*

Tourist Homes in the *B-2* and *B-4 Business Districts* provided such district is located on a state trunk or U.S. numbered highway.

Vehicle Sales, Service, washing and repair stations, garages, taxi stands, and public parking lots, in *all business districts* provided all gas pumps are not less than *thirty (30)* feet from any side or rear lot line and *twenty (20)* feet from any existing or proposed street line.

Any Development within *five hundred (500)* feet of the existing or proposed rights-of-way of freeways, expressways, interstate and controlled access traffic-ways, and within *fifteen hundred (1,500)* feet of their existing or proposed interchange or turning lane rights-of-way shall be deemed to be conditional uses; and no structures shall be erected closer than *one hundred (100)* feet to their rights-of-way.

4.7 Industrial and Agricultural Uses

The following industrial and agricultural uses shall be conditional uses and may be permitted as specified:

Animal Hospitals in the *A-1 Agricultural, M-1* and *M-2 Industrial Districts* provided the lot area is not less than *three (3)* acres, and all principal structures and uses are not less than *one hundred (100)* feet from any residential district.

Dumps, Disposal Areas, Incinerators, and sewage disposal plants in the *A-1 Agricultural* and the *M-1* and *M-2 Industrial Districts.* Municipal earth and sanitary land fill operations may be permitted in any district.

Commercial Raising, propagation, boarding, or butchering of animals, such as dogs, mink, rabbits, foxes, goats, and pigs; the commercial production of eggs; and the hatching, raising, fattening, or butchering of fowl in the *A-1 Agricultural District.* Pea vineries, creameries, and condenseries in the *A-1 Agricultural* or *M-1* and *M-2 Industrial Districts.*

Manufacture and Processing of abrasives, acetylene, acid, alkalies, and ammonia, asbestos, asphalt, batteries, bedding, bleach, bone, cabbage, candle, carpeting, celluloid, cement, cereals, charcoal, chemicals, chlorine, coal tar, coffee, coke, cordage, creosote, dextrine, disinfectant, dye, excelsior, felt, fish, fuel, furs, gelatin, glucose, gyp-sum, hair products, ice, ink, insecticide, lampblack, lime, lime products, linoleum, matches, meat, oil cloth, paint, paper, peas, perfume, pickle, plaster of Paris, plastics, poison, polish, potash, pulp, pyroxylin, radium, rope, rubber, sausage, shoddy, shoe and lampblacking, size, starch, stove polish, textiles, and varnish. Manufacturing, processing, and storage of building materials, explosives, dry ice, fat, fertilizer, flammables, gasoline, glue, grains, grease, lard, plastics, radioactive materials, shellac, soap, turpentine, vinegar, and yeast. Manufacture and bottling of alcoholic beverages, bag cleaning, bleacheries, canneries, cold storage warehouses; electric and steam generating plants; electroplating; enameling; forges; foundries; garbage; incinerators; lacquering; lithographing; offal, rubbish, or animal reduction; oil coal, and bone distillation; refineries; road test facilities; slaughterhouses; smelting; stockyards; tanneries; and weaving in the *M-2 Heavy Industrial District* and shall be at least *six hundred (600)* feet from residential and public and semipublic districts.

Outside Storage and Manufacturing Areas in the *M-2 Heavy Industrial District.* Wrecking, junk, demolition and scrap yards shall be surrounded by a solid fence or evergreen planting screen completely preventing a view from any other property or public right-of-way and shall be at least *six hundred (600)* feet from residential, public, and semipublic districts.

Commercial Service Facilities, such as restaurants and fueling stations, in the *M-1 and M-2 Industrial Districts* provided all such services are physically and saleswise oriented toward industrial district users and employees and other users are only incidental customers.

4.8 Mineral Extraction

Mineral extraction operations including washing, crushing, or other processing are conditional uses and may be permitted in the *M-3 Quarrying District* provided:

The Application for the conditional use permit shall include: an adequate description of the operation; a list of equipment, machinery, and structures to be used; the source, quantity, and disposition of water to be used; a topographic map of the site showing existing contours with minimum vertical contour interval of *five (5)* feet, trees, proposed and existing access roads, the depth of all existing and proposed excavations; and a restoration plan.

The Restoration Plan provided by the applicant shall contain proposed contours after filling, depth of the restored topsoil, type of fill, planting or reforestation, restoration commencement, and completion dates. The applicant shall furnish the necessary fees to provide for the *Village*'s inspection and administrative costs and the necessary sureties that will enable the *Village* to perform the planned restoration of the site in event of default by the applicant. The amount of such sureties shall be based upon cost estimates prepared by the *Village* Engineer, and the form and type of such sureties shall be approved by the *Village Attorney.*

The Conditional Use Permit shall be in effect for a period not to exceed *two (2)* years and may be renewed upon application for a period not to exceed *two (2)* years. Modifications or additional conditions may be imposed upon application for renewal.

The *Village Plan Commission* shall particularly consider the effect of the proposed operation upon existing streets, neighboring development, proposed land use, drainage, water supply, soil erosion, natural beauty, character, and land value of locality and shall also consider the practicality of the proposed restoration of the site.

4.9 Recreational Uses

The following public recreational facilities shall be conditional uses and may be permitted as specified: archery ranges, bathhouses, beaches, boating, camps, conservatories, driving ranges, firearm ranges, golf courses, gymnasiums, hunting, ice boating, marinas, music halls, polo fields, pools, riding academies, skating rinks, sport fields, stadiums, swimming pools, and zoological and botanical gardens in the *P-1 Public and Semipublic District* provided that the lot area is not less than *three (3)* acres and all structures are not less than *fifty (50)* feet from any district boundary.

Commercial Recreation Facilities, such as arcades, bowling alleys, clubs, dance halls, driving ranges, gymnasiums, lodges, miniature golf, physical culture, pool and billiard halls, racetracks, rifle ranges, Turkish baths, skating rinks, and theaters are conditional uses and may be permitted in the *B-2, B-3,* or *B-4 Business Districts.*

SECTION 5.0 TRAFFIC, PARKING, AND ACCESS

5.1 Traffic Visibility

No obstructions, such as structures, parking or vegetation, shall be permitted in any district between the heights of *two and one-half (2½)* feet and *ten (10)* feet above the plane through the mean curb-grades within the triangular space formed by any two existing or proposed intersecting street or alley right-of-way lines and a line joining points on such lines located a minimum of *fifteen (15)* feet from their intersection.

In the Case of Arterial Streets intersecting with other arterial streets or railways, the corner cutoff distances establishing the triangular vision clearance space shall be increased to *fifty (50)* feet.

5.2 Loading Requirements

In all districts adequate loading areas shall be provided so that all vehicles loading, maneuvering, or unloading are completely off the public ways and so that all vehicles need not back onto any public way.

5.3 Parking Requirements

In all districts and in connection with every use, there shall be provided at the time any use or building is erected, enlarged, extended, or increased off-street parking stalls for all vehicles in accordance with the following:

Adequate Access to a public street shall be provided for each parking space, and driveways shall be at least *ten (10)* feet wide for one- and two-family dwellings and minimum of *twenty-four (24)* feet for all other uses.

Size of each parking space shall be not less than *one hundred and eighty (180)* square feet exclusive of the space required for ingress and egress.

Location to be on the same lot as the principal use or not over *four hundred (400)* feet from the principal use. No parking stall or driveway except in residential districts shall be closer than *twenty-five (25)* feet to a residential district lot line or a street line opposite a residential district.

Surfacing. All off-street parking areas shall be graded and surfaced so as to be dust free and properly drained. Any parking area for more than *five (5)* vehicles shall have the aisles and spaces clearly marked.

Curbs or Barriers shall be installed so as to prevent the parked vehicles from extending over any lot lines.

Number of Parking Stalls Required:

Single-family dwellings and mobile homes	2 stalls for each dwelling unit
Multi-family dwellings	1.5 stalls for each dwelling unit
Hotels, Motels	1 stall or each guest room plus 1 stall for every 3 employees
Hospitals, clubs, lodges, sororities, dormitories, lodging and boardinghouses	1 stall for every 5 beds plus 1 stall for every 3 employees
Medical and dental clinics	3 stalls for each doctor
Churches, theaters, auditoriums, community centers, vocational and night schools, and other places of public assembly	1 stall for every 5 seats
Colleges, secondary and elementary schools	1 stall for every 2 employees
Restaurants, bars, places of entertainment, repair shops, retail and service stores	1 stall for every 150 square feet of floor area
Manufacturing and processing plants, laboratories, and warehouses	1 stall for every three employees
Funeral homes	1 stall for every 4 seats
Bowling alleys	5 stalls for each alley

Uses Not Listed. In the case of structures or uses not mentioned, the provision for a use that is similar shall apply.

Combinations of any of the above uses shall provide the total of the number of stalls required for each individual use.

5.4 Driveways

All driveways installed, altered, changed, replaced, or extended after the effective date of this Ordinance shall meet the following requirements:

Islands between driveway openings shall be provided with a minimum of *twelve (12)* feet between all driveways and *six (6)* feet at all lot lines.

Openings for vehicular ingress and egress shall not exceed *twenty-four (24)* feet at the street line and *thirty (30)* feet at the roadway.

Vehicular Entrances and Exits to drive-in theaters, banks, and restaurants; motels; funeral homes; vehicular sales, service, washing and repair stations; garages; or public parking lots shall be not less than *two hundred (200)* feet from any pedestrian entrance or exit to a school, college, university, church, hospital, park, playground, library, public emergency shelter, or other place of public assembly.

5.5 Highway Access

No direct private access shall be permitted to the existing or proposed rights-of-way of expressways; nor to any controlled access arterial street without permission of the highway agency that has access control jurisdiction.

No direct public or private access shall be permitted to the existing or proposed rights-of-way of the following:

Freeways, Interstate Highways, and their interchanges or turning lanes nor to intersecting or interchanging streets within *fifteen hundred (1,500)* feet of the most remote end of the taper of the turning lanes.

Arterial Streets intersecting another arterial street within *one hundred (100)* feet of the intersection of the right-of-way lines.

Streets intersecting an arterial street within *fifty (50)* feet of the intersection of the right-of-way lines.

Access barriers, such as curbing, fencing, ditching, landscaping, or other topographic barriers, shall be erected to prevent unauthorized vehicular ingress or egress to the above specified streets or highways.

Temporary Access to the above rights-of-way may be granted by the *Village Plan Commission* after review and recommendation by the highway agencies having jurisdiction. Such access permit shall be temporary, revocable, and subject to any conditions required and shall be issued for a period not to exceed *twelve (12)* months.

SECTION 6.0 EXCEPTIONS

6.1 Height

The district height limitations stipulated elsewhere in this Ordinance may be exceeded, but such modification shall be in accord with the following:

Architectural Projections, such as spires, belfries, parapet walls, cupolas, domes, flues, and chimneys, are exempt from the height limitations of this Ordinance.
Special Structures, such as elevator penthouses, gas tanks, grain elevators, scenery lofts, radio and television receiving antennas, manufacturing equipment and necessary mechanical appurtenances, cooling towers, fire towers, substations, and smoke stacks, are exempt from the height limitations of this Ordinance.
Essential Services, utilities, water towers, electric power, and communication transmission lines are exempt from the height limitations of the Ordinance.
Communication Structures, such as radio and television transmission and relay towers, aerials, and observation towers, shall not exceed in height *three (3)* times their distance from the nearest lot line.
Agricultural Structures, such as barns, silos, and windmills, shall not exceed in height twice their distance from the nearest lot line.

Public or Semipublic Facilities, such as schools, churches, hospitals, monuments, sanitariums, libraries, governmental offices and stations, may be erected to a height of *sixty (60)* feet, provided all required yards are increased not less than *one (1)* foot for each foot the structure exceeds the district's maximum height requirement.

6.2 Yards

The yard requirements stipulated elsewhere in this Ordinance may be modified as follows:

Uncovered Stairs, landings, and fire escapes may project into any yard but not to exceed *six (6)* feet and not closer than *three (3)* feet to any lot line.
Architectural Projections, such as chimneys, flues, sills, eaves, belt courses, and ornaments, may project into any required yard; but such projection shall not exceed *two (2)* feet.
Residential Fences are permitted on the property lines in residential districts but shall not in any case exceed a height of *six (6)* feet; shall not exceed a height of *four (4)* feet in the street yard and shall not be closer than *two (2)* feet to any public right-of-way.
Security Fences are permitted on the property lines in all districts except residential districts but shall not exceed *ten (10)* feet in height and shall be of an open type similar to woven wire or wrought iron fencing.

Accessory Uses and detached accessory structures are permitted in the rear yard only; they shall not be closer than *ten (10)* feet to the principal structure, shall not exceed *fifteen (15)* feet in height, shall not occupy more than twenty (20) percent of the rear yard area, and shall not be closer than *three (3)* feet to any lot line nor *five (5)* feet to an alley line.

Off-Street Parking is permitted in all yards of the *B-3* and *B-4 Business Districts* but shall not be closer than *twenty-five (25)* feet to any public right-of-way.

Essential Services, utilities, electric power and communication transmission lines are exempt from the yard and distance requirements of this Ordinance.

Landscaping and vegetation are exempt from the yard requirements of this Ordinance.

6.3 Additions

Additions in the street yard of existing structures shall not project beyond the *average* of the existing street yards on the abutting lots or parcels.

6.4 Average Street Yards

The required street yards may be decreased in any residential or business districts to the average of the existing street yards of the abutting structures on each side but in no case less than *fifteen (15)* feet in any residential district and *five (5)* feet in any business district.

6.5 Noise

Sirens, whistles, and bells that are maintained and utilized solely to serve a public purpose are exempt from the sound level standards of this Ordinance.

SECTION 7.0 SIGNS

7.1 Permit Required

No sign shall hereafter be located, erected, moved, reconstructed, extended, enlarged, converted, or structurally altered without a zoning permit except those signs excepted in Section 7.2 and without being in conformity with the provisions of this Ordinance. The sign shall also meet all the structural requirements of the Building Code.

7.2 Signs Excepted

All signs are prohibited in all *Residential, Agricultural, Conservancy, Flood, Public and Semipublic Districts* except the following:

Signs Over Show Windows or Doors of a nonconforming business establishment announcing without display or elaboration only the name and occupation of the proprietor and not to exceed *two (2)* feet in height and *ten (10)* feet in length.

Real Estate Signs not to exceed *eight (8)* square feet in area which advertise the sale, rental, or lease of the premises upon which said signs are temporarily located.

Name, Occupation, and Warning Signs not to exceed *two (2)* square feet located on the premises.

Bulletin Boards for public, charitable or religious institutions not to exceed *eight (8)* square feet in area located on the premises.

Memorial Signs, tablets, names of buildings, and date of erection when cut into any masonry surface or when constructed of metal and affixed flat against a structure.

Official Signs, such as traffic control, parking restrictions, information, and notices.

Temporary Signs or banners when authorized by the *Board of Zoning Appeals.*

7.3 Signs Permitted

Signs are permitted in all *Business* and *Industrial Districts* subject to the following restrictions:

Wall Signs placed against the exterior walls of buildings shall not extend more than *six (6)* inches outside of a building's wall surface, shall not exceed *five hundred (500)* square feet in area for any one premises, and shall not exceed *twenty (20)* feet in height above the mean centerline street grade.

Projecting Signs fastened to, suspended from, or supported by structures shall not exceed *one hundred (100)* square feet in area for any one premises; shall not extend more than six (6) feet into any required yard; shall not extend more than *three (3)* feet into any public right-of-way; shall not be less than ten (10) feet from all side lot lines; shall not exceed a height of *twenty (20)* feet above the mean centerline street grade; and shall not be less than *ten (10)* feet above the sidewalk nor *fifteen (15)* feet above a driveway or an alley.

Ground Signs shall not exceed twenty (20) feet in height above the mean centerline street grade, shall meet all yard requirements for the district in which it is located, shall not exceed *one hundred (100)* square feet on one side nor *two hundred (200)* square feet on all sides for any one premises.

Roof Signs shall not exceed *ten (10)* feet in height above the roof, shall meet all the yard and height requirements for the district in which it is located, and shall not exceed *three hundred (300)* square feet on all sides for any one premises.

Window Signs shall be placed only on the inside of commercial buildings and shall not exceed *twenty-five (25)* percent of the glass area of the pane upon which the sign is displayed.

Combinations of any of the above signs shall meet all the requirements for the individual sign.

7.4 Facing

No sign except those permitted in Section 7.2 shall be permitted to face a *Residential* or *Public and Semi-public District* within *one hundred (100)* feet of such district boundary.

7.5 Traffic

Signs shall not resemble, imitate, or approximate the shape, size, form, or color of railroad or traffic signs, signals, or devices. Signs shall not obstruct or interfere with the effectiveness of railroad or traffic signs, signals, or devices. No sign shall be erected, relocated, or maintained so as to prevent free ingress to or egress from any door, window, or fire escape; and no sign shall be attached to a standpipe or fire escape. No sign shall be placed so as to obstruct or interfere with traffic visibility.

7.6 Existing Signs

Signs lawfully existing at the time of the adoption or amendment of this Ordinance may be continued although the use, size, or location does not conform with the provisions of this Ordinance. However, it shall be deemed a nonconforming use or structure; and the provisions of Section 8.0 shall apply.

7.7 Bonds

Ever applicant for a zoning permit for a sign shall, before the permit is granted, execute a surety bond in a sum to be fixed by the Zoning Administrator, but not to exceed *Twenty-five Thousand Dollars ($25,000)*; and it shall be of a form and type approved by the *Village Attorney*, indemnifying the *municipality* against all loss cost damages or expense incurred or sustained by or recovered against the *municipality* by reason of the erection, construction, or maintenance of such sign. A liability insurance policy issued by an insurance company authorized to do business in the State of Wisconsin, and conforming to the requirements of this section, may be permitted by the *Village Attorney* in lieu of a bond.

SECTION 8.0 NONCONFORMING USES

8.1 Existing Nonconforming Uses

The lawful nonconforming use of a structure, land, or water existing at the time of the adoption or amendment of this Ordinance may be continued although the use does not conform with the provisions of this Ordinance; however:

Only That Portion of land or water in actual use may be so continued and the structure may not be extended, enlarged, reconstructed, substituted, moved, or structurally altered except when required to do so by law or order or so as to comply with the provisions of this Ordinance.

Total Lifetime Structural Repairs or alterations shall not exceed *fifty (50)* percent of the *Village* assessed value of the structure at the time of its becoming a nonconforming use unless it is permanently changed to conform to the use provisions of this Ordinance.

Substitution of New Equipment may be permitted by the *Board of Zoning Appeals* if such equipment will reduce the incompatibility of the nonconforming use with the neighboring uses.

8.2 Abolishment or Replacement

If such nonconforming use is discontinued or terminated for a period of *twelve (12)* months, any future use of the structure, land, or water shall conform to the provisions of this Ordinance. When a non-conforming use or structure is damaged by fire, explosion, flood, the public enemy, or other calamity, to the extent of more than *fifty (50)* percent of its current assessed value, it shall not be restored except so as to comply with the use provisions of this Ordinance.

A Current File of all nonconforming uses shall be maintained by the *Zoning Administrator* listing the following: owner's name and address; use of the structure, land, or water; and assessed value at the time of its becoming a nonconforming use.

8.3 Existing Nonconforming Structures

The lawful nonconforming structure existing at the time of the adoption or amendment of this Ordinance may be continued although its size or location does not conform with the lot width, lot area, yard, height, parking and loading, and access provisions of this Ordinance; however, it shall not be extended, enlarged, reconstructed, moved, or structurally altered except when required to do so by law or order or so as to comply with the provisions of this Ordinance.

8.4 Changes and Substitutions

Once a nonconforming use or structure has been changed to conform, it shall not revert back to a nonconforming use or structure. Once the *Board of Zoning Appeals* has permitted the substitution of a more restrictive nonconforming use for an existing nonconforming use, the substituted use shall lose its status as a legal nonconforming use and become subject to all the conditions required by the *Board of Zoning Appeals.*

8.5 Substandard Lots

In any residential district, a one-family detached dwelling and its accessory structures may be erected on any legal lot or parcel of record in the County Register of Deeds office before the effective date or amendment of this Ordinance:

Such Lot or Parcel shall be in separate ownership from abutting lands. If abutting lands and the substandard lot are owned by the same owner the substandard lot shall not be sold or used without full compliance with the provisions of this Ordinance. If in separate ownership, all the district requirements shall be complied with insofar as practical but shall not be less than the following:

Lot	Width	Minimum *30* ft.
	Area	Minimum *4,000* sq. ft.
Building	Area	Minimum *1,000* sq. ft.
	Height	Maximum *30* ft.

Yards	Street	Minimum *25* ft.; the second street yard on corner lots shall be not less than *10* ft.
	Rear	Minimum *25* ft.
	Side	Minimum *16* percent of the frontage, but not less than *5* ft.

SECTION 9.0 BOARD OF ZONING APPEALS

9.1 Establishment

There is hereby established a *Board of Zoning Appeals* for the *Village of* ________ for the purpose of hearing appeals and applications, and granting variances and exceptions to the provisions of this Zoning Ordinance in harmony with the purpose and intent of the Zoning Ordinance.

9.2 Membership

The *Board of Zoning Appeals* shall consist of *five (5)* members appointed by the *Village President* and confirmed by the *Village Board.*

Terms shall be staggered three-year periods.

Chairman shall be designated by the *Village President.*

An Alternate Member may be appointed by the *Village President* for a term of three (3) years and shall act only when a regular member is absent or refuses to vote because of interest.

One Member shall be a Village Plan Commissioner and one member shall be a registered architect, registered professional engineer, builder, or real estate appraiser.

Secretary shall be the *Village* Clerk.

Zoning Inspector shall attend all meetings for the purpose of providing technical assistance when requested by the Board.

Official Oaths shall be taken by all members in accordance with Section 19.01 of the Wisconsin Statutes within *ten (10)* days of receiving notice of their appointment.

Vacancies shall be filled for the unexpired term in the same manner as appointments for a full term.

9.3 Organization

The *Board of Zoning Appeals* shall organize and adopt rules of procedure for its own government in accordance with the provisions of this Ordinance.

Meetings shall be held at the call of the chairman and shall be open to the public.

Minutes of the proceedings and a record of all actions shall be kept by the secretary, showing the vote of each member upon each question, the reasons for the Board's determination, and its finding of facts. These records shall be immediately filed in the office of the Board and shall be a public record.

The Concurring Vote of four (4) members of the Board shall be necessary to correct an error; grant a variance; make an interpretation; and permit a utility, temporary, unclassified, or substituted use.

9.4 Powers

The *Board of Zoning Appeals* shall have the following powers:

Errors. To hear and decide appeals where it is alleged there is error in any order, requirement, decision, or determination made by the *Zoning Inspector or Architectural Board.*

Variances. To hear and grant appeals for variances as will not be contrary to the public interest, where, owing to special conditions, a literal enforcement will result in practical difficulty or unnecessary hardship, so that the spirit and purposes of this Ordinance shall be observed and the public safety, welfare, and justice secured. Use variances shall not be granted.

Interpretations. To hear and decide application for interpretations of the zoning regulations and the boundaries of the zoning districts after the *Village Plan Commission* has made a review and recommendation.

Substitutions. To hear and grant applications for substitution of more restrictive nonconforming uses for existing nonconforming uses provided no structural alterations are to be made and the *Village Plan Commission* has made a review and recommendation. Whenever the Board permits such a substitution, the use may not thereafter be changed without application.

Unclassified Uses. To hear and grant applications for unclassified and unspecified uses provided that such uses are similar in character to the principal uses permitted in the district and the *Village Plan Commission* has made a review and recommendation.

Temporary Uses. To hear and grant applications for temporary uses, in any district provided that such uses are of a temporary nature, do not involve the erection of a substantial structure, and are compatible with the neighboring uses and the *Village Plan Commission* has made a review and recommendation. The permit shall be temporary, revocable, subject to any conditions required by the *Board of Zoning Appeals*, and shall be issued for a period not to exceed *twelve (12)* months. Compliance with all other provisions of this Ordinance shall be required.

Permits. The Board may reverse, affirm wholly or partly, modify the requirements appealed from, and may issue or direct the issue of a permit.

Assistance. The Board may request assistance from other *Village officers*, departments, commissions, and boards.

Oaths. The chairman may administer oaths and compel the attendance of witnesses.

9.5 Appeals and Applications

Appeals from the decision of the *Zoning Administrator* or the *Architectural Board* concerning the literal enforcement of this Ordinance may be made by any person aggrieved or by any officer, department, board, or bureau of the *Village*. Such appeals shall be filed with the secretary within *thirty (30)* days after the date of written notice of the decision or order of the *Zoning Administrator* or *Architectural Board.* Applications may be made by the owner or lessee of the structure, land, or water to be affected at any time and shall be filed with the secretary. Such appeals and application shall include the following:

Name and Address of the appellant or applicant and all abutting and opposite property owners of record.

Plat of Survey prepared by a registered land surveyor showing all of the information required under Section 2.3 for a *Zoning* Permit.

Additional Information required by the *Village Plan Commission, Village Engineer, Board of Zoning Appeals, or Zoning Administrator.*

9.6 Hearings

The *Board of Zoning Appeals* shall fix a reasonable time and place for the hearing, give public notice thereof at least *ten (10)* days prior, and shall give due notice to the parties in interest, the *Zoning Administrator*, and the *Village Plan Commission.* At the hearing the appellant or applicant may appear in person, by agent, or by attorney.

9.7 Findings

No variance to the provisions of this Ordinance shall be granted by the Board unless it finds beyond a reasonable doubt that all the following facts and conditions exist and so indicates in the minutes of its proceedings.

Exceptional Circumstances. There must be exceptional, extraordinary, or unusual circumstances or conditions applying to the lot or parcel, structure, use, or intended use that do not apply generally to other properties or uses in the same district and the granting of the variance would not be of so general or recurrent nature as to suggest that the Zoning Ordinance should be changed.

Preservation of Property Rights. That such variance is necessary for the preservation and enjoyment of substantial property rights possessed by other properties in the same district and same vicinity.

Absence of Detriment. That the variance will not create substantial detriment to adjacent property and will not materially impair or be contrary to the purpose and spirit of this Ordinance or the public interest.

9.8 Decision

The *Board of Zoning Appeals* shall decide all appeals and applications within *thirty (30)* days after the final hearing and shall transmit a signed copy of the Board's decision to the appellant or applicant, *Zoning Administrator*, and *Village Plan Commission.*

Conditions may be placed upon any zoning permit ordered or authorized by this Board.

Variances, Substitutions, or Use Permits granted by the Board shall expire within *six (6)* months unless substantial work has commenced pursuant to such grant.

9.9 Review by Court of Record

Any person or persons aggrieved by any decision of the *Board of Zoning Appeals* may present to the court of record a petition duly verified setting forth that such decision is illegal and specifying the grounds of the illegality. Such petition shall be presented to the court within *thirty (30)* days after the filing of the decision in the office of the *Board of Zoning Appeals.*

SECTION 10.0 AMENDMENTS

10.1 Authority

Whenever the public necessity, convenience, general welfare, or good zoning practice require, the *Village Board of Trustees* may, by ordinance, change the district boundaries or amend, change or supplement the regulations established by this Ordinance or amendments thereto.

Such Change or Amendment shall be subject to the review and recommendation of the *Village Plan Commission and the appropriate Joint Extraterritorial Zoning Committee.*

10.2 Initiation

A change or amendment may be initiated by the *Village Board*, *Village Plan Commission*, or by a petition of one or more of the owners or lessees of property within the area proposed to be changed.

10.3 Petitions

Petitions for any change to the district boundaries or amendments to the regulations shall be filed with the *Village Clerk*, describe the premises to be rezoned or the regulations to be amended, list the reasons justifying the petition, specify the proposed use and have attached the following:

Plot Plan drawn to a scale of 1 inch equals 100 feet showing the area proposed to be rezoned, its location, its dimensions, the location and classification of adjacent zoning districts, and the location and existing use of all properties within *two hundred (200)* feet to the area proposed to be rezoned.

Owners' Names and Addresses of all properties lying within *two hundred (200)* feet of the area proposed to be rezoned.

Additional Information required by the *Village Plan Commission*, *Joint Extraterritorial Zoning Committee*, or *Village Board.*

Fee Receipt from the *Village* Treasurer in the amount of *Twenty-five Dollars ($25).*

10.4 Recommendations

The *Village Plan Commission* shall review all proposed changes and amendments within the corporate limits and shall recommend that the petition be granted as requested, modified, or denied. The recommendation shall be made at a meeting subsequent to the meeting at which the petition is first submitted and shall be made in writing to the *Village Board.*

10.5 Hearings

The *Village Board* shall hold a public hearing upon each recommendation, giving at least *ten (10)* days' prior notice by publication at least *three (3)* times during the preceding *thirty (30)* days listing the time, place, and the changes or amendments proposed. The *Village Board* shall also give at least *ten (10)* days' prior written notice to the clerk of any municipality within *one thousand (1,000)* feet of any land to be affected by the proposed change or amendment.

10.6 Village Board's Action

Following such hearing and after careful consideration of the *Village Plan Commission*'s recommendations, the *Village Board* shall vote on the passage of the proposed change of amendment.

The Village Plan Commission's Recommendations may only be overruled by three-fourths (¾) of the full Village Board's membership.

10.7 Protes

In the event of a protest against such district change or amendment to the regulations of this Ordinance, duly signed and acknowledged by the owners of *twenty (20)* percent or more either of the areas of the land included in such proposed change, or by the owners of *twenty (20)* percent or more of the land immediately adjacent extending *one hundred (100)* feet therefrom, or by the owners of *twenty (20)* percent or more of the land directly opposite thereto extending *one hundred (100)* feet from the street frontage of such opposite land, such changes or amendments shall not become effective except by the favorable vote of *three-fourths (¾)* of the full *Village Board* membership.

SECTION 11.0 DEFINITIONS

For the purposes of this Ordinance, the following definitions shall be used. Words used in the present tense include the future; the singular number includes the plural number; and the plural number includes the singular number. The word "shall" is mandatory and not directory.

Accessory Use or Structure

A use or detached structure subordinate to the principal use of a structure, land, or water and located on the same lot or parcel serving a purpose customarily incidental to the principal use or the principal structures.

Alley A special public right-of-way affording only secondary access to abutting properties.

Arterial Street A public street or highway used or intended to be used primarily for fast or heavy through traffic. Arterial streets and highways shall include freeways and expressways as well as arterial streets, highways, and parkways.

Basement That portion of any structure located partly below the average adjoining lot grade.

Boardinghouse A building other than a hotel or restaurant where meals or lodging are regularly furnished by prearrangement for compensation for *four (4)* or more persons not members of a family, but not exceeding *twelve (12)* persons and not open to transient customers.

Building Any structure having a roof supported by columns or walls used or intended to be used for the shelter or enclosure of persons, animals, equipment, machinery, or materials.

Building Area The total living area bounded by the exterior walls of a building at the floor levels, but not including basement, utility rooms, garages, porches, breezeways, and unfinished attics.

Building Height The vertical distance measured from the mean elevation of the finished lot grade along the street yard face of the structure to the highest point of flat roofs; to the mean height level between the eaves and ridges of gable, gambrel, hip, and pitch roofs; or to the deck line of mansard roofs.

Clothing Repair Shops Shops where clothing is repaired, such as shoe repair shops, seamstress, tailor shops, shoe shine shops, clothes pressing shops, but none employing over *five (5)* persons.

Clothing Stores Retail stores where clothing is sold, such as department stores, dry goods and shoe stores, dress, hosiery, and millinery shops.

Conditional Uses Uses of a special nature as to make impractical their predetermination as a principal use in a district.

Corner Lot A lot abutting two or more streets at their intersection provided that the corner of such intersection shall have an angle of *one hundred and thirty-five (135)* degrees or less, measured on the lot side.

Dwelling A detached building designed or used exclusively as a residence or sleeping place, but does not include boarding or lodging houses, motels, hotels, tents, cabins, or mobile homes.

Efficiency A dwelling unit consisting of one principal room with no separate sleeping rooms.

Emergency Shelter Public or private enclosures designed to protect people from aerial, radiological, biological, or chemical warfare; fire, flood, windstorm, riots, and invasions.

Essential Services Services provided by public and private utilities, necessary for the exercise of the principal use or service of the principal structure. These services include underground, surface, or overhead gas, electrical, steam, water, sanitary sewerage, storm water drainage, and communication systems and accessories thereto, such as poles, towers, wires, mains, drains, vaults, culverts, laterals, sewers, pipes, catch basins, water storage tanks, conduits, cables, fire alarm boxes, police call boxes, traffic signals, pumps, lift stations, and hydrants, but not including buildings.

Expressway A divided arterial street or highway with full or partial control of access and with or without grade-separated intersections.

Family Any number of persons related by blood, adoption, or marriage, or not to exceed *four (4)* persons not so related, living together in one dwelling as a single housekeeping entity.

Freeway An expressway with full control of access and with fully grade-separated intersections.

Frontage The smallest dimension of a lot abutting a public street measured along the street line.

Gift Stores Retail stores where items such as art, antiques, jewelry, books, and notions are sold.

Hardware Stores Retail stores where items such as plumbing, heating, and electrical supplies, sporting goods, and paints are sold.

Household Occupation Any occupation for gain or support conducted entirely within buildings by resident occupants which is customarily incidental to the principal use of the premises, does not exceed *twenty-five (25)* percent of the area of any floor, uses only household equipment, and no stock in trade is kept or sold except that made on the premises. A household occupation includes uses such as babysitting, millinery, dressmaking, canning, laundering, and crafts, but does not include the display of any goods nor such occupations as barbering, beauty shops, dance schools, real estate brokerage, or photographic studios.

Interchange A grade-separated intersection with one or more turning lanes for travel between intersection legs.

Joint Extraterritorial Zoning Committee Any Zoning committee established in accordance with Section 62.23(7a) of the Wisconsin Statutes (Chapter 241, Laws of 1963).

Living Rooms All rooms within a dwelling except closets, foyers, storage areas, utility rooms, and bathrooms.

Loading Area A completely off-street space or berth on the same lot for the loading or unloading of freight carriers, having adequate ingress and egress to a public street or alley.

Lot A parcel of land having frontage on a public street, occupied or intended to be occupied by a principal structure or use and sufficient in size to meet the lot width, lot frontage, lot area, yard, parking area, and other open space provisions of this Ordinance.

Lot Lines and Area The peripheral boundaries of a parcel of land and the total area lying within such boundaries.

Lot Width The width of a parcel of land measured at the rear of the specified street yard.

Machine Shops Shops where lathes, presses, grinders, shapers, and other wood and metal working machines are used, such as blacksmith, tinsmith, welding, and sheet metal shops; plumbing, heating, and electrical repair and overhaul shops.

Minor Structures Any small, movable accessory erection or construction such as birdhouses; tool houses; pet houses; play equipment; arbors; and walls and fences under *four (4)* feet in height.

Motel A series of attached, semi-attached, or detached sleeping units for the accommodation of transient guests.

Nonconforming Uses or Structures Any structure, land, or water lawfully used, occupied, or erected at the time of the effective date of this Ordinance or amendments thereto that does not conform to the regulations of this Ordinance or amendments thereto. Any such structure conforming in respect to use but not in respect to frontage, width, height, area, yard, parking, loading, or distance requirements shall be considered a nonconforming structure and not a nonconforming use.

Parking Lot A structure or premises containing *ten (10)* or more parking spaces open to the public for rent or a fee.

Parking Space A graded and surfaced area of not less than *one hundred and eighty (180)* square feet in area either enclosed or open for the parking of a motor vehicle, having adequate ingress and egress to a public street or alley.

Parties in Interest Includes all abutting property owners, all property owners within *one hundred (100)* feet, and all property owners of opposite frontages.

Professional Home Offices Residences of doctors of medicine, practitioners, dentists, clergymen, architects, landscape architects, professional engineers, registered land surveyors, lawyers, artists, teachers, authors, musicians, or other recognized professions used to conduct their professions where the office does not exceed *one-half (½)* the area of only one floor of the residence and only one nonresident person is employed.

Rear Yard A yard extending across the full width of the lot, the depth of which shall be the minimum horizontal distance between the rear lot line and a line parallel thereto through the nearest point of the principal structure. This yard shall be opposite the street yard or one of the street yards on a corner lot.

Side Yard A yard extending from the street yard to the rear yard of the lot, the width of which shall be the minimum horizontal distance between the side lot line and a line parallel thereto through the nearest point of the principal street.

Signs Any words, letters, figures, numerals, phrases, sentences, emblems, devices, designs, trade names, or trade marks by which anything is made known and which are used to advertise or promote an individual, firm, association, corporation, profession, business, commodity or product and which is visible from any public street or highway.

Smoke Unit The number obtained when the smoke density in Ringelmann number is multiplied by the time of emission in minutes.

Street A public right-of-way not less than *fifty (50)* feet wide providing primary access to abutting properties.

Street Yard A yard extending across the full width of the lot, the depth of which shall be the minimum horizontal distance between the existing or proposed street or highway line and a line parallel thereto through the nearest point of the principal structure. Corner lots shall have two such yards.

Structure Any erection or construction, such as buildings, towers, masts, poles, booms, signs, decorations, carports, machinery, and equipment.

Structural Alterations Any change in the supporting members of a structure, such as foundations, bearing walls, columns, beams, or girders.

Turning Lanes An existing or proposed connecting roadway between two arterial streets or between an arterial street and any other street. Turning lanes include grade-separated interchange ramps.

Utilities Public and private facilities such as water wells, water and sewage pumping stations, water storage tanks, power and communication transmission lines and towers, electrical power substations, static transformer stations, telephone and telegraph exchanges, microwave radio relays, and gas regulation stations, but not including sewage disposal plants, municipal incinerators, warehouses, shops, and storage yards.

Yard An open space on the same lot with a structure, unoccupied and unobstructed from the ground upward except for vegetation. The street and rear yards extend the full width of the lot.

Index

F

G

H

I

J

L

M

N

O

P

V

W

Z

For Product Safety Concerns and Information please contact our
EU representative GPSR@taylorandfrancis.com Taylor & Francis
Verlag GmbH, Kaufingerstraße 24, 80331 München, Germany